METHODS IN MOLECULAR BIOLOGY

For further volumes:
http://www.springer.com/series/7651

For over 35 years, biological scientists have come to rely on the research protocols and methodologies in the critically acclaimed *Methods in Molecular Biology* series. The series was the first to introduce the step-by-step protocols approach that has become the standard in all biomedical protocol publishing. Each protocol is provided in readily-reproducible step-by step fashion, opening with an introductory overview, a list of the materials and reagents needed to complete the experiment, and followed by a detailed procedure that is supported with a helpful notes section offering tips and tricks of the trade as well as troubleshooting advice. These hallmark features were introduced by series editor Dr. John Walker and constitute the key ingredient in each and every volume of the *Methods in Molecular Biology* series. Tested and trusted, comprehensive and reliable, all protocols from the series are indexed in PubMed.

Mammalian Synthetic Systems

Edited by

Francesca Ceroni and Karen Polizzi

Department of Chemical Engineering and Imperial College Centre for Synthetic Biology, Imperial College London, London, UK

Humana Press

Editors
Francesca Ceroni
Department of Chemical Engineering
and Imperial College Centre
for Synthetic Biology
Imperial College London
London, UK

Karen Polizzi
Department of Chemical Engineering
and Imperial College Centre
for Synthetic Biology
Imperial College London
London, UK

ISSN 1064-3745 ISSN 1940-6029 (electronic)
Methods in Molecular Biology
ISBN 978-1-0716-3717-3 ISBN 978-1-0716-3718-0 (eBook)
https://doi.org/10.1007/978-1-0716-3718-0

This Humana imprint is published by the registered company Springer Science+Business Media, LLC, part of Springer Nature.
The registered company address is: 1 New York Plaza, New York, NY 10004, U.S.A.

Paper in this product is recyclable.

Preface

We are currently sailing inside the third decade of synthetic biology and witnessing the breadth of possibilities and applications it has already enabled. Advancement in synthetic biology start from the growing number of genetic components and constructs available for cell engineering and goes through a plethora of chassis and tools at our disposal for basic research and applications, in therapy as well as in bioproduction and beyond. In this collection, we would like to feature examples of the advancements enabled by synthetic biology in mammalian cells, from construct design and assembly to computational analysis and circuitry with desired functionalities.

We start by considering technological advances and the development of automated and parallel gene construct design and testing that have accelerated assembly and validation. Two examples are provided in Chaps. 1 and 2. In Chap. 1, authors adopt CIS display, an *in vitro*-directed evolution protocol, for high-throughput selection of DNA-binding candidates from pools of non-binding proteins. In Chap. 2, authors present an example of a Golden Gate-based high-throughput construct assembly, the Modular Protein Expression Toolbox (MOPET), that enables highly efficient DNA assembly of pre-defined, standardized functional DNA modules effecting protein expression with a focus to minimize the cloning burden in coding regions and facilitating generation of genetic vectors coding for difficult-to-express proteins.

Thanks to automation and parallel high-throughput assembly, a growing number of genetic components have so far been developed, widening the toolbox at the disposal of the community for synthetic construct design. Chapters 3, 4, and 5 exemplify some of building blocks synthetic biologists can now select from when in the search for cell engineering components. Chapter 3 provides example of such libraries, describing a designed set of coiled-coiled (CC) motifs useful for mammalian cell engineering and providing a protocol for the construction of CC-mediated logic circuits in mammalian cells. In Chaps. 4 and 5 authors present a mammalian-based synthetic biology toolbox to engineer membrane-membrane interactions and a set of protease-responsive RNA Binding Proteins, to widen the applications of synthetic biology circuitry to cell-to-cell communication and RNA-based regulation, respectively.

The rational engineering of synthetic circuits and systems has benefitted from computational tools and characterization standards. Indeed, mathematical modeling plays a vital role in mammalian synthetic biology by providing a framework to design and optimize design circuits, predict their behavior, and guide experimental design. In Chap. 6, authors review recent models used in the literature, considering mathematical frameworks at the molecular, cellular, and system levels. Reporting key challenges in the field, and discussing opportunities for genome-scale models, machine learning and cybergenetics to expand the capabilities of model-driven mammalian cell biodesign, the authors open to next steps in the field. Chapters 7 and 8 go instead into the detail of network architectures useful to increase the robustness and performance of gene circuits. One example of such architectures is provided by integral feedback control, pivotal in minimizing disturbances (Chap. 7). Chapter 8 presents a systematic approach based on predictive mathematical modeling to guide the design and construction of gene activity-based biosensors. Mathematical modeling is useful not only to support genetic design but also to take into account the context of cellular growth and integrate cell and process engineering, providing guidance on the performance of engineered cells. This is especially important in the context of bioproduction applications, where specific growth and nutrient conditions are usually adopted. Flux balance analysis (FBA) is a computational methodology to analyze and model the metabolic behavior of cells. In Chap. 9, the authors break down the key steps for formulating flux balance analysis models and other FBA-derived methodologies in the context of mammalian cell biology, including strain design, cell line–specific models, and metabolic flux sampling, directly addressing the need to optimize cell and process conditions for best performance in bioprocessing. One specific aspect of mammalian cell engineering and experimentation is the adoption of transient transfections for the insertion of plasmid DNA in episomal form within the cells. These enable faster assessment than the more laborious workflow required for genome integration. However, transient transfections suffer from great variability of plasmid internalization and may lead to challenging data interpretation. Over the last decades, novel approaches to enable faster integration of synthetic systems were developed, which could surpass the variability of transient transfection but also the time-consuming workflows of traditional cell-line genome engineering. One such example is presented in Chap. 10 where authors provide a detailed protocol describing how to create and utilize landing pad cells for large-scale, library-based genetic experiments. The experimental template can be easily adapted to explore libraries of proteins or protein variants within mammalian cells and allow fast and high-throughput construct analysis in these systems. Both transiently and stably engineered mammalian cells are often engineered with constructs bearing fluorescent proteins as easy-to-assessed reporters of construct functionality. Construct performance is usually assessed by flow cytometry a few days after transfection has been performed, in the case of transient transfection, or when a desired condition is reached, in the case of stable systems. Synthetic biology approaches have been developed to aid in quantitative and reproducible flow cytometry data analysis. Chapter 11 focuses on methods developed in the context of transient transfections in mammalian cells and focuses on the process controls necessary for effective calibration and the analysis of flow cytometry datasets. The chapter shows how these methods can be adopted to develop, model, refine, and predict the behavior of a variety of genetic devices. Moving from these foundational tools, both experimental and computational, available for mammalian system design, our collection then focuses on the plethora of synthetic systems available as of today, from gene expression regulation to the engineering of cell-to-cell communication and biosensing. For regulatory systems, we showcase examples of CRISPR-based control in Chaps. 12, 13 and 14, biotherapeutics in Chap. 15, and biosensors for the detection of key biomolecular properties and biomarkers in Chaps. 16, 17 and 18. Finally, we close the overview of mammalian synthetic systems by featuring two approaches relevant in the application of mammalian synthetic biology toward application, specifically considering how synthetic biology can be effective in supporting cell fate engineering and the control of differentiation in Chap. 19 and the design of chromobodies in Chap. 20, i.e., the fusion of nanobodies and fluorescent proteins, for characterization and detection of binder/antigen interactions, pivotal for research and therapeutic applications. While we appreciate the collection presented here is only covering a small portion of what is currently available, we hope it will serve as a useful dive into the area of mammalian synthetic systems and how they can be designed and engineered for transforming our approach to mammalian cell research and applications.

London, UK *Francesca Ceroni*
London, UK *Karen Polizzi*

Contents

Contributors

ROCHELLE AW • *Department of Chemical Engineering and Imperial College Centre for Synthetic Biology, Imperial College London, London, UK; Department of Bioengineering, Stanford University, Stanford, CA, USA*

UGNE BARONAITE • *Centre for Engineering Biology, School of Biological Sciences, The University of Edinburgh, Edinburgh, UK; Institute for Quantitative Biology, Biochemistry and Biotechnology, School of Biological Sciences, The University of Edinburgh, Edinburgh, UK*

JACOB BEAL • *Raytheon BBN Technologies, Cambridge, MA, USA*

EMILY BENNETT • *Department of Life Sciences, Imperial College London, London, UK*

JAMES W. BRYSON • *Biotech Research and Innovation Centre, Faculty of Health and Medical Sciences, University of Copenhagen, Copenhagen, Denmark*

ELISE CACHAT • *Centre for Engineering Biology, School of Biological Sciences, The University of Edinburgh, Edinburgh, UK; Institute for Quantitative Biology, Biochemistry and Biotechnology, School of Biological Sciences, The University of Edinburgh, Edinburgh, UK*

FABIANA CALANDRA • *Synthetic and Systems Biology lab for Biomedicine, Istituto Italiano di Tecnologia-IIT, Naples, Italy*

FRANCESCA CERONI • *Department of Chemical Engineering and Imperial College Centre for Synthetic Biology, Imperial College London, London, UK*

YIN HOON CHEW • *School of Mathematics, University of Birmingham, Birmingham, UK*

TARA L. DEANS • *Department of Biomedical Engineering, University of Utah, Salt Lake City, UT, USA*

ROBERTO DI BLASI • *Department of Chemical Engineering and Imperial College Centre for Synthetic Biology, Imperial College London, London, UK*

TINA FINK • *National Institute of Chemistry, Department of Synthetic Biology and Immunology, Ljubljana, Slovenia*

TIMOTHY FREI • *Department of Biosystems Science and Engineering, Eidgenössische Technische Hochschule (ETH) Zürich, Basel, Switzerland*

MARTIN FUSSENEGGER • *Department of Biosystems Science and Engineering, ETH Zurich, Basel, Switzerland; Faculty of Science, University of Basel, Basel, Switzerland*

CAROLIN MAJA GEBHARDT • *Centre for Synthetic Biology, Department of Biology, Technical University Darmstadt, Darmstadt, Germany*

MOE HIROSAWA • *Department of Life Science Frontiers, Center for iPS Cell Research and Application (CiRA), Kyoto University, Kyoto, Japan*

MARK ISALAN • *Department of Life Sciences, Imperial College London, London, UK*

ROMAN JERALA • *National Institute of Chemistry, Department of Synthetic Biology and Immunology, Ljubljana, Slovenia; EN-FIST Centre of Excellence, Ljubljana, Slovenia*

NISHA D. KAMATH • *Department of Pathology, Case Western Reserve University School of Medicine, Cleveland, OH, USA*

MUSTAFA KHAMMASH • *Department of Biosystems Science and Engineering, Eidgenössische Technische Hochschule (ETH) Zürich, Basel, Switzerland*

CLEO KONTORAVDI • *Department of Chemical Engineering, Imperial College London, London, UK; Department of Chemical Engineering and Imperial College Centre for Synthetic Biology, Imperial College London, London, UK*

MITCHELL R. LEWIS • *Department of Biomedical Engineering, University of Utah, Salt Lake City, UT, USA*
HA EUN LIM • *Department of Bioengineering, Rice University, Houston, TX, USA*
ALLEN P. LIU • *Department of Mechanical Engineering, University of Michigan, Ann Arbor, MI, USA; Department of Biomedical Engineering, University of Michigan, Ann Arbor, MI, USA; Cellular and Molecular Biology Program, University of Michigan, Ann Arbor, MI, USA; Department of Biophysics, University of Michigan, Ann Arbor, MI, USA*
CARLOS D. LLANOS • *Systems, Synthetic, and Physical Biology, Rice University, Houston, TX, USA*
MAYSAM MANSOURI • *Department of Biosystems Science and Engineering, ETH Zurich, Basel, Switzerland*
LUCIA MARUCCI • *Department of Engineering Mathematics, University of Bristol, Bristol, UK; School of Cellular and Molecular Medicine, University of Bristol, Bristol, UK*
KENNETH A. MATREYEK • *Department of Pathology, Case Western Reserve University School of Medicine, Cleveland, OH, USA*
HOSSEIN MOGHIMIANAVVAL • *Department of Mechanical Engineering, University of Michigan, Ann Arbor, MI, USA*
SONISILPA MOHAPATRA • *Department of Biophysics and Biophysical Chemistry, Johns Hopkins University, Baltimore, MD, USA*
JARED D. MOON • *Department of Psychiatry, Columbia University, New York, NY, USA; Research Foundation for Mental Hygiene, Menands, NY, USA; New York State Psychiatric Institute, New York, NY, USA*
JAMES MORRISSEY • *Department of Chemical Engineering, Imperial College London, London, UK*
IGNACIO MOYA-RAMÍREZ • *Department of Chemical Engineering and Imperial College Centre for Synthetic Biology, Imperial College London, London, UK; Departamento de Ingeniería Química, Universidad de Granada, Granada, Spain*
LINDSEY NASSIMOS • *Department of Chemistry, Binghamton University, Binghamton, NY, USA*
DOMINIK NIOPEK • *Institute of Pharmacy and Molecular Biotechnology (IPMB), Faculty of Engineering Sciences, Heidelberg University, Heidelberg, Germany*
KAREN POLIZZI • *Department of Chemical Engineering and Imperial College Centre for Synthetic Biology, Imperial College London, London, UK*
PRATIMA POUDEL • *Department of Chemistry, Chemical and Biomedical Engineering, Tagliatela College of Engineering, University of New Haven, West Haven, CT, USA*
LIN QI • *Department of Life Sciences, Imperial College London, London, UK*
CHILEAB REDWOOD-SAWYERR • *Department of Chemical Engineering and Imperial College Centre for Synthetic Biology, Imperial College London, London, UK*
ERIK RIHTAR • *National Institute of Chemistry, Department of Synthetic Biology and Immunology, Ljubljana, Slovenia*
SUSAN J. ROSSER • *Department of Quantitative Biology, Biochemistry and Biotechnology, University of Edinburgh, Edinburgh, UK; Centre for Synthetic and Systems Biology and UK Centre for Mammalian Synthetic Biology, School of Biological Sciences, University of Edinburgh, Edinburgh, UK*
HIROHIDE SAITO • *Department of Life Science Frontiers, Center for iPS Cell Research and Application (CiRA), Kyoto University, Kyoto, Japan*
LAURA SEGATORI • *Systems, Synthetic, and Physical Biology, Rice University, Houston, TX, USA; Department of Bioengineering, Rice University, Houston, TX, USA; Department of*

Chemical and Biochemical Engineering, Rice University, Houston, TX, USA; Department of Biosciences, Rice University, Houston, TX, USA

VELIA SICILIANO • *Synthetic and Systems Biology lab for Biomedicine, Istituto Italiano di Tecnologia-IIT, Naples, Italy*

BENJAMIN STRAIN • *Department of Chemical Engineering, Imperial College London, London, UK*

SHUE WANG • *Department of Chemistry, Chemical and Biomedical Engineering, Tagliatela College of Engineering, University of New Haven, West Haven, CT, USA*

ERNST WEBER • *Molecular Design & Engineering, Biologics Research, Bayer AG, Wuppertal, Germany*

JIAHUI WU • *Department of Chemistry, Binghamton University, Binghamton, NY, USA; Department of Chemistry, University of Massachusetts, Amherst, MA, USA*

TIANYI XIE • *Department of Bioengineering, Rice University, Houston, TX, USA*

KEVIN YUSKO • *Department of Chemistry, Binghamton University, Binghamton, NY, USA*

YUWEN ZHAO • *Department of Chemistry, Chemical and Biomedical Engineering, Tagliatela College of Engineering, University of New Haven, West Haven, CT, USA; Department of Bioengineering, Lehigh University, Bethlehem, PA, USA*

Chapter 1

A Directed Evolution Protocol for Engineering Minimal Transcription Factors, Based on CIS Display

Lin Qi, Emily Bennett, and Mark Isalan

Abstract

Directed evolution is an efficient strategy for obtaining desired biomolecules. Since the 1990s, the emergence of display techniques has enabled high-throughput screening of functional proteins. However, classical methods require library construction by plasmid cloning and are limited by transformation efficiencies, typically limiting library sizes to ~10^6–10^7 variants. More recently, in vitro techniques have emerged that avoid cloning, allowing library sizes of >10^{12} members. One of these, CIS display, is a DNA-based display technique which allows high-throughput selection of biomolecules in vitro. CIS display creates the genotype–phenotype link required for selection by a DNA replication initiator protein, RepA, that binds exclusively to the template from which it has been expressed. This method has been successfully used to evolve new protein–protein interactions but has not been used before to select DNA-binding proteins, which are major components in mammalian synthetic biology. In this chapter, we describe a directed evolution method using CIS display to efficiently select functional DNA-binding proteins from pools of nonbinding proteins. The method is illustrated by enriching the minimal transcription factor Cro from a low starting frequency (1 in 10^9). This protocol is also applicable to engineering other DNA-binding proteins or transcription factors from combinatorial libraries.

Key words DNA-binding proteins, High-throughput selection, Transcription factors (TF), Directed evolution, Protein engineering, CIS display

1 Introduction

Transcription factors (TF) are crucial in the regulation of gene expression. Typically, transcription factors regulate gene expression by binding to a specific DNA sequence, either alone or with other proteins as a complex [1]. The expressed gene can be activated or repressed, depending on whether transcription factors promote or block the recruitment of cellular RNA polymerases [2, 3]. To achieve robust regulation, transcription factors normally need strong sequence-specific binding affinities with their target DNA sequences (typically nanomolar dissociation constants). Directed evolution and selection technologies are a powerful way to improve

Francesca Ceroni and Karen Polizzi (eds.), *Mammalian Synthetic Systems*, Methods in Molecular Biology, vol. 2774,
https://doi.org/10.1007/978-1-0716-3718-0_1,

TF affinities and alter target recognition sequences and have hence been used in many biotechnological applications, including the engineering of mammalian transcription factors [4]. These have the potential to be used in a variety of synthetic gene regulation applications, including human gene therapy [5–8].

Directed evolution simulates the natural selection process: a protein with better function can be screened from a mutant combinatorial library through iterative rounds of selection. High-throughput screening methods such as phage display [9] have been widely applied in recent years to obtain proteins with new structures and functions. Typically, display techniques require an evolvable protein with a specific binding target: biomolecules that have strong binding affinity with their targets can be recovered, whereas biomolecules with weak binding affinity can be removed or washed away. Desired biomolecules can therefore be efficiently enriched using rounds of selection. For instance, phage display can create a genotype and phenotype link between a gene variant and the protein it codes for, by inserting foreign DNA fragments into a phage coat protein gene [9]. Phage display has now successfully been used in the selection of numerous biomolecules, including antibody fragments and protease substrates [10, 11].

Another example of a display technology is yeast display, which has advantages in supporting eukaryotic posttranslational modifications [12]. Yeast display has been used in the selection of human antibodies [13].

Critically, the library sizes of cell-dependent display techniques, such as phage and yeast display, are limited by plasmid transformation efficiency (typically resulting in ~10^6 to 10^7 variants per library). Importantly, a larger library has a higher chance to isolate biomolecules with higher affinity and so restricted library sizes are a problem [14]. Because of this issue, other techniques without transformation efficiency limitations were proposed. For example, mRNA display, one of the most prevalent cell-free display techniques, uses 3′ puromycin to create a genotype and phenotype link [15]. Consequently, the resulting larger libraries (~10^{12}) can theoretically isolate biomolecules such as protein-binding peptides with high binding affinities (binding at a low nanomolar levels) [16]. Similarly, another in vitro high-throughput screening technique, ribosome display [17], which creates the genotype and phenotype link by preventing the dissociation of ribosome–mRNA–protein complex, has also been used to engineer functional proteins, such as designed ankyrin repeat proteins (DARPins) [18]. Although powerful, these techniques are quite fragile and carry the risk of losing genes of interest because of degradation of the RNA fragments in selection complexes. In this chapter, we therefore use an alternative DNA-based cell-free display technique, CIS display [19], which overcomes the limitations of other display techniques to engineer proteins of interest.

CIS display is a DNA-based high-throughput selection technique. The genotype–phenotype link is generated by a DNA replication initiator protein, RepA. Due to the closest *ori* sequence being in *cis* on the same DNA template, a faithful link is created between the template and expressed polypeptide—the same link that is required for all display techniques [19].

CIS display has been used for many applications, including selecting 18-mer peptides to bind antibodies and lysozyme [19], peptides to inhibit proteases [20], and peptides to inhibit human vascular endothelial growth factor receptor isoform 2 (VEGFR-2) [21]. To date, all these applications of CIS display have been for protein–protein interactions. To complement this technique, we have developed protocols to enable CIS display for protein–DNA interactions, to enable transcription factor engineering.

Here, we illustrate the method by demonstrating that it can select bacteriophage lambda Cro out of a large excess of non-DNA-binding protein. Cro is a minimal 66-amino acid transcription factor [22] which can bind to its target DNA with high binding affinity (~0.1 nM) [23]. Through this method, we show that Cro can be enriched from a billion-fold excess of non-DNA-binding proteins such as GFP, with ~60-fold of enrichment per round. By substituting Cro with a mutagenized library of a DNA-binding protein of interest, this method can be used to establish a CIS display system to engineer other DNA-binding proteins to bind to desired target DNA sequences (Fig. 1).

2 Materials

2.1 DNA Construction and Preparation of Biotinylated Target DNA

1. *Ptac-Cro-RepA-CIS-ori* and *Ptac-GFP-RepA-CIS-ori* sequence obtained by gene synthesis (GenScript; Table 1; *see* **Note 1**).
2. CIS display construct amplification primer (Table 2).
3. Gibson Assembly® Master Mix (NEB). Store at −20 °C.
4. KOD hot-start polymerase (Sigma-Aldrich). Store at −20 °C.
5. 25 mM $MgSO_4$ stock solution.
6. 2 mM dNTP stock solution.
7. 10 × KOD hot-start polymerase buffer.
8. Monarch® DNA gel extraction kit (NEB).
9. Agarose.
10. Cro target and negative control target (Table 3).
11. Annealing buffer: 10 mM Tris–HCl, 1 mM EDTA, 0.1 M NaCl, pH 8.0.
12. Nuclease-free water.
13. Q5® High-Fidelity 2× Master Mix (NEB). Store at −20 °C.

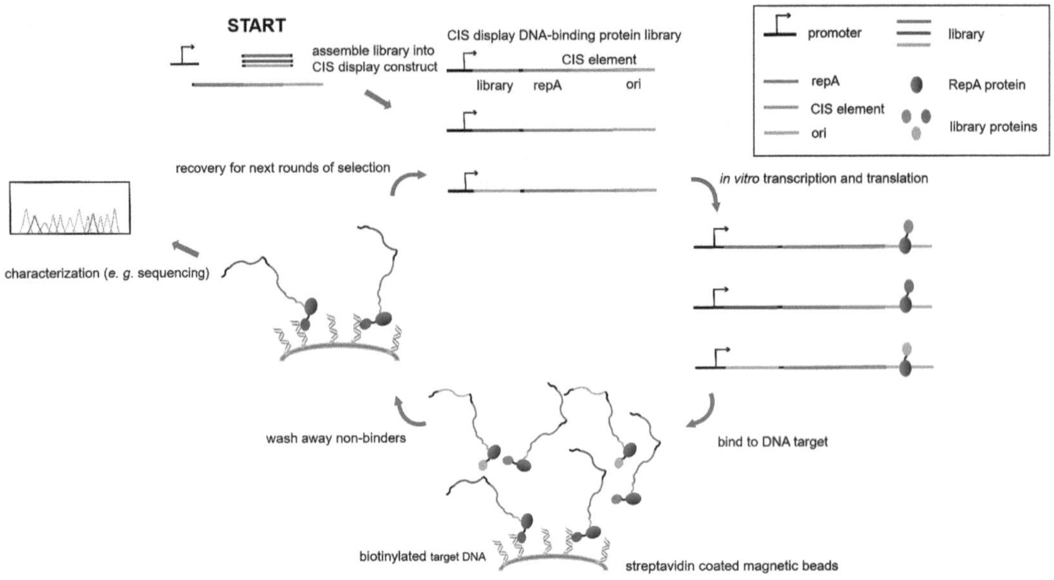

Fig. 1 Schematic figure of DNA-binding protein engineering through CIS display. A combinatorially randomized library of a DNA-binding protein of interest [24, 25] is introduced upstream of RepA, typically using gene synthesis, ligation, and/or PCR. The linear DNA fragment library is expressed in vitro in a transcription–translation reaction. A faithful link between each variant protein and its DNA template is generated by RepA having the property of binding in *cis* to the DNA molecule from which it was expressed. DNA-binding protein variants in the library that have high binding affinity with a user-provided biotinylated target DNA can be recovered with streptavidin-coated magnetic beads. Conversely, proteins with low binding affinity can be washed away with an appropriate buffer. After a suitable number of cycles or rounds of enrichment, the selected proteins can finally be recovered by PCR of their associated genes (orange). Candidates can subsequently be characterized by sequencing and the proteins verified with an appropriate assay, such as ELISA or EMSA

14. QIAquick PCR purification kit (QIAGEN).
15. 1 × Tris acetate–EDTA buffer.
16. SYBR™ Safe DNA Gel Stain (Invitrogen).
17. Purple DNA gel loading dye (NEB).

2.2 Preparation of Streptavidin Magnetic Beads

1. Dynabeads™ M-280 Streptavidin (Invitrogen) or equivalent (*see* **Note 2**).
2. Phosphate-buffered saline (PBS), pH 7.4.
3. Blocking buffer: 2% (w/v) of bovine serum albumin (BSA), 0.1 mg/mL herring sperm DNA in PBS.

2.3 In Vitro Transcription and Translation

1. *E. coli* S30 extract for linear templates kit (Promega). Store at −80 °C (*see* **Note 3**).
2. Nuclease-free water.

Table 1
Sequences of *Ptac-Cro-RepA-CIS-ori* and *Ptac-GFP-RepA-CIS-ori*

Sequence name	Sequence (5′ to 3′))	Annotation
Upstream sequence	cccagtcacgacgttgtaaaacgacggccagtgaattcgagctcggtacctcgcgaatgcatctagat ccccatcccc	For PCR recovery of CIS display construct
Ptac	ctgttgacaattaatcatggctcgtataatgtgtggaattgtgagcggataacaatttcacac	Promoter
Ribosome binding site	aggaaa	Allows ribosome binding
5′ untranslated region	aggaaacaggatctacc	
Cro	atggaacaacgcataaccctgaaagattatgcaatgcgctttgggcaaaccaagacagctaaagatc tcggcgtatatcaaag cgcgatcaacaaggccattcatgcaggccgaaagatttttttaactataaacgctgatggaagcgt ttatgcggaagaggtaaagcccttcccgagtaacaaaaaaacaacagca	Transcription factor: mediates specific binding with target DNA sequence
GFP	atgcgtaaaggagaagaacttttcactggagttgtcccaattcttgttgaattagatggtgatgtt aatgggcacaaattttctgtcagtggagagggtgaaggtgatgcaacatacgga aaacttacccttaaatttatttgcactactggaaaactacctgttccatggccaacacttgtcactactttcggt tatggtgttcaatgctttgcgagatacccagatcatatgaaacagcatgactttttcaagagtgccatgcc cgaaggttatgtacaggaaagaactatatttttcaaagatgacgggaactacaagacacgtgctgaa gtcaagtttgaaggtgatacccttgttaatagaatcgagttaaaaggtattgattttaaagaagatggaaa cattcttggacacaaattggaatacaactataactcacacaatgtatacatcatggcag acaaacaaaagaatggaatcaaagttaacttcaaaattagacacaacattgaagatggaa gcgttcaactagcagaccattatcaacaaaatactccaattggcgatggccctgtccttttacc agacaaccattacctgtccacacaatctgccctttcgaaagatcccaacgaaaagagagaccaca tggtccttcttgagtttgtaacagctgctgggattacacatggcatggatgaactatacaaa	Non-DNA-binding protein: negative control for nonspecific interactions
linker	gggggaggaggatcagcggcccca	Flexible GGGGS linker

(continued)

Table 1
(continued)

Sequence name	Sequence (5′ to 3′))	Annotation
RepA	actgatcttcaccaaacgtattaccgccaggtaaagaaccc gaatccggtgttcactccccgtgaaggtgccggaacgccgaagttccgcgaaaaaccgatg gaaaaggcggtgggcctcacctcccgttttgatttcgccattcatgtggcgcatgcccgttcccgtggt ctgcgtcggcgcatgccaccggtgctgcgtcgacgggctattgatgcgctgctgcaggggctgtg tttccactatgacccgctggccaaccgcgtccagtgttccatcaccacactggccattgagtgcgga ctggcgacagagtccggtgcaggaaaactctccatcacccgtgccacccgggccctgacgttcct gtcagagctgggactgattacctaccagacggaatatgacccgcttatcgggtgcta cattccgaccgacatcacgttcacactggctctgtttgctgcccttgatgtgtctgaggatg cagtggcagctgcgcgccgcagtcgtgttgaatgggaaaacaaacagcgcaaaaag caggggctggataccctgggtatggatgagctgatagcgaaagcctggcgttttgtgcgtga gcgtttccgcagttaccagacagagcttcagtcccgtggaataaaacgtgcccgtgcgcgtcgtg atgcgaacagagaacgtcaggatatcgtcaccctagtgaaacggcagctgacgcgtgaaatctcggaaggacgcttcac tgctaatggtgaggcggtaaaacgcgaagtggagcgtcgtgtgaaggagcgcatgattctgtcacg taaccgcaattacagccggctggccacagcttctccctga	Binding to *ori* sequence: provides genotype and phenotype linkage
CIS	aagtgatctcctcagaataatccggcctgcgccggagg catccgcacgcctgaagcccgccggtgcacaaaaaaacagcgtcgcatgcaaaaaacaatct catcatccaccttctggagcatccgattccccctgtttttaatacaaaatacgcctcagcgacggggaattt	Allows transient interaction between DNA and expressed peptide
ori	tgcttatccacatttaactgcaagggacttccccataaggttac aaccgttcatgtcataaagcgccagccgccagtcttacagggtgcaatgtatcttttaaaca cctgtttatatctcctttaaactacttaattacattcatttaaaaagaaaacctattcactgcctgt cctgtggacagacagatatgcacctcccaccgcaagcggcgggccccgaccggagccacttta gttacaacacacaaaaacaacctccagaaaaaccccggtccagcgcagaa ccgaaaccacaaagccc	Allows expressed RepA fusion protein to bind; creates genotype and phenotype linkage
Downstream sequence	atcggatcccgggcccgtcgactgcagaggcctgcatg caagcttggcgtaatcatggtcatagctgtttcctg	For PCR recovery of CIS display construct

Table 2
Primers used for PCR recovery of CIS display constructs

Sequence name	Sequence (5′ to 3′))	Annotation
FAR-1-F	cccagtcacgacgttgtaaaacg	Round 1 recovery
FAR-2-F	acggccagtgaattcgagctc	Round 2 recovery
TAC-1-F	ccccatccccctgttgacaattaatc	Round 3 recovery
TAC-2-F	cccctgttgacaattaatcatggc	Round 4 recovery
TAC-N-F	gtgtggaattgtgagcggataac	Round 5 recovery
RepA-N-R	atcggatgctccagaaggtg	Round 5 recovery
ORI-2-R	tgcatatctgtctgtccacagg	Round 4 recovery
ORI-1-R	gggctttgtggtttcggttc	Round 3 recovery
FAR-2-R	ccaagcttgcatgcaggcc	Round 2 recovery
FAR-1-R	caggaaacagctatgaccatgattacg	Round 1 recovery

Table 3
Primers used for target DNA preparation

Sequence name	Sequence (5′ to 3′))	Annotation
Cro target Forward	GCAACCATTATCACCGCCGGTGATAAAATAG TCAACACCGGCGGTGATAGATATTTCACAGTCAG TCCACACGTC	To make Cro target
Cro target Reverse	GACGTGTGGACTGACTGTGAAATATCTA TCACCGCCGGTGTTGACTATTTTATCACCGGCGG TGATAATGGTTGC	To make Cro target
Negative control Forward	GCAACCATTATAAATAGTGGTGATAAAATAGTTA TCACCACTATTTATAGATATTTCACAGTCAG TCCACACGTC	To make negative control target
Negative control Reverse	GACGTGTGGACTGACTGTGAAATATCTATAAATAG TGGTGATAACTATTTTATCACCACTATTTATAA TGGTTGC	To make negative control target
Cro target amplification Forward	gtgtggaattgtgagcggataac	To make biotinylated Cro target
Negative control amplification Forward	GCAACCATTATAAATAGTGGTGAT	To make biotinylated negative control target
Target amplification Reverse	Biotin-GACGTGTGGACTGACTGTGA	To make biotinylated Cro/negative control target

2.4 Affinity Selection

1. Washing buffer: 0.1–1% of Tween 20 in PBS.
2. Magnets.
3. ThermoPol® Reaction Buffer (NEB).
4. Rotary mixer.

3 Methods

3.1 Preparation of DNA Templates and Target DNA

1. Amplify the CIS display constructs (*Ptac-Cro-RepA-CIS-ori* and *Ptac-GFP-RepA-CIS-ori*) using primers (Table 2) with the following PCR mixture: 3 μL 10 μM each primer, 4 μL of 25 mM $MgSO_4$, 5 μL of 2 mM each dNTP, 5 μL of 10 × buffer, 1 ng of template, 1U KOD hot-start DNA polymerase, and nuclease-free water to a final volume of 50 μL. Set up the following PCR protocol: initial denaturation at 95 °C for 2 min, followed by 25–35 cycles at 95 °C, 20 s; at 65 °C, 30 s; at 70 °C, 50 s, followed by a final extension at 70 °C for 2 min (*see* **Note 4**).
2. Analyze the results by running on a 1% agarose gel and purify the CIS display DNA templates by gel purification with a Monarch DNA gel extraction kit (*see* **Note 5**).
3. Anneal Cro target and negative control target with the following PCR mixtures: 5 μL of 100 μM each primer and 40 μL of annealing buffer. Set up the following PCR protocol: one cycle at 95 °C for 5 min, cool down slowly to 50 °C (−1 °C/cycle), 1 min per cycle.
4. Dilute annealed DNA to appropriate concentration and amplify with the biotinylated primer with the following PCR mixture: 5 μL of 10 μM each primer, 25 μL of Q5 High-Fidelity 2× Master Mix (NEB), 0.1 ng of template, and nuclease-free water to a final volume of 50 μL. Set up the following PCR protocol: initial denaturation at 98 °C for 30 s, followed by 30 cycles at 98 °C, 10 s; at 65 °C, 12 s; at 72 °C, 5 s, followed by a final extension at 72 °C for 10 s.
5. Analyze the results by running on a 2.5% agarose gel. Purify the biotinylated target DNA with a QIAquick PCR purification kit (*see* **Note 6**).

3.2 In Vitro Transcription and Translation

1. Dilute *Ptac-Cro-RepA-CIS-ori* DNA with nuclease-free water to reach a final molar ratio of $1{:}10^9$ to *Ptac-GFP-RepA-CIS-ori*. Add 3–4 μg of mixed DNA templates into *E. coli* S30 extract to a total reaction volume of 50 μL. Incubate the in vitro TnT (transcription and translation) reaction at 37 °C for 90 min (*see* **Note 7**).
2. Stop the TnT reaction by cooling it on ice for 10 min.

3.3 Affinity Selection

1. Wash 25 μL M-280 streptavidin-coated magnetic beads in 200 μL PBS for by pipetting. Place the mixture on a strong magnet for 2 minutes, allowing the beads to become attracted to the magnet. Repeat this step once. Block the beads with 200 μL of blocking buffer for an hour at room temperature on a rotary mixer (for gentle suspension).
2. Dilute the TnT reaction with blocking buffer and incubate for 10 min at room temperature on a rotary mixer.
3. Incubate the prepared biotinylated target DNA with diluted TnT reaction for an hour at room temperature on a rotary mixer, to allow protein–target–DNA interactions to occur.
4. Incubate blocked beads with TnT reaction mix from **step 3**, for 15 min at room temperature, on a rotary mixer.
5. Wash beads 4–12 times with 0.1% Tween 20 in PBS, as previously described in **step 1**, followed by two washes with PBS (*see* **Note 8**).
6. Resuspend the beads with 50 μL of ThermoPol® Reaction Buffer (NEB) after washing (elution of DNA from beads is optional). Recover the enriched DNA sequence, coding for the desired DNA-binding protein, with the following PCR mixture: 3 μL 10 μM each primer, 4 μL of 25 mM $MgSO_4$, 5 μL of 2 mM each dNTP, 5 μL of 10 × buffer, 1–5 μL of templates (beads), 1U KOD hot-start DNA polymerase, and nuclease-free water to a final volume of 50 μL. Set up the following PCR protocol: initial denaturation at 95 °C for 2 min, followed by 25–35 cycles at 95 °C, 20 s; at 65 °C, 30 s; at 70 °C, 50 s, followed by a final extension at 70 °C for 2 min (*see* **Note 9**).
7. Analyze the results by running on a 1% agarose gel (Fig. 2). Purify the selection results with a QIAquick PCR purification kit (*see* **Note 10**); purified PCR products can be used in the next round of selection or characterization (Table 4).

4 Notes

1. To demonstrate enrichment of a desired DNA-binding protein (bacteriophage lambda Cro) from low starting frequencies of DNA-binding proteins (as would be found in a combinatorial library), any non-DNA-binding "dummy" proteins (e.g., GFP, RFP) can be mixed with Cro. The "dummy" protein CIS display construct can be obtained by either gene synthesis or cloning (e.g., Gibson assembly).
2. Here we recommend streptavidin-coated magnetic beads and strong magnets for efficient washing and selection. Other streptavidin-coated systems such as Pierce™ streptavidin-coated high-capacity plates (Thermo Scientific) can also be applied.

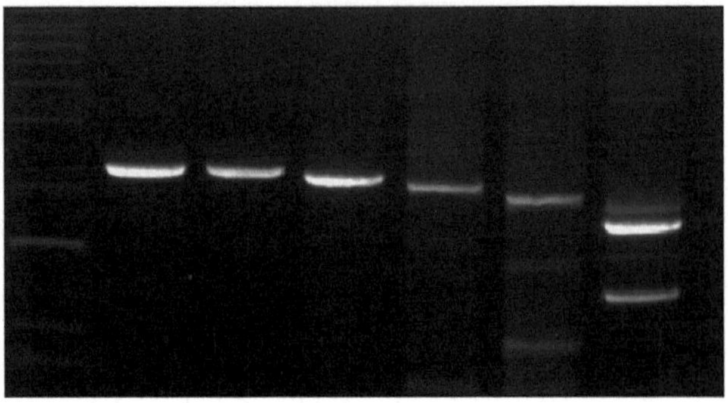

Fig. 2 Typical results of CIS display of DNA-binding proteins. The enrichment of Cro from a billion-fold excess of a non-DNA-binding protein, GFP (1:10^9 dilution of Cro/GFP). A Cro band (1254 bp; shown by arrow) demonstrated enrichment after five rounds of selection on this agarose DNA gel. The Cro band was gel purified and verified by DNA sequencing (Eurofins). After each round of selection, internal nested primers were applied to recover the full CIS constructs and to avoid generating PCR artifacts from primer reuse. Therefore, both GFP and Cro bands shorten in size from round to round: GFP (round 0, 2300 bp; round 1, 2300 bp; round 2, 2250 bp; round 3, 2158 bp; round 4, 2043 bp; round 5, 1770 bp) and Cro (round 0, 1784 bp; round1, 1784 bp; round 2, 1734 bp; round 3, 1642 bp; round 4, 1527 bp; round 5, 1254 bp)

Table 4
Cro enrichment conditions

Selection round	Streptavidin magnetic beads	Target DNA	Blocking buffer	Washing condition
Round 1	25 μL	20 pmol	2% BSA in PBS, 0.1 mg/mL herring sperm	4 washes with 0.1% Tween 20 in PBS, 2 washes with PBS
Round 2	20 μL	16 pmol	2% BSA in PBS, 0.1 mg/mL herring sperm	6 washes with 0.1% Tween 20 in PBS, 2 washes with PBS
Round 3	20 μL	12 pmol	2% BSA in PBS, 0.1 mg/mL herring sperm	8 washes with 0.1% Tween 20 in PBS, 2 washes with PBS
Round 4	15 μL	8 pmol	2% BSA in PBS, 0.1 mg/mL herring sperm	10 washes with 0.1% Tween 20 in PBS, 2 washes with PBS
Round 5	15 μL	4 pmol	2% BSA in PBS, 0.1 mg/mL herring sperm	12 washes with 0.1% Tween 20 in PBS, 2 washes with PBS

3. *E. coli* S30 extract for linear templates kits (Promega) can be used to initiate the in vitro transcription and translation reaction. Alternatively, S30 extract for linear templates can be prepared by *E. coli* SL119, which is a derivative of *E. coli* BL21 [26, 27]. *n.b. recD* genes are knocked out for preventing degradation of linear DNA fragments in *E. coli* SL119. For obtaining a desired expression level, supercoil-insensitive promoters such as *Ptac* are recommended [28].

4. Here, we aimed to enrich a low frequency of DNA-binding protein Cro from "dummy" protein GFP. We therefore did a serial dilution of Cro CIS display construct and mixed it with excess GFP CIS display construct such that a final molar ratio of $1:10^9$ was reached. If one wishes to use a combinatorial library of any chosen transcription factor, this step may be replaced with PCR amplification of a codon-diversified library (replacing Cro within this CIS display construct). To optimize the selection process more easily, we also recommend initiating model selections with a higher frequency of target-binding protein, for example, diluting one target-binding protein in a thousand-fold excess of "dummy" protein. This allows the target-binding protein to be enriched with fewer rounds of selection, enabling selection conditions to be identified more rapidly.
5. A standard TnT reaction has a reaction volume of 50 μL. Dilution of a TnT reaction mixture may make the reaction less efficient or even stop the reaction. Therefore, concentrated DNA templates are recommended for adequate protein expression. To obtain higher concentrations in DNA purification, using a smaller volume of pre-warmed elution buffer, or other yield-increasing steps, is highly recommended.
6. In this protocol, QIAquick PCR purification kits were used. For shorter DNA targets, other appropriate kits such as QIAquick nucleotide removal kits can be applied. 2.5% agarose gels were prepared because they showed a better resolution for DNA with low molecular weight. Also, negative control target DNA and beads without any biotinylated target should be used as controls; these control experiments verify that only target DNA can enrich the desired biomolecules.
7. One advantage of using negative control (non-DNA-binding) fluorescent proteins in an enrichment experiment is that the in vitro TnT reaction can be monitored by a fluorescence plate reader (e.g., Tecan). The expression time and conditions (e.g., temperature) can be varied according to observed expression.
8. Selection stringency can be tuned by many factors in CIS display, for instance, blocking buffer composition, Tween 20 concentration, number of washes, and amount of target DNA used. Generally, to enrich low frequencies of biomolecules, selections should start with low stringency, and the stringency can be gradually increased in the following selection rounds.

9. We note that it is essential to use nested primers, to provide higher-quality PCR products without amplification artifacts, over multiple selection rounds.
10. In this case, PCR purification is used because Cro and GFP have different molecular weights. The selection results should be determined by sequencing. According to specific applications, gel purification can also be applied.

References

1. Latchman DS (1997) Transcription factors: an overview. Int J Biochem Cell Biol 29(12): 1305–1312
2. Roeder RG (1996) The role of general initiation factors in transcription by RNA polymerase II. Trends Biochem Sci 21(9):327–335
3. Nikolov D, Burley S (1997) RNA polymerase II transcription initiation: a structural view. Proc Natl Acad Sci 94(1):15–22
4. Pabo CO, Peisach E, Grant RA (2001) Design and selection of novel Cys2His2 zinc finger proteins. Annu Rev Biochem 70(1):313–340
5. Garriga-Canut M, Agustín-Pavón C, Herrmann F, Sánchez A, Dierssen M, Fillat C et al (2012) Synthetic zinc finger repressors reduce mutant huntingtin expression in the brain of R6/2 mice. Proc Natl Acad Sci 109(45):E3136–E3E45
6. Agustín-Pavón C, Mielcarek M, Garriga-Canut M, Isalan M (2016) Deimmunization for gene therapy: host matching of synthetic zinc finger constructs enables long-term mutant Huntingtin repression in mice. Mol Neurodegener 11(1):1–16
7. Bailus BJ, Pyles B, McAlister MM, O'geen H, Lockwood SH, Adams AN et al (2016) Protein delivery of an artificial transcription factor restores widespread Ube3a expression in an Angelman syndrome mouse brain. Mol Ther 24(3):548–555
8. Pyles B, Bailus BJ, O'Geen H, Segal DJ (2018) Purified protein delivery to activate an epigenetically silenced allele in mouse brain. Epigenome Editing. Springer, p. 227–239
9. Smith GP (1985) Filamentous fusion phage: novel expression vectors that display cloned antigens on the virion surface. Science 228(4705):1315–1317
10. Hoogenboom HR (2005) Selecting and screening recombinant antibody libraries. Nat Biotechnol 23(9):1105–1116
11. Diamond SL (2007) Methods for mapping protease specificity. Curr Opin Chem Biol 11(1):46–51
12. Boder ET, Midelfort KS, Wittrup KD (2000) Directed evolution of antibody fragments with monovalent femtomolar antigen-binding affinity. Proc Natl Acad Sci 97(20):10701–10705
13. Chao G, Lau WL, Hackel BJ, Sazinsky SL, Lippow SM, Wittrup KD (2006) Isolating and engineering human antibodies using yeast surface display. Nat Protoc 1(2):755–768
14. Perelson AS, Oster GF (1979) Theoretical studies of clonal selection: minimal antibody repertoire size and reliability of self-non-self discrimination. J Theor Biol 81(4):645–670
15. Roberts RW, Szostak JW (1997) RNA-peptide fusions for the in vitro selection of peptides and proteins. Proc Natl Acad Sci 94(23): 12297–12302
16. Wilson DS, Keefe AD, Szostak JW (2001) The use of mRNA display to select high-affinity protein-binding peptides. Proc Natl Acad Sci 98(7):3750–3755
17. Hanes J, Plückthun A (1997) In vitro selection and evolution of functional proteins by using ribosome display. Proc Natl Acad Sci 94(10): 4937–4942
18. Binz HK, Amstutz P, Kohl A, Stumpp MT, Briand C, Forrer P et al (2004) High-affinity binders selected from designed ankyrin repeat protein libraries. Nat Biotechnol 22(5): 575–582
19. Odegrip R, Coomber D, Eldridge B, Hederer R, Kuhlman PA, Ullman C et al (2004) CIS display: in vitro selection of peptides from libraries of protein–DNA complexes. Proc Natl Acad Sci 101(9):2806–2810
20. Eldridge B, Cooley RN, Odegrip R, McGregor DP, FitzGerald KJ, Ullman CG (2009) An in vitro selection strategy for conferring protease resistance to ligand binding peptides. Protein Eng Des Sel 22(11):691–698
21. Patel S, Mathonet P, Jaulent AM, Ullman CG (2013) Selection of a high-affinity WW domain against the extracellular region of VEGF receptor isoform-2 from a combinatorial library using CIS display. Protein Eng Des Sel 26(4): 307–315

22. Takeda Y, Folkmanis A, Echols H (1977) Cro regulatory protein specified by bacteriophage lambda. Structure, DNA-binding, and repression of RNA synthesis. J Biol Chem 252(17): 6177–6183
23. Kim JG, Takeda Y, Matthews BW, Anderson WF (1987) Kinetic studies on Cro repressor-operator DNA interaction. J Mol Biol 196(1): 149–158
24. Isalan M, Choo Y (2001) Engineering nucleic acid-binding proteins by phage display. DNA-protein interactions. Springer, pp 417–429
25. Isalan M (2006) Construction of semi-randomized gene libraries with weighted oligonucleotide synthesis and PCR. Nat Protoc 1(1):468–475
26. Lesley SA (1995) Preparation and use of E. Coli S-30 extracts. In: Vitro transcription and translation protocols. Springer, pp 265–278
27. Lesley SA, Brow M, Burgess RR (1991) Use of in vitro protein synthesis from polymerase chain reaction-generated templates to study interaction of Escherichia coli transcription factors with core RNA polymerase and for epitope mapping of monoclonal antibodies. J Biol Chem 266(4):2632–2638
28. Chen H-Z, Zubay G (1983) Prokaryotic coupled transcription—translation. Methods in Enzymology. Elsevier, p. 674–690

Chapter 2

Setup and Applications of Modular Protein Expression Toolboxes (MoPET) for Mammalian Systems

Ernst Weber

Abstract

The design and generation of an optimal protein expression construct is the first and essential step in the characterization of any protein of interest. However, the exchange and modification of the coding and/or noncoding elements to analyze their effect on protein function or generating the optimal result can be a tedious and time-consuming process using standard molecular biology cloning methods. To streamline the process to generate defined expression constructs or libraries of otherwise difficult to express proteins, the *Mo*dular *P*rotein *E*xpression *T*oolbox (MoPET) has been developed (Weber E, PloS One 12(5):e0176314, 2017). The system applies Golden Gate cloning as an assembly method and follows the standardized modular cloning (MoClo) principle (Weber E, PloS One 6(2):e16765, 2011). This cloning platform allows highly efficient DNA assembly of pre-defined, standardized functional DNA modules effecting protein expression with a focus on minimizing the cloning burden in coding regions. The original MoPET system consists of 53 defined DNA modules divided into eight functional main classes and can be flexibly expanded dependent on the need of the experimenter and expression host. However, already with a limited set of only 53 modules, 792,000 different constructs can be rationally designed or used to generate combinatorial expression optimization libraries. We provide here a detailed protocol for the (1) design and generation of level 0 basic parts, (2) generation of defined expressions constructs, and (3) generation of combinatorial expression libraries.

Key words Golden Gate cloning, Modular cloning , Library construction, High-throughput cloning, Synthetic biology, Expression optimization

1 Introduction

Production of soluble and active recombinant protein in sufficient amounts is central for structural, functional, and biochemical protein characterization. However, the path to an optimal expression construct for a gene of interest is often a tedious trial and error process. Multiple variables can influence expression and function. The expression host, expression conditions, expression vector backbone, promoters, terminators, signal peptides, different N- and/or C-terminal protein tags, linkers, and combinations thereof must be

Francesca Ceroni and Karen Polizzi (eds.), *Mammalian Synthetic Systems*, Methods in Molecular Biology, vol. 2774,
https://doi.org/10.1007/978-1-0716-3718-0_2,

explored and tested to achieve optimal protein expression constructs tailored for an application of interest. Modifications and adaptations of a given expression construct, like exchanging a protein tag or replacing the promoter, generally require individual cloning strategies which are time-consuming and resource-intensive and may introduce unwanted additional sequences between the DNA fragments.

To allow a fast and streamlined design–built–test–learn cycle to identify the best DNA construct, two key engineering principles had to be applied to the field of molecular biology [3]. First the modularization and standardization of basic parts and second the highly efficient and standardized assembly process of multiple DNA fragments in a single reaction.

For highly efficient simultaneous DNA assembly of multiple fragments, the key methodologies include homology-based assembly methods and type IIS-based assembly methods relying on Golden Gate cloning, which permits directional and seamless assembly of multiple DNA fragments in a one-tube one-step reaction by the concurrent use of type IIS restriction endonucleases and DNA ligase [4, 5]. This is based on the special ability of type IIS enzymes to cleave outside their recognition site in a defined distance independently of the target sequence. When these recognition sites are placed to the far 5′ and 3′ end of any DNA fragment in inverse orientation, they are removed by the restriction enzyme in the cleavage process. Two DNA fragments flanked by compatible sequence overhangs resulting from type IIS cleavage, termed fusion sites, can be ligated seamlessly. Since type IIS sites can be designed to have different fusion site sequences, directional assembly of multiple fragments is feasible (Fig. 1). The method proved to be highly robust and efficient in DNA shuffling applications and for assembly of up to 52 DNA fragments and/or highly repetitive DNA sequences [6–9].

The unique feature of Golden Gate cloning, to require only a single amino acid for the fusion of two functional DNA parts, is the key differentiator to other widely used cloning technologies. Methods like Gibson assembly [10], SLIC [11], and related technologies reviewed by Casini et al. [12], which are also able to combine multiple DNA fragments with high efficiency, need longer identical sequence stretches between the single parts (15–30 nucleotides). The longer sequences have restrictions when used in modular toolboxes: (i) they have the potential to insert larger stretches of unwanted amino acids, (ii) highly repetitive sequences like GS linkers are problematic in efficient assembly, (iii) very short functional parts like tags are not possible as independent single entities, and (iv) a simple expansion by adding new functional parts may require major adjustments in the cloning system. This becomes even more prominent when multiple functional parts are assembled in a linear fashion together with multiple fusion sites in the coding region.

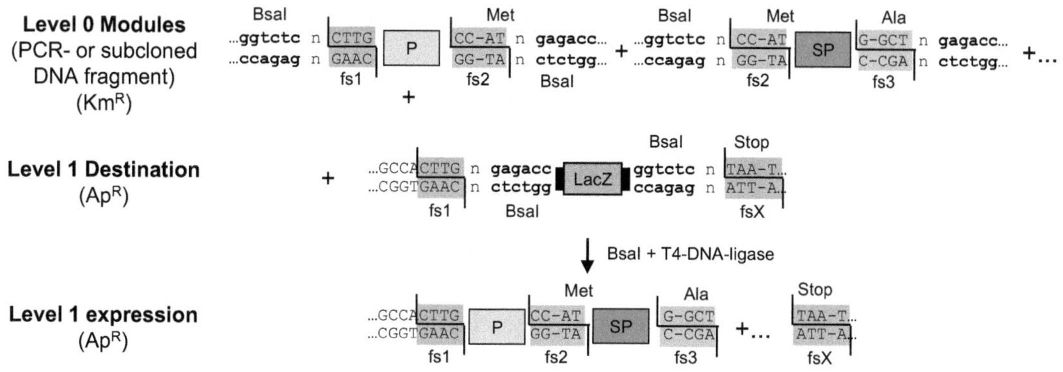

Fig. 1 Golden Gate assembly used in the MoPET system. A detailed overview of the organization and orientation of the type IIS restriction sites and the fusion sites at the different levels of the MoPET system is shown. Level 0 modules are flanked by BsaI recognition sites and module-specific fusion sites (fs) are highlighted with color. Promoter (P) and signal peptide (SP) are shown as an example. The level 0 promoter module and the other level 0 modules required to form a complete expression construct (not shown) are then assembled via *BsaI* into a level 1 destination vector, creating the final level 1 expression construct

Based on Golden Gate cloning as an assembly method, several standardized modular cloning systems have been set up ranging from the prototype modular cloning systems (MoClo) and GoldenBraid focusing on plants [2, 13–15] to a wide range of host organisms including bacteria [16, 17], yeasts [18, 19], chloroplasts [20], cyanobacteria [21], and mammalian cells [22, 23].

The here described modular protein expression toolbox (MoPET) was developed for standardized assembly of expression constructs and protein expression optimization libraries in mammalian expression hosts [1]. MoPET has a strong focus on the special requirements when working with coding regions and is designed to minimize cloning burden like additional amino acids between the functional parts which may influence expression and/or function of the protein of interest (POI).

2 Materials

2.1 Cloning

1. DNA analysis and editing software. Any suitable sequence analysis and editing software package can be used. Popular examples include Geneious Prime (www.geneious.com) and SnapGene (www.snapgene.com).
2. 10 U/μL *Bsa*I-HFv2 (NEB, New England Biolabs Inc., Ipswich, MA, USA).
3. 10 U/μL *Bpi*I (Thermo Fisher Scientific, Waltham, MA USA).
4. 20 U/μL T4 DNA ligase (HC) with 10× ligation buffer (300 mM Tris–HCl pH 7.8, 100 mM MgCl2, 100 mM

DTT, 10 mM ATP) (Promega Corporation, Madison, WI, USA).

5. 5 High-Fidelity PCR Kit, e.g., Novagen® KOD Polymerase PCR Systems.
6. QIAquick PCR Purification Kit.
7. Spectrophotometer for measuring DNA concentration, e.g., Nanodrop® Spectrophotometer ND-1000 (Peqlab, Erlangen, Germany).
8. Thermocycler (e.g., Bio-Rad) to perform PCR reactions and Golden Gate cloning reactions.
9. Luria–Bertani (LB) medium: 10 g tryptone, 5 g yeast extract, 10 g NaCl in 1 L water, autoclave. For agar plates, add 1.5% agar.
10. Antibiotics: kanamycin and ampicillin—filter-sterilized stock solutions of 50 mg/mL in H_2O stored at −20 °C and diluted 1:1000 for a final concentration of 50 μg/mL were added to a liquid or solid medium.
11. 5-Bromo-4-chloro-3-indolyl-β-D-galactopyranoside (X-gal): stock solution of 20 mg/mL in dimethylformamide. The stock is diluted 1:500 (final concentration: 40 μg/mL) in an appropriate amount of LB agar after autoclaving/melting and cooling down.
12. Electrocompetent cells, e.g., Top10 (Thermo Fisher Scientific, Waltham, MA USA).
13. Electroporator: Gene Pulser Electroporation System (Bio-Rad).
14. Eppendorf ThermoMixer®.

2.2 Screening for Correct Assembled Constructs

1. Plasmid DNA preparation: QIAprep Spin Miniprep Kit for single constructs or QIAprep 96 Turbo Miniprep Kit for 96 preparations (Qiagen, Hilden, Germany).
2. GeneRuler™ 1 kb DNA Ladder Plus.
3. 50× TAE buffer: 242.0 g Tris, 57.1 mL of acetic acid, and 100 mL of 0.5 M EDTA, pH 8.0, in 1 L of deionized water. Running buffer for agarose gels is 1× TAE.
4. Agarose solution (0.8–1.5%).
5. Ethidium bromide (stock concentration of 10 mg/mL).
6. MCE Membrane MF-Millipore, 0.025 μm.
7. 2.5 mL deep-well plate (HJ-Bioanalytics GmbH).
8. Gels are checked visually using a gel imager.

3 Methods

3.1 Conceptual Design and Consideration of MoPET

The key first step is the definition and standardization of the functional DNA parts. We defined eight basic module types which were named in accordance with the MoClo system syntax as level 0 modules [2] (*see* **Note 1**). They cover the main variables in expression construct design and optimization, such as promoter (P), signal peptide (SP), N-terminal tag (N-TAG), N-terminal linker (N-Li), the core of the protein to be expressed (CP), C-terminal linker (C-Li), and C-terminal tag (C-TAG) (Fig. 2a) and are flanked by *Bsa*I restriction sites in opposite directions. All are provided as Km^R storage vectors. The expression plasmid backbone itself is also regarded as a module, providing additional expression level determining functions like origin of replication and the 3′-UTR. To allow highly efficient cloning and selection, these backbones provide an Ap^R resistance and a *LacZ* cassette for blue/white screening.

To enable assembly by Golden Gate technology and to assure that all modules of a defined type are interchangeable with each other, we defined specific fusion sites flanking each module type (Fig. 2a). Fusion sites overlapping with coding sequences were chosen to minimize changes to the encoded proteins. The fusion site between promoter and the start of the protein was chosen to be CC-AT. The two cytosine residues represent the last two nucleotides of an optimal Kozak sequence for mammalian expression, which are followed by the two first nucleotides of a start AT-G. To allow creation of the original N-terminus of any given protein to express, the fusion site between the signal peptide and the N-terminal tag is located in the coding sequence of the signal peptide. As alanine is one of the consensus residues for the signal peptidase and present in many mammalian signal peptides at the C-terminus, the fusion site here reads G-GCT [24]. Definition of two linker positions (N-Linker and C-Linker) flanking the core protein allowed to design fusion sites that were located inside the linker coding region. As most linker sequences used in literature typically contain glycine, and a single additional glycine in front of the linkers is regarded as minor disturbance of the expression construct, all four glycine codons were used (GGA-T, C-GGG, GGT-G, A-GGC). As fusion site to connect the POI to the backbone, the TAA-T stop codon was chosen. The last remaining fusion site is in the non-translated region upstream of the promoter module and was selected with the only requirements as to be unique and non-palindromic to allow efficient assembly (CTTG).

A

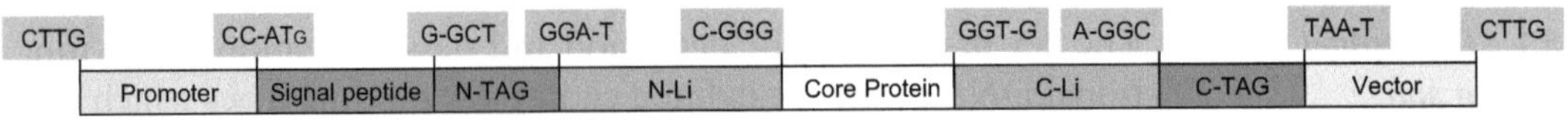

B

Function (Module Name)	5' extension for direct ordering	3' extension for direct ordering
Promoter (P)	GGTCTCACTTG	CC-ATAGAGACC
Signal peptide (SP)	GGTCTCTCC-**ATG** (**Met**)	XXG-**GCT**AGAGACC (X/**Ala**)
N-terminal tag (N-TAG)	GGTCTCTG-**GCT** (**Ala**)	**GGA**-TAGAGACC (**Gly**)
N-terminal linker (N-Li)	GGTCTCT**GGA-TCC** (**Gly/Ser**)	**TCC-GGG**AGAGACC (**Ser/Gly**)
Protein/domain in focus (CP)	GGTCTCTC-**GGG** (**Gly**)	**GGT**-GAGAGACC (**Gly**)
C-terminal linker (C-Li)	GGTCTCT**GGT-G**XX (**Gly**)	**TCA-GGC**AGAGACC (**Ser/Gly**)
C-terminal tag (C-TAG)	GGTCTCTA-**GGC** (**Gly**)	**TAA**-TAGAGACC (**Stop**)

C

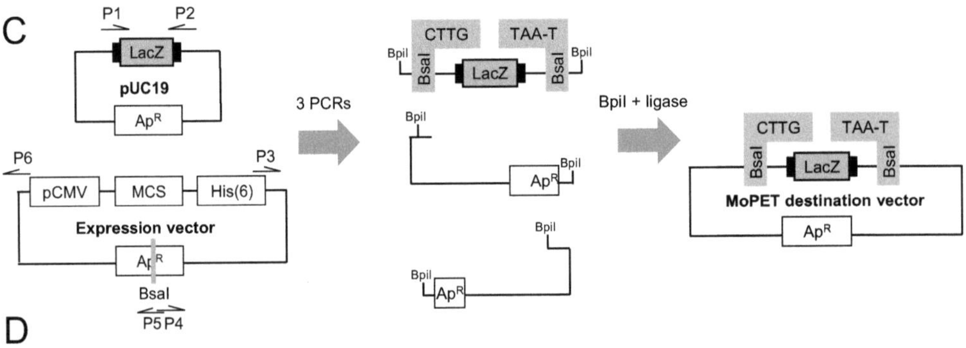

D

P1	ttt*gaagac*ttCTTGt**gagacc**gcagctggcacgacaggtttc
P2	ttt*gaagac*ttATTAt**gagacc**gtcacagcttgtctgtaagcg
P3	ttt*gaagac*ttTAATgcggccgctcgaggccggcaag
P4	ttt*gaagac*ttCGTGacccacgctcaccggctccag
P5	ttt*gaagac*ttCACGcggtatcattgcagcactggg
P6	ttt*gaagac*ttCAAGcggtcatattggacatgagcc

Fig. 2 MoPET module design overview. (**a**) Modular structure of the MoPET system consisting of the eight basic module types: promoter, signal peptide, N-TAG (N-terminal tag), N-Li (N-terminal linker), core protein, C-Li (C-terminal linker), C-TAG (C-terminal tag), and the vector. Boxes show the fusion sites separating the modules and indicating the reading frame. (**b**) Detailed overview of 5′ and 3′ standard DNA sequences containing the *Bsa*I restriction site and the respective fusion sites that must be added for direct ordering of level 0 modules. (**c**) Process of generating MoPET-compatible level 1 expression plasmids. Relevant functional elements are PCR amplified from both starting vectors. The PCR fragments are assembled by a Golden Gate cloning reaction using the *Bpi*I restriction sites added via the PCR primers. Unwanted parts like here the original promoter, MCS, and purification tag, as well additionally present *Bsa*I site in the Ap^R marker, are removed in the process. Ap^R, ampicillin resistance marker; His(6), histidine tag; *LacZ*, marker for blue/white

3.2 Design, Generation, and Domestication of Level 0 Functional Parts

With constantly dropping prices for gene synthesis and the fact that most coding parts like signal peptides, linkers, and tags are short and often cover only few amino acids, a straightforward and fast option is to design level 0 functional parts completely in silico. They can then be ordered at a gene synthesis provider already subcloned in a storage vector with the appropriate antibiotic resistance (kanamycin for MoPET level 0 functional parts).

1. Remove any internal *Bsa*I type IIS recognition site(s) from the fragment of interest as they will interfere with the Golden Gate cloning reaction. Dependent on later planned expansion to higher-order platforms, it may also be advisable to remove additional frequently used sites, e.g., *Bpi*I, *Lgu*I, and *Bsm*BI, at this stage (*see* **Note 2**). In case of noncoding regions like promoters, a functional validation of the altered modules might be required. In case of coding sequences, they can also be codon optimized for the respective expression organism at this step.
2. Dependent on the functional part type, add *Bsa*I restriction sites and respective module-specific fusion sites to the fragment of interest as 5′ and 3′ extension (*see* Fig. 2b). The complete sequence can then be ordered as subcloned DNA fragment in a storage vector. The final complete vector sequence should be rechecked before finally ordering if indeed the storage vector backbone of the provider does not contain the respective type IIS restriction sites.
3. Dependent on the chosen DNA synthesis provider, the amount and the quality of the DNA will vary. To ensure high quality and cloning robustness, retransform and miniprep all modules with the identical miniprep kit.

3.3 Design, Generation, and Domestication of Level 1 Expression Plasmids

The level 0 modules will be assembled to complete transcriptional units in level 1 expression plasmids which must fulfill several specific requirements in contrast to the level 0 plasmids: (i) confer resistance to ampicillin (or any other marker not present in the level 0 plasmids than kanamycin for counterselection), (ii) a *LacZ* fragment for later blue/white screening flanked by *Bsa*I restriction sites generating the fusion sites CTTG/TAAT, and (iii) absence of additional type IIS restriction sites in the backbone sequence. A general example of how to convert any given plasmid into a MoPET-compatible level 1 destination plasmid is provided (Fig. 2c). Here, the required vector fragments and the *LacZ*

Fig. 2 (continued) screening; MCS; multicloning site. (**d**) Primers used for vector creation. *BpiI* restriction sites are shown in *italics*, *BsaI* restriction sites in bold, and fusion sites generated by *BpiI* in red. *BpiI* sites (gaagac) are removed during vector assembly, whereas the *BsaI* sites (gagacc) remain in the vector for subsequent cloning of the desired DNA fragments

selection cassette are generated by PCR. In addition, *Bsa*I sites present in the vector are removed using primers designed to insert single-nucleotide mutations in the DNA recognition sequences. For PCR fragment assembly, *Bpi*I as a second type IIS enzyme is used, because *Bsa*I sites will be introduced to allow Golden Gate assembly into the final vector.

1. Primer design for amplification of the *LacZ* fragments and backbone fragments. The sequences of the primers are listed (Fig. 2d). The primers P1 and P2 are designed to amplify the *LacZ* fragment from a commercially available pUC19 as example. Both primers have *Bpi*I (italics) restriction sites at their end, generating the fusion sites CTTG/TAAT. The fusion sites are followed by *Bsa*I restriction sites (bold) in reverse complement orientation. After the assembly via *Bpi*I, the *Bsa*I sites will become part of the destination vector.

   The expression plasmid backbone contains an additional *Bsa*I restriction site in the Ap^R which will be removed by introducing a single silent nucleotide substitution in the used fusion site of P4 and P5. All four primers (P3–P6) have a *Bpi*I site with a fusion site selected to be compatible with the fusion site of the fragment to which it will be ligated.
2. Set up one PCR with P1/P2 on pUC19 and two PCRs with P3/P4 and P5/P6 on the expression plasmid. A proofreading polymerase like KOD should be used to minimize errors. The following conditions are used following the manufacturer's instructions: 1 μL plasmid DNA (5–20 ng/μL), 5 μL of 10× buffer, 3 μL of 25 mM $MgSO_4$, 5 μL of 2 mM dNTPs, 1.5 μL each of 10 μM sense and antisense primers, and 1 μL of KOD Hot Start DNA Polymerase (10 U/μL, final concentration 0.02 U/μL) in a total reaction volume of 50 μL.
3. Analyze a small amount of each PCR reactions by agarose gel electrophoresis. For preparation of agarose gels for electrophoresis, agarose (0.7–1.5%) in 1× TAE is melted in a microwave oven and one drop of a 0.025% (w/v) ethidium bromide solution is added per 100 mL of melted agarose.
4. If a single band can be detected, gel-purify the remaining PCR reaction to remove potential primer dimers and backbone template which may interfere with the *Bpi*I-based Golden Gate reaction.
5. Set up a BpiI restriction–ligation by combining 30 fmol of each purified PCR product, 2.5 μL 10× ligation buffer, 1 μL of BpiI, and 1.5 μL of ligase (final volume of 25 μL) (*see* **Note 3**).
6. Incubate the Golden Gate restriction–ligation mix in a thermal cycler with the following program: 2 min at 37 °C, 5 min at 16 °C, both repeated 50 times, followed by a digestion step

(5 min at **37 °C** for *Bpi*I) and then by heat inactivation for 10 min at 80 °C.

7. Transformation into *E. coli*. Mix 2.5 μL of the reaction with 50 μL electrocompetent cells, transfer to a pre-cooled electroporation cuvette, and transform by electroporation.
8. Add 1 mL LB medium and regenerate cells for 1 h at 37 °C at 600 rpm in an Eppendorf ThermoMixer®.
9. Plate 20 μL and the complete remaining culture on LB plates containing ampicillin and X-gal. Incubate overnight at 37 °C.
10. Correct colonies should be **blue**. Inoculate 4 mL LB medium with ampicillin and incubate overnight at 37 °C in a shaker incubator.
11. Prepare DNA using a miniprep kit.
12. Verification of the complete constructs by DNA sequencing.

3.4 Standard Golden Gate Assembly of Defined Expression Constructs

Level 0 modules will be assembled into complete transcriptional units in level 1 expression plasmids. In contrast to the level 0 module plasmids, level 1 destination plasmids confer resistance to ampicillin to allow selection pressure toward correctly assembled constructs as all other modules in the Golden Gate reaction are kanamycin resistant. For example, dependent on the specific design, a module for each position or only a subset can be used (Fig. 3A).

1. Design the expression construct(s) and select the required functional part plasmids and the destination expression plasmid.
2. Measure the DNA concentrations of all purified plasmids using a spectrophotometer.
3. Set up a Golden Gate restriction–ligation reaction providing 30 fmol (*see* **Note 3**) of each plasmid in a total reaction volume of 25 μL:

   x μL destination expression plasmid

   x μL level 0 plasmids 1–7

   1.5 μL T4 DNA ligase

   μL BsaI-HFv2 enzyme (*see* **Note 4**)

   2.5 μL 10× ligation buffer

   H_2O to final 25 μL
4. Incubate the Golden Gate restriction–ligation mix in a thermal cycler with the following program: 2 min at 37 °C, 5 min at 16 °C, both repeated 50 times, followed by a digestion step (5 min at **50 °C**) and then by heat inactivation for 10 min at 80 °C.

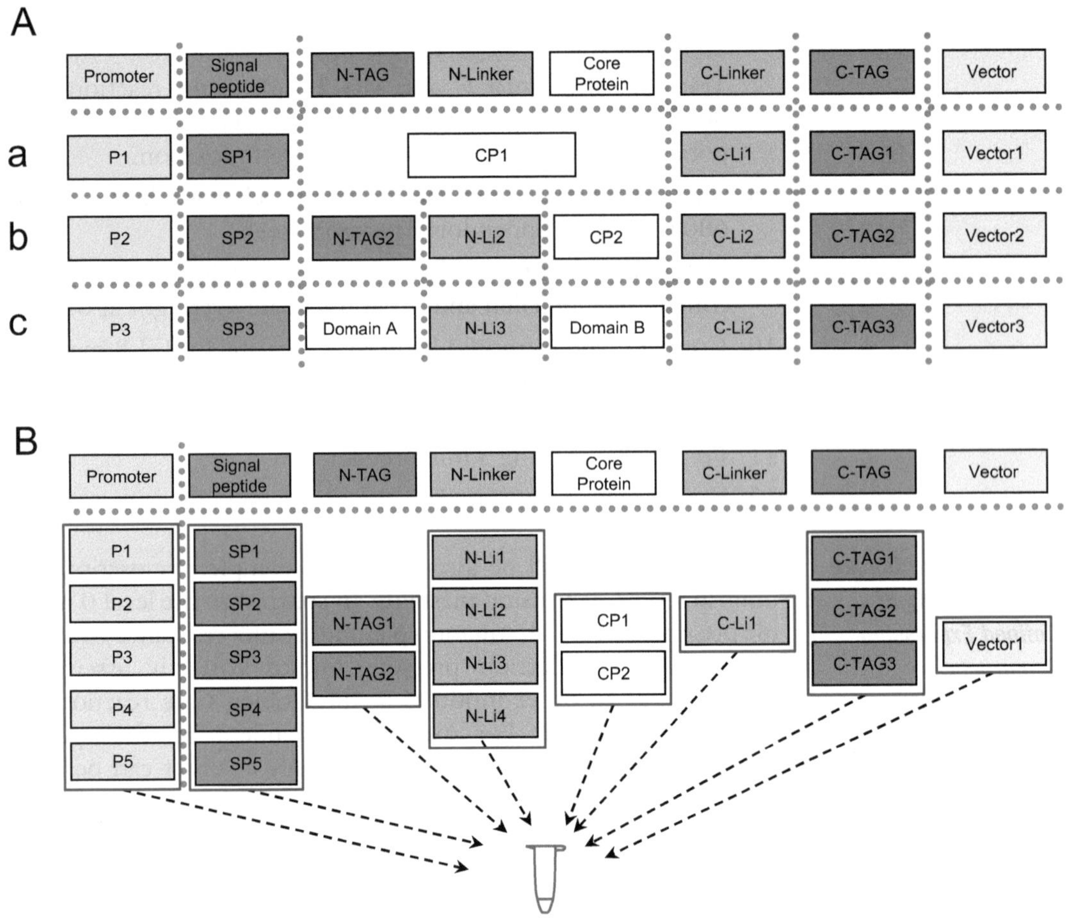

Fig. 3 MoPET design examples for defined expression constructs or library approaches. (**A**) Three examples for the use of MoPET to generate defined expression construct are shown. (**a**) In case no N-terminal tags and linker sequences are required, the core protein (CP) module serves as a flexible adaptor. In this case, the CP variant with the fusion sites (GGCT/GGTT) bridges the N-TAG and N-Li positions. (**b**) All eight functional parts are combined. (**c**) Repurposing of positions. Here the N-terminal tag position contains the first domain of a two-domain protein, so that the dedicated linker modules can be used to separate the two domains. (**B**) In case of expression optimization libraries, multiple level 0 modules from a functional class are combined in a single one-pot reaction

5. Transformation into *E. coli*. Mix 2.5 μL of the reaction with 50 μL electrocompetent *E. coli* cells, transfer to a pre-cooled electroporation cuvette, and transform by electroporation.
6. Add 1 mL LB medium and regenerate cells for 1 h at 37 °C with 600 rpm in an Eppendorf ThermoMixer®.
7. Plate 20 μL on LB plates containing ampicillin and X-gal. Incubate overnight at 37 °C.
8. Correct colonies should be **white**. Picking two white colonies is usually sufficient to identify a correct clone. Inoculate 4 mL

LB medium with ampicillin and incubate overnight at 37 °C in a shaker incubator (*see* **Note 5**).

9. Prepare DNA using a miniprep kit.
10. Verification of constructs by either DNA sequencing or restriction digest.

3.5 Standard Golden Gate Assembly of Expression Libraries

In cases where rational design and generation of defined expression constructs does not lead to suitable constructs, the standardized and modular system also allows the generation of expression optimization libraries from the same set of modules applying an identical assembly strategy. In contrast to the standard assembly procedure, where one selected member of each module class is combined, for a library approach module sets for each position can be selected and combined in a single one-pot reaction. The workflow is like the one for the assembly of defined constructs but with a few special considerations.

1. Design the expression optimization library and select the functional parts and the destination expression plasmid. Calculate the theoretical complexity of the library
2. Measure the DNA concentrations of all purified plasmids using a spectrophotometer.
3. Set up a Golden Gate restriction–ligation reaction providing 30fmol *total amount for each position*. In case five different signal peptides have been selected for the library (Fig. 3B), 6 fmol of each selected module is used. In case the pipetting volume is too low, the sample should be diluted with tenfold water and a tenfold higher volume has to be added to the reaction:

   x μL destination expression plasmid

   x μL level 0 modules of positions 1–7

   1.5 μL T4 DNA ligase

   μL BsaI-HFv2 enzyme

   2.5 μL 10× ligation buffer

   H_2O to final 25 μL
4. Incubate the Golden Gate restriction–ligation mix in a thermal cycler with the following program: 2 min at 37 °C, 5 min at 16 °C, both repeated 50 times, followed by a digestion step (5 min at 50 °C) and then by heat inactivation for 10 min at 80 °C.
5. Dependent on the library complexity and number of variants that will be tested, transform the **whole** reaction into electrocompetent *E. coli* cells.

6. Drop dialysis of the library restriction–ligation reaction. Fill the bottom of a Petri dish with distilled water and float an MCE filter disc (13 mm, 0.025 μm; Millipore) on top. Add the 25 μL reaction to the filter disc and place a tight-fitting lid on the Petri dish to prevent evaporation. Most samples are dialyzed in 30 min.
7. Mix the whole dialyzed restriction–ligation reaction with 50 μL electrocompetent *E.coli* cells and transfer to a pre-cooled electroporation cuvette. Transform by electroporation.
8. Add 1 mL LB medium and regenerate cells for 1 h at 37 °C with 600 rpm in an Eppendorf ThermoMixer®.
9. Plate 20 μL, 100 μL, and the remaining culture on LB plates containing ampicillin and X-gal. Incubate overnight at 37 °C.
10. Pick the required number of white colonies in 2.5 mL 96 deep-well plates prefilled with 1.2 mL LB medium and appropriate antibiotic. Grow overnight and perform DNA preps with, e.g., QIAprep 96 Turbo Miniprep Kit.
11. Send the prepped DNA for sequencing with backbone-specific primers and evaluate for each clone the module composition which has been assembled. In certain cases, also sequencing primers located in the core protein region may be required to cover the whole construct.

4 Notes

1. The flexibility can be further increased by understanding that the number of positions and their nominal assignment to a function only sets an initial frame for the system. There are multiple options to flexibly extend and adjust the system without losing the advantages. For example, the module positions can be freely allocated to other functions, like in case of a two-domain protein the first domain of the POI using the position designated for N-TAG, separated by the N-Linker, and putting the second domain on the original core protein position (Fig. 3c). Also, the number of modules is not fixed to eight and can be adapted to the specific needs of a given project. In cases where not all module positions are needed, skipped positions can be bridged by a modified POI module. In case no N-terminal tags and linkers are required, the POI would start with G-GCT and ends with GGTT (Fig. 3a). When additional module positions are needed, however, an existing module can be split into two new module positions separated by a newly defined, unique fusion site. By keeping the outer fusion sites of the original module, the compatibility to MoPET is ensured at the same time. However, the design and

selection of additional fusion sites to, e.g., expand or adapt the system to a specific project can be a challenge even for trained users as the addition of new fusion sites will interfere with the system. An extremely helpful online tool from New England Biolabs helps here, where all selected fusions sites can be checked for assembly efficiency (http://ggtools.neb.com/viewset/run.cgi).

2. If an additional *Bsa*I type IIS restriction sites is present in a DNA fragment that should be assembled and could not be removed by introducing silent point mutations to disrupt these sites without functional impact, the assembly will still occur (since the digest/ligation is a reversible process), but the efficiency may be decreased. To compensate for this, an additional ligation step should be performed by adding 1 μL T4-DNA ligase for 1 h after the Golden Gate assembly reaction.

3. To calculate the concentration, the following formula can be used: 1 μg of a 1000 bp DNA fragment corresponds to 1.52 pmol. Therefore, the volume of DNA to pipet (in μL) to have 30 fmol is given by the following equation: 30 (fmol) x size (bp) of the DNA fragment/(concentration (ng/μL) × 1520). With standard miniprep DNA, often the volume is far below <μL, so to avoid pipetting errors, the DNA must be diluted. NEB provides a helpful online tool under https://nebiocalculator.neb.com/.

4. There have been several *Bsa*I variants from NEB. In addition, isoschizomers of *Bsa*I (enzymes with the same recognition site sequence) are available from other vendors like *Eco*31I from Thermo Fisher Scientific. With the conditions used here, we achieved optimal results with BsaI-HFv2 which was also used in a recent example to assemble up to 52 fragments [9].

5. The expectation in case of an optimal restriction–ligation reaction with an optimally designed modular toolbox is that after the reaction, a high number of white colonies and a low blue/white ratio is observed. In case that only a low number of white colonies and/or the appearance of a high number of blue colonies is observed, a possibility to increase the chance of finding correctly assembled single constructs can be the dialysis of the restriction–ligation reaction and transformation of the whole reaction into electrocompetent *E. coli* cells to maximize the number of colonies (described in Subheading 3.5, **step 6**). However, this is not a solution for a library screening approach. The main reasons and how to troubleshoot are listed below.

   The most important category is an unbalanced amount of the different DNA modules in the reaction. This can result from simple pipetting errors to the fact that a module was "forgotten." The first step is to simply repeat the reaction. Another case is that one or more of the inserts are degraded, which cannot be detected by the photometric concentration determination with, e.g., the Nanodrop. In parallel with a simple repetition, all samples should be visually inspected on an agarose gel to troubleshoot in a most efficient way. The second larger group of reasons the reaction is not efficiently working are design mistakes. Of course, validated standardized systems like MoPET have a key advantage here, but in case new module types are integrated or during the domestication of the expression destination plasmid errors in the fusion site design can happen, e.g., using palindromic sequences, fusion sites that are too similar to others already used in the system. Here the tool https://ggtools.neb.com/viewset/run.cgi can be used. Also unwanted/unexpected additional type IIS restriction sites can be present in the construct from the gene synthesis provider. So, they should be rechecked and tested by restriction digest.

Acknowledgments

This work was funded by Bayer AG.

References

1. Weber E et al (2017) Modular Protein Expression Toolbox (MoPET), a standardized assembly system for defined expression constructs and expression optimization libraries. PLoS One 12(5):e0176314. https://doi.org/10.1371/journal.pone.0176314
2. Weber E et al (2011) A modular cloning system for standardized assembly of multigene constructs. PLoS One 6(2):e16765. https://doi.org/10.1371/journal.pone.0016765
3. Hughes RA, Ellington AD (2017) Synthetic DNA synthesis and assembly: putting the synthetic in synthetic biology. Cold Spring Harb Perspect Biol 9(1). https://doi.org/10.1101/cshperspect.a023812
4. Engler C, Gruetzner R, Kandzia R, Marillonnet S (2009) Golden gate shuffling: a one-pot DNA shuffling method based on type IIs restriction enzymes. PLoS One 4(5):e5553. https://doi.org/10.1371/journal.pone.0005553
5. Engler C, Kandzia R, Marillonnet S (2008) A one pot, one step, precision cloning method with high throughput capability. PLoS One 3(11):e3647. https://doi.org/10.1371/journal.pone.0003647
6. Zhang F et al (2011) Efficient construction of sequence-specific TAL effectors for modulating mammalian transcription. Nat Biotechnol 29(2):149–153. https://doi.org/10.1038/nbt.1775
7. Cermak T et al (2011) Efficient design and assembly of custom TALEN and other TAL effector-based constructs for DNA targeting. Nucleic Acids Res 39(12):e82. https://doi.org/10.1093/nar/gkr218
8. Weber E et al (2011) Assembly of designer TAL effectors by Golden Gate cloning. PLoS One 6(5):e19722. https://doi.org/10.1371/journal.pone.0019722
9. Pryor JM et al (2022) Rapid 40 kb genome construction from 52 parts through data-optimized assembly design. ACS Synth Biol 11(6):2036–2042. https://doi.org/10.1021/acssynbio.1c00525
10. Gibson DG et al (2009) Enzymatic assembly of DNA molecules up to several hundred kilobases. Nat Methods 6(5):343–345. https://doi.org/10.1038/nmeth.1318
11. Li MZ, Elledge SJ (2007) Harnessing homologous recombination in vitro to generate recombinant DNA via SLIC. Nat Methods

4(3):251–256. https://doi.org/10.1038/nmeth1010

12. Casini A, Storch M, Baldwin GS, Ellis T (2015) Bricks and blueprints: methods and standards for DNA assembly. Nat Rev Mol Cell Biol 16(9):568–576. https://doi.org/10.1038/nrm4014
13. Sarrion-Perdigones A et al (2011) GoldenBraid: an iterative cloning system for standardized assembly of reusable genetic modules. PLoS One 6(7):e21622. https://doi.org/10.1371/journal.pone.0021622
14. Engler C et al (2014) A golden gate modular cloning toolbox for plants. ACS Synth Biol 3(11):839–843. https://doi.org/10.1021/sb4001504
15. Werner S et al (2012) Fast track assembly of multigene constructs using Golden Gate cloning and the MoClo system. Bioeng Bugs 3(1):38–43. https://doi.org/10.4161/bbug.3.1.18223
16. Moore SJ et al (2016) EcoFlex: a multifunctional MoClo kit for E. Coli synthetic biology. ACS Synth Biol 5(10):1059–1069. https://doi.org/10.1021/acssynbio.6b00031
17. Iverson SV, Haddock TL, Beal J, Densmore DM (2016) CIDAR MoClo: improved MoClo assembly standard and new E. coli part library enable rapid combinatorial design for synthetic and traditional biology. ACS Synth Biol 5(1):99–103. https://doi.org/10.1021/acssynbio.5b00124
18. Lee ME, DeLoache WC, Cervantes B, Dueber JE (2015) A highly characterized yeast toolkit for modular, multipart assembly. ACS Synth Biol 4(9):975–986. https://doi.org/10.1021/sb500366v
19. Mitchell LA et al (2015) Versatile genetic assembly system (VEGAS) to assemble pathways for expression in S. Cerevisiae. Nucleic Acids Res 43(13):6620–6630. https://doi.org/10.1093/nar/gkv466
20. Occhialini A et al (2019) MoChlo: a versatile, modular cloning toolbox for chloroplast biotechnology. Plant Physiol 179(3):943–957. https://doi.org/10.1104/pp.18.01220
21. Vasudevan R et al (2019) CyanoGate: a modular cloning suite for engineering cyanobacteria based on the plant MoClo syntax. Plant Physiol 180(1):39–55. https://doi.org/10.1104/pp.18.01401
22. Fonseca JP et al (2019) A toolkit for rapid modular construction of biological circuits in mammalian cells. ACS Synth Biol 8(11):2593–2606. https://doi.org/10.1021/acssynbio.9b00322
23. Martella A et al (2017) EMMA: an extensible mammalian modular assembly toolkit for the rapid design and production of diverse expression vectors. ACS Synth Biol 6(7):1380–1392. https://doi.org/10.1021/acssynbio.7b00016
24. Nielsen H, Engelbrecht J, Brunak S, von Heijne G (1997) Identification of prokaryotic and eukaryotic signal peptides and prediction of their cleavage sites. Protein Eng 10(1):1–6. https://doi.org/10.1093/protein/10.1.1

Chapter 3

Coiled-Coil Interaction Toolbox for Engineering Mammalian Cells

Erik Rihtar, Tina Fink, and Roman Jerala

Abstract

Protein interactions play a crucial role in a variety of biological processes. Therefore, regulation of these interactions has received considerable attention in terms of synthetic biology tool development. Of those, a toolbox of small peptides known as coiled coils (CCs) represents a unique effective tool for mediating protein–protein interactions because their binding specificity and affinity can be designed and controlled. CC peptides have been used as a building module for designing synthetic regulatory circuits in mammalian cells, construction of fast response to a signal, amplification of the response, and localization and regulation of function of diverse proteins. In this chapter, we describe a designed set of CCs used for mammalian cell engineering and provide a protocol for the construction of CC-mediated logic circuits in mammalian cells. Ultimately, these tools could be used for diverse biotechnological and therapeutic applications.

Key words Coiled-coil peptides, Protein–protein interactions, Logic gates, Proteases, Orthogonal coiled coils

1 Introduction

Protein–protein interactions (PPIs) regulate virtually all molecular processes within a living cell. Synthetic biology has established powerful tools to control PPIs in order to study and interrogate biological processes in mammalian cells. An attractive approach for mediating PPIs is the use of small peptide interaction domains called coiled coils (CCs). CCs are protein structural motifs composed of two or more α-helices that twist around one another in a superhelical fashion [1]. A consensus CC sequence consists of multiple seven amino acid heptad repeats, labeled $(abcdefg)_n$, where n is the number of repeats. The interstrand interaction between two CC dimers is based on hydrophobic and electrostatic interactions between the amino acids at the interfacial positions *a*, *d*, and *e*, *g* of the heptad repeat, respectively [2]. The amino acid sequence of the repeats determines interaction selectivity

Francesca Ceroni and Karen Polizzi (eds.), *Mammalian Synthetic Systems*, Methods in Molecular Biology, vol. 2774,
https://doi.org/10.1007/978-1-0716-3718-0_3,

(homo- vs. hetero-oligomers), affinity, oligomeric state (e.g., dimer, trimer, or tetramer), and helix orientation (parallel vs. antiparallel) between the CCs. Moreover, the simplicity of the structure and assembly of CCs provides a multitude of opportunities for the rational computational design of numerous CCs de novo [3, 4]. This has led to the design of a highly versatile synthetic CC toolkit with defined binding affinity, specificity, helix orientation, and oligomeric state. These efforts have significantly improved the utility of CC for a variety of applications, making them a highly useful building block for synthetic biology.

The designed CCs have been widely used as modular building blocks for in vitro assembly of protein nanostructures [3, 5, 6]. In addition to in vitro assembly, the versatile CC protein motif has also shown great promise for numerous in vivo applications in mammalian cells [3, 7–11]. CCs have been used to induce interactions between proteins to regulate a diverse set of cellular processes, including reconstitution of split enzymes [7, 9], multiplex protein subcellular localization [8], transcriptional regulation [8, 9], and genome engineering [11]. CC peptides were also used for the recruitment of extracellular recognition domain (scFv) of therapeutic CAR T cells [12] and to mediate specific cell–cell interactions [13]. In synthetic biology, the construction of modular molecular logic gates is essential for building complex logic devices in living cells. Using orthogonal CC peptides, novel protein-only circuits have been implemented in mammalian cells, termed split protease-cleavable orthogonal-CC-based (SPOC) logic circuits [7]. The SPOC system is based on proteolysis-responsive orthogonal CCs and split proteases able to implement Boolean logic gates in mammalian cells that could respond to an external chemical signal within 15 min. The principle of SPOC logic circuits has also been used to regulate protein secretion in mammalian cells [10]. Recently, CC-responsive allosteric ON/OFF switches (INSRTRs) were constructed via the insertion of a short CC peptide into a diverse set of proteins. In this design, an intrinsically unstructured CC peptide segment is inserted into a target protein without altering its function; however, when the corresponding CC peptide pair is added, the resulting CC interaction leads to the disruption of target protein function in an allosteric fashion. The INSRTR protein switches can be used for the regulation of diverse proteins and functions [14]. All of the above examples show that CC interactions can be regulated by proteolysis [7], phosphorylation [15], metal ions [16], and competitive binding [9] and highlight the utility of the CC toolbox for mammalian cell engineering as well as the usefulness of the CC tools for diverse medical and biotechnological applications.

In this chapter, we outline the current collection of orthogonal synthetic CC peptide pairs that have been engineered and characterized by our group. This set of CCs has been routinely used for

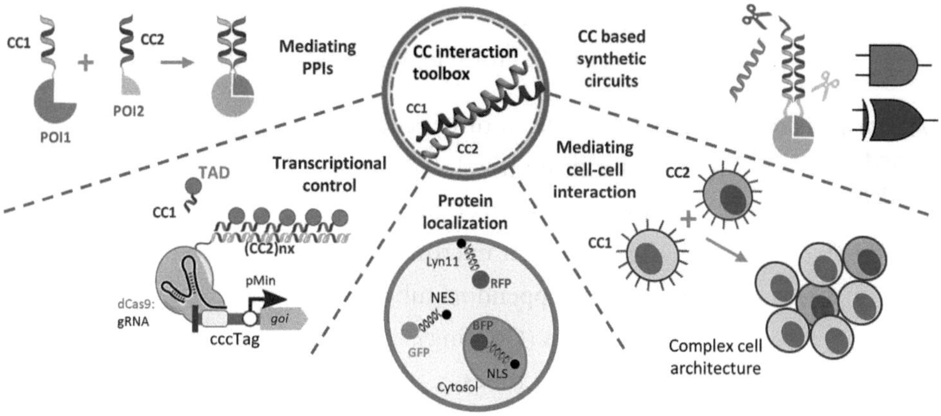

Fig. 1 Coiled-coil interaction toolbox for regulation of diverse cellular processes in mammalian cells. CC toolbox has been systematically exploited for inducing protein interactions and has been shown to regulate split protein reassembly, transcription, subcellular localization, cell–cell aggregation, and construction of logic gates in mammalian cells

mammalian cell engineering. We demonstrate two applications of the CC toolbox, including mediating PPIs (reassembly of split firefly luciferase (fLuc)) and construction of CC-mediated logic gates in mammalian cells (exemplified by AND gate). These methods could be generalized to other applications in mammalian cell engineering using CCs to induce and regulate PPIs (Fig. 1).

2 Materials

2.1 Plasmids

2.1.1 Equipment

1. PCR thermocycle.
2. Nanodrop.
3. Agarose electrophoresis system.
4. Camera for gel imaging.
5. Shaker incubator for bacterial cell culture.

2.1.2 Reagents

1. Plasmids used for SPOC logic: protease encoding plasmids and plasmids encoding CC peptides genetically fused to split firefly luciferase were reported previously [7] and are used as examples in this protocol.
2. Transfection control plasmids: pcDNA3.1 (plasmid without an insert) and pGL4.16/TK_rLuc (Promega) (constitutive expression of the *Renilla* luciferase, driven from the TK promoter) (*see* **Note 1**).
3. Plasmid miniprep kit.
4. PCR purification kit.
5. Gel purification kit.

6. DNA oligos.
7. KAPA HiFi HotStart ReadyMix for PCR (Roche).

2.2 Mammalian Cell Culture and Transfection

2.2.1 Equipment

1. CO_2 tissue culture incubator.
2. Cell culture hood.
3. Automated cell counter.
4. Inverted light microscope.
5. 1.5 mL Eppendorf tubes.
6. T75 flasks (Corning).
7. 50 mL test tubes.

2.2.2 Reagents

1. Immortalized human embryonic kidney 293 T cells—HEK293T (American Type Culture Collection) (*see* **Note 2**).
2. Sterile white bottom 96-well culture plates.
3. Cell growth medium: Dulbecco's Modified Eagle Medium (DMEM), low glucose (1 g/L) with GlutaMAX™ + 10% fetal bovine serum (FBS).
4. Phosphate-buffered saline (PBS).
5. Trypsin–EDTA solution.
6. 150 mM NaCl.
7. 1 mg/mL polyethylenimine (PEI; MW 40,000; PolyScience) stock solution.

2.3 Luciferase Reporter Assay

1. Microplate luminometer.
2. Dual-Luciferase® Reporter Assay System (Promega).

3 Methods

3.1 Selection and Preparation of CC Encoding Plasmids for Inducing PPIs

Optimal utilization of the synthetic CC interaction toolbox for mammalian cell engineering applications demands knowledge of their interaction affinities, orientation (parallel vs. antiparallel), and specificity (homo- vs heterodimer) (*see* **Note 3**). For simplicity, we have listed the characteristics and amino acid sequences of our synthetic CC peptides in Table 1. Here we describe the design and preparation of CC encoding plasmids for the construction of synthetic circuits in mammalian cells.

1. Genes encoding split firefly luciferase reporter domains (nLuc and cLuc) and orthogonal proteases (TEV and PPV) can be obtained from the literature [7] and ordered as synthetic genes (*see* **Note 4**).
2. The chosen CC segments are ordered as long synthetic DNA oligonucleotides and introduced into the constructs using

Table 1
Amino acid sequences of synthetic heterodimeric CC peptides used in mammalian cells

Peptide name	Amino acid sequence Basic NIC CC toolbox	Comments	Interacting partners	Reference
	gabcdef gabcdef gabcdef gabcdef			
P1	SPED EIQALEE ENAQLEQ ENAALEE EIAQLEY G	Part of NCIP set of CC	P2	[8]
P2	SPED KIAQLKE KNAALKE KNQQLKE KIQALKY G	Part of NCIP set of CC	P1	[8]
P3	SPED EIQQLEE EIAQLEQ KNAALKE KNQALKY G	Part of NCIP set of CC	P4, P4mS, AP4, AP4mS	[7, 8]
P4	SPED KIAQLKQ KIQALKQ ENQQLEE ENAALEY G	Part of NCIP set of CC	P3, P3mS	[7, 8]
P5	SPED ENAALEE KIAQLKQ KNAALKE EIQALEY G	Part of NCIP set of CC	N6, P6, P6A, P6SN	[8, 9]
P6	SPED KNAALKE EIQALEE ENQALEE KIAQLKY G	Part of NCIP set of CC	N5, P5, P5A, P5SN	[8, 9]
P7	SPED EIQALEE KNAQLKQ EIAALEE KNQALKY G	Part of NCIP set of CC	N8, P8, P8A, P8SN	[8, 9]
P8	SPED KIAQLKE ENQQLEQ KIQALKE ENAALEY G	Part of NCIP set of CC	N7, P7 P7A, P7SN	[8, 9]
P9	SPED ENQALEQ KNAQLKQ EIAALEQ EIAQLEY G	Part of NCIP set of CC	P10, AP10	[8, 9]
P10	SPED KNAQLKE ENAALEE KIQQLKE KIQALKY G	Part of NCIP set of CC	P9	[8, 9]

(continued)

Table 1 (continued)

Peptide name	Amino acid sequence Basic NIC CC toolbox	Comments	Interacting partners	Reference
P11	SPED ENQALEQ EIAQLEQ EIAALEQ KNAQLKY G	Part of NCIP set of CC	P12	[8]
P12	SPED KNAQLKE KIAALKE KIQQLKE ENQALEY G	Part of NCIP set of CC	P11	[8]
CC segments with different affinities				
P3mS	SPED EIQQLEE EISQLEQ KNSQLKE KNQQLKY G	Derived from P3 (weaker affinity)	P4, P4mS, AP4, AP4mS	[7]
P4mS	SPED KISQLKQ KIQQLKQ ENQQLEE ENSQLEY G	Derived from P4 (weaker affinity)	P3, P3mS	[7]
P9mS	SPED KLAQIKE KLQQIKE ELAANEE KLQANKY G	Derived from P9 (weaker affinity)	P10, AP10	[7]
N5	Y EIAALEA KIAALKA KNAALKA EIAALEA GC	Derived from P5 (higher affinity, low degree of homodimerization at higher concentration)	N6, P6, P6A, P6SN	[9]
N6	Y KIAALKA EIAALEA ENAALEA KIAALKA GC	Derived from P6 (higher affinity, low degree of homodimerization at higher concentration)	N5, P5, P5A, P5SN	[9]
N7	Y EIAALEA KNAALKA EIAALEA KIAALKA GC	Derived from P7 (higher affinity, low degree of homodimerization at higher concentration)	N8, P8, P8A, P8SN	[9]
N8	Y KIAALKA ENAALEA KIAALKA EIAALEA GC	Derived from P8 (higher affinity, low degree of homodimerization at higher concentration)	N7, P7 P7A, P7SN	[9]
P5A	YG ENAALEA KIAALKA KNAALKA EIAALEA GC	Derived from P5 (higher affinity, no homodimer formation, weaker reconstitution of split luciferase)	N6, P5, P6A, P6SN	[9]

(continued)

Table 1
(continued)

Peptide name	Amino acid sequence Basic NIC CC toolbox	Comments	Interacting partners	Reference
P6A	YG KNAALKA EIAALEA ENAALEA KIAALKA GGC	Derived from P6 (higher affinity, no homodimer formation, weaker reconstitution of split luciferase)	N5, P5, P5A, P5SN	[9]
P7A	YG EIAALEA KNAALKA EIAALEA KNAALKA GC	Derived from P7 (higher affinity, no homodimer formation, weaker reconstitution of split luciferase)	N8, P8, P8A, P8SN	[9]
P8A	YG KIAALKA ENAALEA KIAALKA ENAALEA GGC	Derived from P8 (higher affinity, no homodimer formation, weaker reconstitution of split luciferase)	N7, P7, P7A, P7SN	[9]
P5SN	ENSQLEE KISQLKQ KNSELKE EIQQLEY G	Derived from P5 (weaker association with N6)	N6, P6, P6A, P6SN	[9]
P6SN	KNSELKE EIQQLEE ENQQLEE KISELKY G	Derived from P6 (weaker association with N6)	N5, P5, P5A, P5SN	[9]
P7SN	EIQQLEE KNSQLKQ EISQLEE KNQELKY G	Derived from P7 (weaker association with N8)	N8, P8, P8A, P8SN	[9]
P8SN	KISELKE ENQQLEQ KIQQLKE ENSQLEY G	Derived from P8 (weaker association with N8)	N7, P7, P7A, P7SN	[9]
Antiparallel CC segments				
AP4	SPED KLAQIKE KLQQIKE ELAANEE KLQANKY G	Derived from P4 (antiparallel CC)	P3, P3mS	[7]
AP10	SPED KLAQIKE KLQQIKE ELAANEE KLQANKY G	Derived from P10 (antiparallel CC)	P9	[7]
AP4mS	SPED ELQSNEE ELQQNEQ KLQQIKQ KLQSIKY G	Derived from AP4 (weaker affinity)	P3, P3mS	[7]

PCR extension at either N- or C-terminus of the protein domains (*see* **Note 5**). Usually, short flexible linker sequences are placed between the linked CC sequences and the protein domains if the spacing is required (*see* **Note 6**). Additionally, in the case of SPOC logic constructs, protease cleavage sites were strategically introduced into the constructs via PCR (TEV site: ENLYFQS and PPV site NVVVHQS).

3. The genes are amplified with PCR and inserted into a pcDNA3.1 mammalian expression vector using the Gibson assembly method [17]. The constructs are cloned downstream of the strong constitutive pCMV promoter and should include a consensus Kozak sequence (GCCACC), adjacent to the start codon (ATG). The coding sequence should end with a stop codon (TAA, TAG, or TGA).

3.2 Mammalian Cell Culture and Transfection

1. HEK293T cells are maintained at standard conditions at 37 °C and 5% CO_2 in the cell culture incubator. Cells are split every 3–4 days at a ratio of 1:5 to 1:10. All work with cell cultures should be performed in sterile conditions in a cell culture hood.
2. On day 0 (at 90% confluence), trypsinize and count the cells using an automated cell counter. Seed HEK293T cells by pipetting 100 μL of the cell suspension in each well of a white 96-well plate at a density of 2×10^4 cells per well.
3. Incubate the inoculated plates in a cell culture incubator for 16–24 h or until cells reach a confluency of 60–90% which is optimal for transfection.
4. Prior to transfection, dilute plasmid DNA to 50 ng/μL in 150 mM NaCl.
5. On day 1, prepare DNA transfection mixes in 1.5 mL Eppendorf tubes inside the cell culture hood (*see* **Note 7**). Referring to Table 2, add appropriate volume of 150 mM NaCl and plasmids. In a separate Eppendorf tube, prepare mastermix for PEI transfection solution (3 μL of PEI stock solution per 1000 ng of DNA) (*see* **Note 8**).
6. Add PEI solution to transfection mixtures at equal volume (10 μL PEI solution + 10 μL DNA mixture). Mix the transfection mixtures by gently tapping the side of the 1.5 mL Eppendorf tube. Incubate the transfection mixtures at room temperature for 15–20 min to allow the DNA to complex with PEI transfection reagent (the incubation time might affect the transfection efficiency).
7. Slowly add 20 μL of transfection mixture to the HEK293T cells in the designated wells of the 96-well plate (*see* **Note 9**).
8. Incubate the transfected cells for 24–48 h in a cell culture incubator.

Table 2
Optimized plasmid amounts for the construction of SPOC logic circuits (AND gate) in HEK293T cells

A	B	Input plasmids	(ng)	Logic function plasmids	(ng)
SPOC-based logic functions					
0	0	/	/	nLuc_AP4_TEVs_P3mSAP4_PPV_P3_cLuc	1010
1	0	pCMV_TEVp	90		
0	1	pCMV_PPVp	90		
1	1	pCMV_TEVp, pCMV_PPVp	90; 90		
AND logic function protease titration					
pCMV_TEVp pCMV_PPVp			0;10;25;500; 10; 25; 50	nLuc_AP4_TEVs_P3mS AP4mS_PPVs_P3_cLu	1010

The listed plasmid amounts are used for the transfection of one well of a 96-well plate (*see* **Note 10**)

3.3 Determination of Firefly Luciferase Activity

Here we describe the protocol for measuring firefly luciferase (fLuc) activity, which is a functional output in the exemplified SPOC logic circuits. Depending on the intended application, fLuc can be switched out with a different protein (e.g., split fluorescent protein, recombinase, transcriptional factor, etc.) in which case different downstream analysis should be performed (e.g., flow cytometry, Western blotting, real-time PCR, etc.).

1. Harvest the cells by carefully removing the media from the wells with a vacuum pump and adding 30 μL of 1× passive lysis buffer (Promega). For complete lysis of the cells, freeze and thaw the plate before proceeding to the next step.
2. Measure luciferase activity in cell lysates according to the Dual-Luciferase® Reporter Assay System (Promega) on a microplate luminometer.
3. Calculate relative luciferase units (RLU) by normalizing each sample's firefly luciferase activity to the constitutive *Renilla* luciferase activity determined in the same sample.

4 Notes

1. phRL-TK/TK_rLuc drives constitutive expression of *Renilla* luciferase. It is used as a transfection control in luciferase experiments and to calculate relative luciferase units by calculating the firefly luciferase activity of each well to the constitutive *Renilla* luciferase activity measured in the same well. Empty pcDNA3 plasmid is used to equalize total amount of transfected DNA.

2. The HEK293T cells are easy to culture and transfect with PEI. Nevertheless, other cell lines can be used (e.g., HeLa, MCF-7, Neuro2A, etc.), but optimization of transfection conditions might be required.
3. Selection of CC peptide pairs depends on the application and complexity of the designed system. For example, interaction between proteins can be simply modulated by the correct choice of CC peptide pairs with different affinities, orientation, and specificity, where heterodimerizing CCs enable more complex engineering by bringing together different protein partners. While CC-mediated homo- or heterodimerization of target proteins is the simplest example of this strategy, designing more complex circuits requires additional mutually orthogonal CC modules (i.e., CC segments that interact only with their designated binding partners) with fine-tuned affinities. The latter was illustrated in engineering SPOC signaling pathway in mammalian cells in which a processing module was constructed using proteolysis-responsive orthogonal CC pairs with an autoinhibitory CC segment. The affinity of the autoinhibitory and target CC should be sufficiently strong to inhibit reconstitution of the target protein, but at the same time sufficiently weak enough that upon proteolytic cleavage, the displacer CC can effectively displace the autoinhibitory CC.
4. Alternatively, many effector domains can also be obtained from Addgene where plasmids are provided in the form of bacterial stab (e.g., Addgene plasmid numbers: 118966–118970; 119182; 119211–119214; 119299; 119300; 119302; 119303; 135982–135986).
5. The orientation of CC segments at the N- or C-termini is important for the formation of the desired supramolecular structure or the correct reconstitution of split enzymes. It is beneficial to build a molecular model of the desired reconstituted structure or experimentally test different orientations of CC segments.
6. We prefer to use flexible Gly-Ser (GS) linkers. The optimal linker length is usually determined experimentally—we start with linker lengths of 0, 3, 5, 10, 20, and 30 aa.
7. Transfection mixes (without PEI!) can be prepared 1 day prior to transfection and stored at 4 °C overnight. The PEI transfection solution can be added 15–20 min prior to transfection.
8. The PEI/DNA ratio can vary between batches of PEI. We recommend that you test each batch of PEI by transfecting the cells at different ratios of DNA/PEI and determine the transfection efficiency and cell viability.

9. For reliable results, each experiment should be independently repeated at least two times and include at least three technical repeats (three separate wells on a multiwall plate transfected with the same plasmid mixture) in each individual experiment.
10. For more complex logic functions with several modules/ inputs, each module should be titrated to determine concentration window of transfected DNA where the information flow is optimal.

Acknowledgments

This work was supported by funding from the Slovenian Research Agency grant number P4-0176, Z4-2657, N4-0080.

References

1. Burkhard P, Stetefeld J, Strelkov SV (2001) Coiled coils: a highly versatile protein folding motif. Trends Cell Biol 11:82–88
2. Woolfson DN (2005) The design of coiled-coil structures and assemblies. Adv Protein Chem 70:79–112
3. Ljubetič A, Lapenta F, Gradišar H et al (2017) Design of coiled-coil protein-origami cages that self-assemble in vitro and in vivo. Nat Biotechnol 35:1094–1101
4. Beesley JL, Woolfson DN (2019) The de novo design of α-helical peptides for supramolecular self-assembly. Curr Opin Biotechnol 58:175–182
5. Božič Abram S, Gradišar H, Aupič J et al (2021) Triangular in vivo self-assembling coiled-coil protein origami. ACS Chem Biol 16:310–315
6. Gradišar H, Božič S, Doles T et al (2013) Design of a single-chain polypeptide tetrahedron assembled from coiled-coil segments. Nat Chem Biol 9:362–366
7. Fink T, Lonzarić J, Praznik A et al (2019) Design of fast proteolysis-based signaling and logic circuits in mammalian cells. Nat Chem Biol 15:115–122
8. Lebar T, Lainščček D, Merljak E et al (2020) A tunable orthogonal coiled-coil interaction toolbox for engineering mammalian cells. Nat Chem Biol 16:513–519
9. Plaper T, Aupič J, Dekleva P et al (2021) Coiled-coil heterodimers with increased stability for cellular regulation and sensing SARS-CoV-2 spike protein-mediated cell fusion. Sci Rep 111(11):1–16
10. Praznik A, Fink T, Franko N et al (2022) Regulation of protein secretion through chemical regulation of endoplasmic reticulum retention signal cleavage. Nat Commun 13:1–14
11. Lainščček D, Forstnerič V, Mikolič V et al (2022) Coiled-coil heterodimer-based recruitment of an exonuclease to CRISPR/Cas for enhanced gene editing. Nat Commun 131(13):1–12
12. Cho JH, Collins JJ, Wong WW (2018) Universal chimeric antigen receptors for multiplexed and logical control of T cell responses. Cell 173:1426–1438
13. Chao G, Wannier TM, Gutierrez C et al (2022) helixCAM: a platform for programmable cellular assembly in bacteria and human cells. Cell 185:3551–3567
14. Plaper T, Merljak E, Fink T et al (2022) Designed allosteric protein logic. bioRxiv 2022.06.03.494683. https://doi.org/10.1101/2022.06.03.494683
15. Woodall NB, Weinberg Z, Park J et al (2021) De novo design of tyrosine and serine kinase-driven protein switches. Nat Struct Mol Biol 28:762–770
16. Aupič J, Lapenta F, Strmšek Ž et al (2022) Metal ion–regulated assembly of designed modular protein cages. Sci Adv 8:8243
17. Gibson DG, Young L, Chuang R-Y et al (2009) Enzymatic assembly of DNA molecules up to several hundred kilobases. Nat Methods 6: 343–345

Chapter 4

A Mammalian-Based Synthetic Biology Toolbox to Engineer Membrane–Membrane Interfaces

Hossein Moghimianavval, Sonisilpa Mohapatra, and Allen P. Liu

Abstract

Intercellular membrane–membrane interfaces are compartments with specialized functions and unique biophysical properties that are essential in numerous cellular processes including cell signaling, development, and immunity. Using synthetic biology to engineer or to create novel cellular functions in the intercellular regions has led to an increasing need for a platform that allows generation of functionalized intercellular membrane–membrane interfaces. Here, we present a synthetic biology platform to engineer functional membrane–membrane interfaces using a pair of dimerizing proteins in both cell-free and cellular environments. We envisage this platform to be a helpful tool for synthetic biologists who wish to engineer novel intercellular signaling and communication systems.

Key words Membrane–membrane interfaces, Membrane protein reconstitution, Cell-free expression, Split protein reconstitution, Synthetic biology, SpyTag–SpyCatcher

1 Introduction

Intercellular and intracellular membrane–membrane interfaces play critical roles in various biological processes [1]. For example, in contact-dependent intercellular communication, membrane–membrane interface formation is essential for specific and targeted signal transduction [2]. Due to their unique biochemical and physical traits, cellular interfaces are specific compartments with locally distinct molecular compositions where certain molecules have less 2D diffusional freedom, form microclusters, or become completely excluded [1]. For instance, in immunological synapses, upon contact formation between a T-cell and an antigen-presenting cell (APC), the interfacial region becomes enriched with T-cell receptors (TCR) bound to their ligand. The interface formation also drives CD45 spatial size-dependent segregation and leads to depletion of CD45 from the membrane–membrane interface compartment [3]. In addition to immunological synapses, intercellular

Francesca Ceroni and Karen Polizzi (eds.), *Mammalian Synthetic Systems*, Methods in Molecular Biology, vol. 2774, https://doi.org/10.1007/978-1-0716-3718-0_4,

membrane–membrane interfaces play crucial roles in preventing leakage in endothelial layers [4], forming adhesion in epithelial sheets [5], neuronal signaling [6], and cellular differentiation during development [7]. On the other hand, intracellular membrane–membrane interfaces, also known as membrane contact sites, are essential for cellular processes such as autophagy, lipid transport, and organelle trafficking [8].

Given the importance of membrane–membrane interfaces, there has been an increasing need for in vitro platforms for controllable generation of membrane–membrane interfaces as well as developing probes for detecting membrane–membrane interface formation [9–12]. In top-down cellular systems, overexpression of specific adhesive molecules and receptor and ligand on different cells has been shown to induce cellular interface formation between T-cells and APCs with distinct physicochemical properties [3]. Recently, formation of a phagocytic interface and consequent CD45 depletion has been demonstrated in vitro where macrophages were grown on coverslips micropatterned with IgG [13].

While top-down systems allow probing the biological roles of proteins in physiological contexts, bottom-up platforms offer the unique possibility of creating membrane–membrane interfaces with minimal components. Additionally, such systems permit investigating the role of membrane or specific proteins in determining the biophysical properties of the interface while decoupling the effect of other proteins. For example, in a bottom-up in vitro system, membrane–membrane interface formation induced by affinity of dimerizing GFP proteins, each residing on one membrane, has been demonstrated [11]. Interestingly, size-dependent protein segregation in this minimal system has been observed and is suggested to be regulated by the physical height between two apposing membranes dictated by the size of binding proteins. A similar study has shown an analogous role of claudin-4, a tight junction protein, in driving size-dependent protein organization in membrane–membrane interfaces [12].

Here, we present a bottom-up synthetic biology platform for creating and functionalizing membrane–membrane interfaces. We utilize a peptide–protein pair with high affinity called SpyTag and SpyCatcher to facilitate membrane–membrane interface formation and a split fluorescent cherry (sfCherry) protein to detect successful interface formation and protein reconstitution [14–16]. The SpyTag and SpyCatcher proteins are derived from the fibronectin binding protein (FbaB) of *Streptococcus pyogenes* and are engineered to have high affinity and rapid reaction kinetics [17]. We start from soluble proteins sCatch-GFP and sTag-BFP and utilize a cell-free expression (CFE) system to synthesize them and test their functionality. Next, by leveraging the potential of CFE systems in co-translational translocation of membrane proteins [18, 19], we present a strategy to reconstitute transmembrane proteins

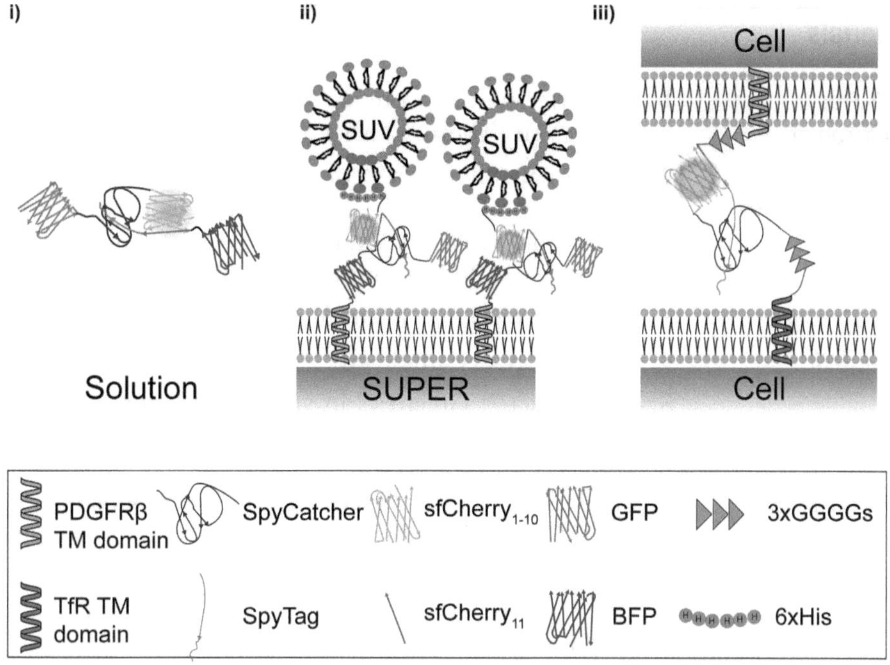

Fig. 1 Schematics of different protein modules that constitute a synthetic biology platform for reconstituting functional membrane–membrane interfaces. These modules include the soluble protein system made by interaction between sCatch-GFP and sTag-BFP proteins (i), the transmembrane system for protein reconstitution between supported lipid bilayers and SUVs (ii), and InterSpy system for intercellular protein complementation (iii). The box on the bottom depicts the domains that each cartoon represents

InterCatch-GFP and InterTag-BFP on supported lipid bilayers and detect the reconstitution of sfCherry on the membrane by supplementing sTag or sCatch, respectively. We next take the advantage of this strategy and demonstrate sfCherry formation in the membrane–membrane interface region between SUPER templates and small liposomes. Finally, we outline an approach to apply this system in a cellular environment for functionalizing intercellular membrane–membrane interfaces and protein complementation using InterTag and InterCatch proteins with different sizes of flexible linkers (Fig. 1).

Given the significance of membrane–membrane interfaces in biology, the platform presented here will be useful for reconstitution of specialized membrane–membrane interfaces for applications in designing synthetic contact-dependent communication pathways between cells or driving protein reorganization in cells or synthetic cells [20, 21].

2 Materials

2.1 Cell-Free Reconstitution of sfCherry in Membrane-Membrane Interfaces

2.1.1 Cell-Free Expression and Protein Dimerization Test

1. One-step human coupled in vitro transcription translation (IVTT) kit (Thermo Fisher Scientific).
2. Individual plasmids containing sCatch-GFP, sTag-BFP, InterCatch-GFP, and InterTag-BFP sequences cloned in a mammalian cell-free expression vector (Addgene #186902, #186903, #186900, and #186901).
3. 96-well conical (V) bottom plate.
4. 4–20% polyacrylamide gel.
5. FluoroTect™ GreenLys in vitro Translation Labeling System (Promega).

2.1.2 SUV and SUPER Template Generation

1. Lipids: 1,2-dioleoyl-sn-glycero-3-phosphatidylcholine (DOPC) and 1,2-dioleoyl-sn-glycero-3-[(N-(5-amino-1-carboxypentyl)iminodiacetic acid)succinyl] (nickel salt) (DGS-NTA(Ni)).
2. 100 nm liposome extruder.
3. 5 M NaCl solution, sterile-filtered.
4. 5 μm and 20 μm silica beads (*see* **Note 1**).
5. 96-well clear flat-bottom plate.

2.1.3 Bacterial Expression and Purification

1. BL21(DE3)pLysS cells.
2. LB broth.
3. 1 M isopropyl β-D-1-thiogalactopyranoside (IPTG).
4. Plasmid DNA containing sCatch-GFP-6xHis or sTag-BFP-6xHis sequence downstream of a bacterial ribosome binding site (RBS) under T7 promoter (Addgene #186904 and #186905).
5. 50 mg/mL kanamycin (*see* **Note 2**).
6. Glucose.
7. Probe sonicator.
8. Lysis buffer: 50 mM Tris–HCl (pH 7.4), 300 mM NaCl, 50 mM imidazole, 1 mM 4-benzenesulfonyl fluoride hydrochloride (AEBSF) protease inhibitor. Store at 4 °C.
9. Washing buffer: 50 mM Tris–HCl (pH 7.4), 300 mM NaCl, 50 mM imidazole. Store at 4 °C.
10. Elution buffer: 50 mM Tris–HCl (pH 7.4), 300 mM NaCl, 300 mM imidazole. Store at 4 °C.
11. Dialysis buffer: 1× PBS.

12. ÄKTA Start fast protein liquid chromatography (FPLC) system.
13. 1 mL His-trap column (Cytiva).

2.1.4 Size Exclusion Chromatography

1. Sepharose 4B resin.
2. 4 mL bed volume liquid chromatography column.

2.2 Intercellular sfCherry Reconstitution

2.2.1 Cell Culturing and Stable Cell Line Generation

1. HEK293T cells.
2. Cell growth medium: Dulbecco's Modified Eagle Medium (DMEM) with high glucose, pyruvate and glutamine, 10% (vol/vol) fetal bovine serum (FBS), 100 μg/mL penicillin–streptomycin, and 0.292 mg/mL glutamine.

   For 500 mL of cell growth medium, mix 445 mL DMEM (includes 4 mM L-glutamine +1 mM sodium pyruvate + 25 mM glucose), 50 mL of FBS, and 5 mL of penicillin–streptomycin (100 mg/mL). Filter-sterilize using a 0.22 μm filter. Prepare aliquots of 40 mL each and store them at 4 °C. This mixture is referred to as DMEM in the sections below, unless otherwise mentioned.
3. 12- and 6-well flat-bottom plate, tissue culture-treated.
4. Transfection reagent Lipofectamine™ 3000 (Thermo Fisher Scientific).
5. Dulbecco's phosphate-buffered solution (DPBS).
6. 2 mg/ mL puromycin.
7. Trypsin–EDTA (0.05%) (Gibco).
8. Nunc™ Lab-Tek™ chambered cover glass (Thermo Fisher Scientific): Coat the wells with a 0.5% solution of fibronectin in DPBS. For coating the wells, prepare a 0.5% solution of fibronectin by adding 25 μL of the fibronectin solution to 5 mL of DPBS. Coat the well with minimal volume to cover the surface and incubate for 10 min at room temperature under sterile conditions. Remove excess solution and air-dry the wells completely before adding further reagents or cells. The remaining fibronectin solution in DPBS can be stored at 4 °C and can be reused for coating wells.
9. Transposase-expressing helper plasmid (Addgene #34879) [22].
10. Plasmids encoding InterTag-3xL or InterCatch-3xL proteins and containing coding DNA sequences for cytosolic fluorescent protein markers compatible with sleeping beauty system (Addgene #186913 and #186909).

3 Methods

3.1 Cell-Free Reconstitution of sfCherry in Membrane–Membrane Interfaces

3.1.1 Cell-Free Expression of sTag and sCatch and Protein Dimerization Test

1. Prechill the aliquoted reaction reagents from IVTT kit on ice prior to assembling reaction.
2. Assemble a master mix for cell-free synthesis of sCatch and sTag proteins by mixing 10 μL lysate with 4 μL reaction mixture and 2 μL accessory proteins.
3. In two separate PCR tubes, assemble 10 μL reactions by mixing 5 μL of the master mix and proper amount of sCatch-GFP or sTag-BFP DNA plasmid for a total concentration of 10 nM DNA. Supply 0.2 μL of GreenLys to the sTag-BFP reaction for fluorescence labeling during SDS–PAGE. Add ultrapure water to bring the reaction volume to 10 μL. Mix the solutions by gently pipetting up and down.
4. Transfer the reactions to a 96-well conical V-bottom plate. Protein synthesis is monitored at 488/528 nm and 400/450 nm excitation/emission for sCatch-GFP and sTag-BFP, respectively, using a fluorescence plate reader set at 30 °C. Figure 2a (left) shows the fluorescence signals from sCatch-GFP and sTag-BFP increase over time, indicating successful synthesis of proteins by the CFE system.
5. After incubating reactions at 30 °C for 4–5 h, recover the reaction solutions from the well plate and store them in separate PCR tubes.
6. Take 5 μL from each reaction and mix them in a separate tube. Keep the remaining 5 μL from each reaction. Transfer the 10 μL mixture to the well plate and incubate in the plate reader at room temperature for 3–4 h while monitoring sfCherry signal at 561/625 nm excitation/emission. Figure 2a (right) shows the rise of sfCherry signal over time indicating the SpyTag-SpyCatcher bond formation and sfCherry reconstitution.
7. Once the incubation is over, recover the 10 μL mixture from the well plate and visualize the sCatch-GFP, sTag-BFP, and their mixture using SDS–PAGE and in-gel fluorescence imaging of GFP and GreenLys (*see* **Note 3**). Figure 2b shows the in-gel imaging of SDS–PAGE indicating individual sCatch-GFP and sTag-BFP and their conjugated product after SpyTag-SpyCatcher interaction.

3.1.2 SUV and SUPER Template Preparation for Direct Membrane Protein Reconstitution

1. Take an appropriate amount of DOPC from stock solution for a final concentration of 5 mM lipid in a final volume of 500 μL aqueous solution and transfer to a clear glass test tube.
2. Dry the lipid film under a gentle stream of argon.

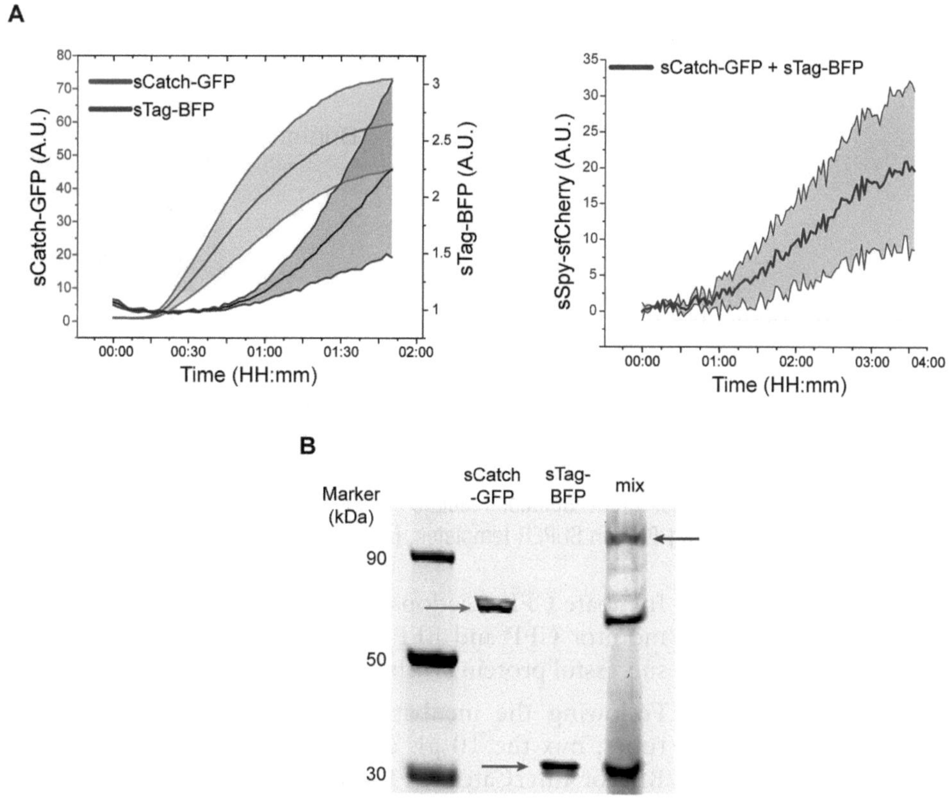

Fig. 2 (**a**) Left: GFP (green) and BFP (blue) fluorescence signal readout during synthesis of proteins sCatch-GFP and sTag-BFP, respectively, by CFE. Data shown as mean ± SD, $n = 3$. (**b**) Right: sfCherry (magenta) fluorescence signal readout after mixing CFE reactions containing sCatch-GFP and sTag-BFP post-translationally. Data shown as mean ± SD ($n = 3$). (Adapted from Moghimianavval et al. [16])

3. To ensure complete organic solvent evaporation, desiccate the dried lipid film for at least 1 h in vacuum.
4. Following desiccation, add 500 μL ultrapure water to the lipid film and vortex until the lipid is completely dissolved in water (*see* **Note 4**).
5. Pass the lipid solution through the liposome extruder with a 100 nm filter 11 times to generate SUVs (*see* **Note 5**). The SUVs are stable at 4 °C up to 2 weeks.
6. Once the SUVs are ready, assemble two CFE reactions expressing InterCatch-GFP and InterTag-BFP by following the procedure described in Subheading 3.1.1 with the modification of adding 2 μL of 5 mM SUV solution instead of water to the reactions.
7. Simultaneously, assemble two more CFE reactions expressing sCatch-GFP and sTag-BFP by repeating **steps 1–5** in Subheading 3.1.1.

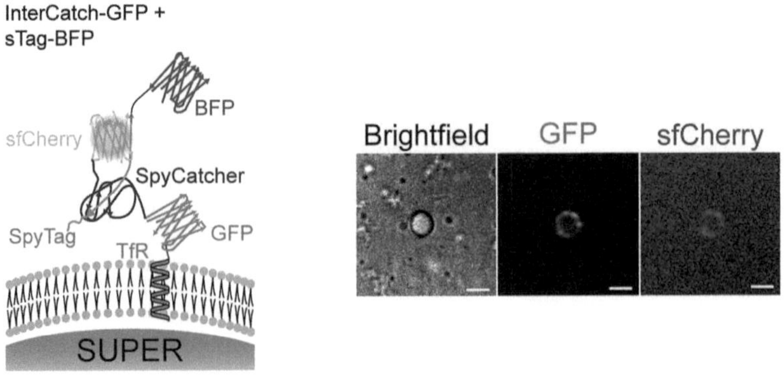

Fig. 3 Representative confocal images of cell-free expressed InterCatch-GFP (green) reconstituted on the 5 μm SUPER templates and sfCherry reconstitution (magenta) on the membrane. Also shown is the schematic representation of sfCherry reconstitution mediated by the interaction between InterCatch-GFP and sTag-BFP on SUPER templates. (Adapted from Moghimianavval et al. [16])

8. Incubate CFE reactions at 30 °C for 4–5 h. If desired, one can monitor GFP and BFP signals using a plate reader to ensure successful protein synthesis.
9. Following the incubation, in two separate microcentrifuge tubes, mix the 10 μL of CFE reactions expressing InterTag-BFP or InterCatch-GFP with 10 μL 5 M NaCl and 3.5 μL of 5 μm silica bead solution. Add ultrapure water to bring the final volume to 50 μL. Incubate the mixture at room temperature for 30 min with occasional gentle flicking.
10. After the incubation, add 1 mL 1× PBS to each mixture and spin down for 5 min at 200 × *g*.
11. For each bead solution, remove 950 μL of the supernatant. Resuspend the beads in the remaining 100 μL solution.
12. Repeat **steps 9–10** twice. The final mixture will be SUPER templates harboring InterCatch-GFP or InterTag-BFP transmembrane proteins.
13. Dilute the SUPER template solutions in a 1:9 ratio and transfer 50 μL of each of the diluted solutions to separate wells in a clear flat-bottom 96-well plate.
14. Add the completed CFE reactions containing sCatch-GFP or sTag-BFP to the wells with SUPER templates harboring InterTag-BFP and InterCatch-GFP, respectively.
15. Incubate the plate at room temperature for 3–4 h before imaging the SUPER templates. Figure 3 shows the reconstitution of InterCatch-GFP on SUPER templates and sfCherry reconstitution on the SUPER template periphery indicating successful split protein reconstitution through the interaction between the SpyCatcher domain residing on the membrane and the SpyTag supplied in soluble form.

3.1.3 Bacterial Expression and Purification of sCatch-GFP-6xHis and sTag-BFP-6xHis

The following steps describe the procedure for purifying sCatch-GFP-6xHis. The purification method for sTag-BFP-6xHis is identical to this.

1. Pick a single colony of overnight grown BL21(DE3)pLysS cells transformed with sCatch-GFP-6xHis. Grow the colony overnight in 5 mL LB broth supplemented with 50 μg/mL kanamycin while shaking at 220 rpm at 37 °C in an orbital shaker.
2. Dilute the culture into a large flask containing 1 L LB broth supplemented with 0.8% w/v glucose and 50 μg/mL kanamycin. Grow the culture at 37 °C while shaking at 220 rpm in the orbital shaker. Start measuring the A_{600} every 20 min once the culture starts to look opaque.
3. When the A_{600} reaches 0.5–0.6, add 420 μL of 1 M IPTG to the culture.
4. Following induction with IPTG, incubate the culture at 30 °C while shaking at 200 rpm for 4–5 h.
5. Once the incubation is over, pellet the culture by centrifuging at 5000 × *g* for 10 min at 4 °C.
6. Discard the supernatant and resuspend the pellet in 30 mL lysis buffer.
7. Using the probe sonicator, lyse the resuspended cells while keeping them on ice (*see* **Note 6**).
8. Centrifuge the lysate at 30,000 x g for 25 min. Recover the supernatant.
9. Load the supernatant onto the His-trap column mounted on the ÄKTA Start FPLC system with a flowrate of 1 mL/min.
10. Wash the column with 15 mL washing buffer.
11. Elute the protein by running elution buffer through the column and collect 1 mL fractions of the eluent.
12. Analyze protein purity and yield in each fraction by SDS–PAGE.
13. Pool the fractions with highest purity and yield and dialyze against 1 L dialysis buffer overnight at 4 °C.
14. Measure the protein concentration (*see* **Note 7**) and concentrate to around 1 mg/mL if desired (*see* **Note 8**).

3.1.4 Size Exclusion Chromatography and Membrane–Membrane Interface Reconstitution

The following steps describe reconstitution of membrane–membrane interfaces between SUPER templates harboring InterCatch-GFP and SUVs carrying sTag-BFP-6xHis. The procedure to reconstitute membrane–membrane interfaces where the proteins reside on the opposite membranes is identical to this.

1. Assemble a CFE reaction for direct reconstitution of InterCatch-GFP on SUPER templates following **steps 1–13**

in Subheading 3.1.2 with the modification of adding 3.5 μL of 20 μm silica bead solution instead of 5 μm bead solution in **step 9**.

2. While the CFE reaction from **step 1** is incubating, prepare 5 mM vesicles of 90% DOPC and 10% DGS-NTA(Ni) following **steps 1–5** in Subheading 3.1.2.
3. In a microcentrifuge tube, mix 50 μL of SUV solution from previous step with 10 μL of stock solution of purified sTag-BFP-6xHis.
4. Incubate the mixture at room temperature for 10 min.
5. In the meantime, pack the liquid chromatography column with ~3 mL of Sepharose-4B resin. Make sure to wash the column with PBS multiple times after packing.
6. Carefully add the SUV–protein mixture from **step 4** to the center of the column and wait until the solution is completely loaded into the column.
7. Carefully add 200 μL PBS to the center of column while collecting the eluent from the bottom of the column.
8. Repeat **step** 7 19 more times.
9. Once all fractions are collected, take 10 μL aliquot from each fraction and transfer to a 96-well conical V-bottom plate.
10. Measure the sTag-BFP-6xHis fluorescence at 400/450 nm excitation/emission and keep the fraction with the highest fluorescence signal that eluted during the first peak in the elution step. This is the fraction that contains the highest concentration of SUV-bound sTag-BFP-6xHis proteins (*see* **Note 9**).
11. Transfer 50 μL of a 1:4 dilution of the 20 μm SUPER templates harboring InterCatch-GFP from **step 1** to a 96-well clear flat-bottom plate.
12. Add 50 μL of the SUV-bound sTag-BFP-6xHis from **step 10** to the SUPER template solution (*see* **Note 10**). Incubate for 3–4 h at room temperature and image. Figure 4 shows sfCherry reconstitution in the intermembrane space between the bilayer of SUPER template harboring InterCatch-GFP and SUVs carrying sTag-BFP-6xHis.

3.2 Intercellular sfCherry Reconstitution

3.2.1 Generation of Stable HEK293 Cell Lines Expressing InterCatch-3xL and InterTag-3xL

The steps below summarize the protocol for generating stable cell lines expressing InterCatch-3xL starting from a frozen tube of HEK293T. The procedure to generate stable cell line expressing InterTag-3xL protein system is identical. We use InterTag-3xL and InterCatch-3xL constructs, each with three repeats of GGGGS linkers, for highest split protein reconstitution efficiency. The steps for generation of membrane–membrane interfaces using other variants of InterTag or InterCatch are the same. All the

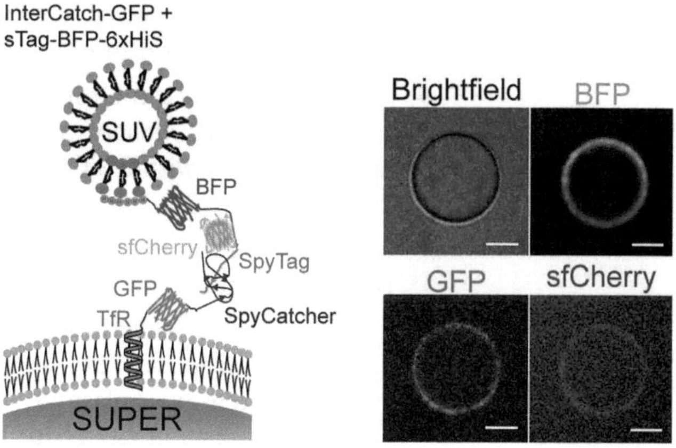

Fig. 4 Representative confocal images of cell-free expressed InterCatch-GFP (green) reconstituted on the 20 μm SUPER templates and sfCherry reconstitution (magenta) in the membrane–membrane interface between SUPER templates and SUVs harboring sTag-BFP-6xHis (cyan). Also shown is the schematic representation of sfCherry reconstitution mediated by the interaction between InterCatch-GFP reconstituted on the SUPER template and SUV-bound sTag-BFP-6xHis. (Adapted from Moghimianavval et al. [16])

steps are performed on a sterile bench, unless otherwise mentioned. DMEM, DPBS, and trypsin–EDTA (0.05%) that will be used during cell passaging and/or transfection are warmed by incubation in a water bath set to 37 °C for 20 min.

1. Mix a thawed aliquot of HEK293T cells with 9 mL DMEM in a 15 mL falcon tube. Mix well by pipetting two to three times (*see* **Note 11**).
2. Transfer the mix to a T75 flask and place the flask in a humidified incubator maintained at 37 °C with 5% CO_2.
3. Once the cells reach 70–80% confluency (determined by counting cells using Trypan blue and a Neubauer cell counter), they are ready for passaging.
4. Gently remove the medium from the T75 flask by using a serological pipette.
5. Add 3 mL of freshly warmed DPBS solution. Swirl the flask gently to wash the cells and then remove the DPBS by pipetting.
6. Add 1 mL warm trypsin–EDTA to detach the cells from the flask surface and incubate the flask in the humidified incubator at 37 °C for 5 min.
7. Mix the solution of trypsin–EDTA with the detached cells by pipetting up and down five to ten times. This is done to ensure that majority of the cells are detached from the flask walls and to avoid cell clumping.

8. Mix 130 μL of trypsinized cells with 13 mL DMEM in a 15 mL falcon tube. Mix thoroughly by pipetting two to three times.
9. Add 1 mL of the mixture in each well of a 12-well plate. Incubate the 12-well plate in a humidified incubator maintained at 37 °C with 5% CO_2.
10. Once the cells reach a confluency of 60%, they are ready to be transfected. Typically, the cells take around 24 h after plating to reach this confluency.
11. Co-transfect the cells in each well of the 12-well plate with 100 ng of transposase-expressing helper plasmid and 1000 ng of InterCatch-3xL plasmid, using Lipofectamine™ 3000 transfection reagent (*see* **Note 12**).
12. After 4–7 h of transfection, replace the cell medium of each well with 1 mL of warm DMEM medium (*see* **Note 13**).
13. After 2 days of transfection, remove the medium from each well by pipetting. Wash with 1 mL of warm DPBS.
14. Add 100 μL of trypsin–EDTA to each well and incubate the well plate in the humidified incubator at 37 °C for 5 min.
15. Mix the content of each well thoroughly by pipetting five to ten times. This step is necessary to ensure that all the cells have been detached from the surface of the well.
16. Mix 2 mL of warm DMEM and 75 μL of trypsinized cells in each well of two 6-well plates.
17. After 24 h of plating, add 2 μL of puromycin (2 mg/mL) (*see* **Note 14**) to each 6-well plate containing 2 mL of DMEM. Isolated islands of cells can be observed the day after. These islands start to become larger over time.
18. Replace the medium in each well with 2 mL of freshly warmed DMEM every other day. Supplement each well with 2 μL of puromycin (2 mg/mL) to select for resistant colonies that have stably integrated InterCatch-3xL constructs introduced through transfection.
19. This is continued until the resistant cell colonies in each well reach a confluency of ~80%. The stable cells in each well are then ready for expansion in a T75 flask.
20. Remove the medium from each well of the 6-well plate by pipetting. Wash with 2 mL of warm DPBS. Add 200 μL of trypsin–EDTA to each well and incubate the well plate in the humidified incubator at 37 °C for 5 min.
21. Add 10 mL of DMEM supplemented with 10 μL puromycin (2 mg/mL) into a T75 flask. Add 125 μL of trypsinized cells into the flask. Mix the contents of the flask well by pipetting several times slowly. Place the flask in a humidified incubator maintained at 37 °C and with 5% CO_2.

22. Once the cells are 80–90% confluent, trypsinize the cells with 1 mL trypsin–EDTA and transfer the trypsinized cells into 10 mL of warm DMEM.
23. Spin the cells down with a benchtop centrifuge at a speed of 500 × *g* (with slow acceleration and deceleration). Remove the supernatant and resuspend the cells in 10 mL DMEM supplemented with 5% DMSO.
24. Prepare 1 mL aliquots of the cells for long-term storage. Store them in −80 °C freezer for 24 h and then transfer them to liquid nitrogen for long-term storage.

3.2.2 Co-Culturing Cells Expressing InterTag-3xL and InterCatch-3xL Protein Systems

1. Culture HEK293T stable cell lines expressing InterTag-3xL and InterCatch-3xL, generated as described in Subheading 3.2.1, separately in the wells of a 6-well plate with 2 mL of DMEM per well.
2. Upon reaching a confluency of ~80%, trypsinize the cells in each well with 200 μL of warm trypsin–EDTA.
3. Add 200 μL of warm DMEM into the wells of the fibronectin-coated 4-well chambered cover glass. In separate wells, add 10 μL of trypsinized InterTag-3xL or InterCatch-3xL expressing cells.
4. The following day, monolayers of InterCatch-3xL or InterTag-3xL cells are formed in the wells. Trypsinize the monolayers with 100 μL of warm trypsin–EDTA per well and resuspend in 500 μL of warm DMEM gently by pipetting several times.
5. For co-culturing cell lines expressing InterTag-3xL and InterCatch-3xL, mix the cell suspensions in a 15 mL tube thoroughly via pipetting.
6. Incubate the tube containing the cell suspension mixture in an orbital shaker placed in the humidified incubator for 45 min shaking at 100 rpm (*see* **Note 15**).
7. Centrifuge and pellet the cell suspension mixture in a benchtop centrifuge at 200 × *g* for 5 min (slow acceleration and deceleration).
8. Aspirate the supernatant and refresh the cell suspension with another 500 μL of warm DMEM. Resuspend the co-cultured cells gently via pipetting.
9. Seed 200 μL of the co-cultured cell suspension onto the fibronectin-coated wells and incubate for 1 h before imaging (*see* **Note 16**). Figure 5 depicts sfCherry reconstitution in the intercellular junctions between InterCatch-3xL and InterTag-3xL expressing cells.

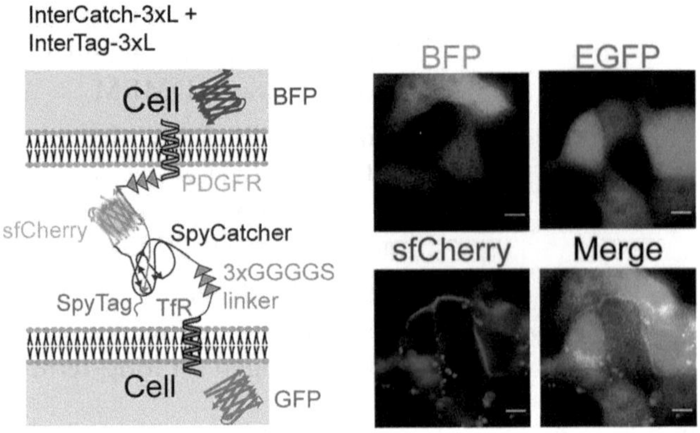

Fig. 5 Representative fluorescence images of co-cultures of HEK293T cells expressing InterCatch-3xL and InterTag-3xL with GFP (green) and BFP (cyan) cytosolic markers, respectively, and reconstituted sfCherry (magenta) in the intercellular interfaces. Also shown is the schematic representation of sfCherry reconstitution mediated by the interaction between InterCatch-3xL and InterTag-3xL at the intercellular interface. (Adapted from Moghimianavval et al. [16])

4 Notes

1. We purchased 5 μm and 20 μm silica beads from Bangs laboratory and Corpuscular, respectively.
2. The choice of antibiotics depends on what antibacterial resistance is encoded in the DNA vector containing sCatch-6xHis or sTag-6xHis.
3. If in-gel imaging of GFP is desired, avoid heating the sample prior to SDS–PAGE as it will cause denaturation of GFP.
4. The solution should look cloudy if the lipid is completely dissolved.
5. We used Avanti polar extruder apparatus, but one can use other possible options such as T&T Scientific extruders.
6. Make sure to use proper settings on the sonicator to avoid overheating lysate. We used 50% duty cycle, 50% output power, and ~eight times sonication cycles of 30 s followed by 2-min incubation on ice.
7. We used a Nanodrop to measure protein concentration and predicted the protein extinction coefficient factor by Expasy. One can use other methods such as Bradford or BCA assay to determine protein concentration.
8. We used MilliporeSigma Amicon ultra-15 centrifugal filter tubes to concentrate proteins.
9. The fraction containing SUVs will look opaque due to the presence of liposomes.

10. Due to the variations in size exclusion chromatography results, make sure that the fluorescence signal from eluted fractions containing SUV-bound sTag-6xHis or sCatch-6xHis matches the signal from CFE reactions in Subheading 3.1.2 for a valid comparison.
11. Always warm the reagents used for passaging/treating the cells to 37 °C by placing them in a water bath. Adding cold reagents to the cells may result in cell death.
12. The presence of endotoxins in purified plasmids can interfere with transfection experiments. Transfection with high-quality plasmids is essential for efficient transfection and obtaining healthy transfected cell lines. We therefore used a DNA miniprep kit (E.Z.N.A.® Endo-Free Plasmid DNA Mini Kit I) that includes an additional endotoxin removal step in the plasmid purification protocol.
13. After 4–7 h of transfection, it is important to replace the medium with freshly warmed DMEM to improve cell viability.
14. The antibiotic choice to impose selection pressure on transfected cells is dependent on the antibiotic resistance encoded in the DNA vectors containing InterTag-3xL or InterCatch-3xL.
15. We co-cultured the trypsinized InterTag-3xL and InterCatch-3xL cell suspension by mixing and transferring them to a humidified orbital shaker and shaking at 100 rpm for 45 min. This is necessary to avoid clumping of cells and for obtaining a monolayer of InterTag-InterCatch co-cultured cell pairs upon plating for imaging.
16. We observed that the efficiency of sfCherry reconstitution is dependent on the incubation time post plating. The various InterTag and InterCatch cell pairs seeded onto fibronectin-coated wells were incubated for 1 h before imaging for maximum observed reconstitution. The efficiency of observed reconstitution saturated at incubation times greater than 1 h. However, this incubation time might vary depending on the type of split protein being reconstituted.

References

1. Belardi B, Son S, Felce JH et al (2020) Cell–cell interfaces as specialized compartments directing cell function. Nat Rev Mol Cell Biol 21: 750–764
2. Yang BA, Westerhof TM, Sabin K et al (2021) Engineered tools to study intercellular communication. Adv Sci (Weinh) 8:2002825
3. James JR, Vale RD (2012) Biophysical mechanism of T-cell receptor triggering in a reconstituted system. Nature 487:64–69
4. Otani T, Furuse M (2020) Tight junction structure and function revisited. Trends Cell Biol 30:805–817
5. Niessen CM (2007) Tight junctions/adherens junctions: basic structure and function. J Invest Dermatol 127:2525–2532
6. Südhof TC, Malenka RC (2008) Understanding synapses: past, present, and future. Neuron 60:469–476

7. Sjöqvist M, Andersson ER (2019) Do as I say, not(ch) as I do: lateral control of cell fate. Dev Biol 447:58–70
8. Prinz WA, Toulmay A, Balla T (2019) The functional universe of membrane contact sites. Nat Rev Mol Cell Biol 21:7–24
9. Feinberg EH, VanHoven MK, Bendesky A et al (2008) GFP Reconstitution Across Synaptic Partners (GRASP) defines cell contacts and synapses in living nervous systems. Neuron 57:353–363
10. Kanadome T, Hayashi K, Seto Y et al (2022) Development of intensiometric indicators for visualizing N-cadherin interaction across cells. Commun Biol 5:1–12
11. Schmid EM, Bakalar MH, Choudhuri K et al (2016) Size-dependent protein segregation at membrane interfaces. Nat Phys 12:704–711
12. Belardi B, Son S, Vahey MD et al (2019) Claudin-4 reconstituted in unilamellar vesicles is sufficient to form tight interfaces that partition membrane proteins. J Cell Sci 132: jcs221556
13. Freeman SA, Goyette J, Furuya W et al (2016) Integrins form an expanding diffusional barrier that coordinates phagocytosis. Cell 164:128–140
14. Zakeri B, Fierer JO, Celik E et al (2012) Peptide tag forming a rapid covalent bond to a protein, through engineering a bacterial adhesin. Proc Natl Acad Sci U S A 109:E690–E697
15. Feng S, Varshney A, Coto Villa D et al (2019) Bright split red fluorescent proteins for the visualization of endogenous proteins and synapses. Communications Biology 2:1–12
16. Moghimianavval H, Patel C, Mohapatra S et al (2022) Engineering functional membrane–membrane interfaces by InterSpy. Small 19: e2202104
17. Keeble AH, Turkki P, Stokes S et al (2019) Approaching infinite affinity through engineering of peptide-protein interaction. Proc Natl Acad Sci U S A 116:26523–26533
18. Moghimianavval H, Hsu YY, Groaz A et al (2022) In vitro reconstitution platforms of mammalian cell-free expressed membrane proteins. Methods Mol Biol 2433:105–120
19. Majumder S, Hsu YY, Moghimianavval H et al (2022) In vitro synthesis and reconstitution using mammalian cell-free lysates enables the systematic study of the regulation of LINC complex assembly. Biochemistry 61:1495–1507
20. Sharma B, Moghimianavval H, Hwang SW et al (2021) Synthetic cell as a platform for understanding membrane-membrane interactions. Membranes (Basel) 11:912
21. Groaz A, Moghimianavval H, Tavella F et al (2021) Engineering spatiotemporal organization and dynamics in synthetic cells. Wiley Interdiscip Rev Nanomed Nanobiotechnol 13. https://doi.org/10.1002/wnan.1685
22. Mátés L, Chuah MKL, Belay E et al (2009) Molecular evolution of a novel hyperactive sleeping beauty transposase enables robust stable gene transfer in vertebrates. Nat Genet 41: 753–761

Chapter 5

Engineered Protease-Responsive RNA-Binding Proteins (RBPs) to Expand the Toolbox of Synthetic Circuits in Mammalian Cells

Fabiana Calandra and Velia Siciliano

Abstract

Genetically encoded sensor–actuator circuits aim at reprogramming cellular functions and are inspired by intracellular networks: from the input signal (sensor) to the desired output response (actuator). In the last years, circuits with posttranscriptional regulation of gene expression have aroused great interest for their potential in the biomedical space. Posttranscriptional modulation can be achieved with ribozymes, riboswitches (simple regulatory elements based on RNA secondary structures), noncoding RNAs, and RNA-binding proteins (RBPs). RBPs are proteins that recognize specific motifs on the mRNA target inducing mRNA decay or translation inhibition. The use of RBPs deriving from different species in mammalian cells has allowed to create sophisticated and multilayered regulatory networks, addressing the previous limitation of regulatory orthogonal parts that can be assembled in synthetic devices. In this chapter, we describe the engineering and tests of protease-responsive RNA-binding proteins (L7Ae and MS2-cNOT7) to expand the toolbox of synthetic circuits in mammalian cells.

Key words Mammalian synthetic biology, Protein–protein interaction, Protein–RNA regulation, RNA-binding proteins, Protein engineering, Tobacco etch virus protease

1 Introduction

Nature has evolved a pool of biological molecules that are logically linked together to form networks, allowing cells to maintain their functions. Spacing from signal transduction to gene regulation, synthetic biology aims at re-engineering these functions using transcriptional or posttranscriptional synthetic circuits (1–3) with robust expression and reduced impact on cellular physiology, ultimately creating systems with tremendous application in biomedicine research and therapeutics (4–6). Recently, RNA-binding proteins (RBPs) became part of the "synthetic biologist's toolbox" in the design of synthetic circuits (7). Their relevance is linked especially to RNA-based devices: when engineering

Francesca Ceroni and Karen Polizzi (eds.), *Mammalian Synthetic Systems*, Methods in Molecular Biology, vol. 2774,
https://doi.org/10.1007/978-1-0716-3718-0_5,

RNA-encoded synthetic networks to program cellular functions, it is paramount to have foundational tools that control RNA activity. In particular, the RBPs L7Ae and MS2-cNOT7 have been recently characterized as efficient mRNA regulators when repurposed in posttranscriptional synthetic networks. However, the construction of RNA-based multilayered networks is limited by the availability of composable and orthogonal regulatory devices: ribozymes and riboswitches, for example, can modulate the translation of the relative output but cannot be interconnected to create modular circuits (8, 9). Here we describe a protein-engineering-based method that was implemented to add further regulation to RNA circuits. In specific, L7Ae and Ms2-cNOT7 sequence was modified to connect their activity to those of proteases. As proof of principle, the cleavage sites of tobacco etch virus protease (TEVp) was inserted in the protein sequence such that its presence should abolish the RBPs' activity. TEVp is widely used in biotechnology and, to date, there are no off-target substrates reported in the human proteome. In other words, by connecting TEV protease to L7Ae and Ms2-cNOT7, we enabled the tuning of output expression at posttranscriptional and translational level in an orthogonal manner (8).

Proteases recognize specific amino acid sequences leading to proteolytic cleavage. Theoretically, these protease-responsive sequences could be transferred to other proteins. Importantly, this engineering step modifies the primary structure of the protein, which however should retain its wild-type function (the tertiary and quaternary structure is not compromised). This is potentially possible due to the availability of protein crystal structures for a large number of proteins, as well as of software that allow to infer protein structure (10, 11), after in silico modification or via homology analysis for native protein sequences (when the structure is not known yet). L7Ae is an archeal protein that recognizes an RNA kink-turn (K-turn) motif located in the 5'UTR of the target mRNA (12). The mammalian synthetic systems were designed such that the K-turn motif is placed in the 5'UTR of the target mRNA. Thus, in the absence of L7Ae, the ribosome can activate the translation of the mRNA, whereas when present, L7Ae interacts with the K-turn forming a s2e conformation which represses the mRNA translation by inhibiting the access to the ribosomes (Fig. 1).

L7Ae was further engineered to include in the protein structure a consensus motif (cleavage site (CS)) that responds to the tobacco etch virus protease (TEVp) (L7Ae-CS). Based on the L7Ae protein structure, three different insertion points were selected to insert the TEVp CS: they were placed after amino acid residue N70 (L7Ae-CS1), P56 (L7Ae-CS2), or K77 (L7Ae-CS3). The rationale behind the choice of the insertion sites was:

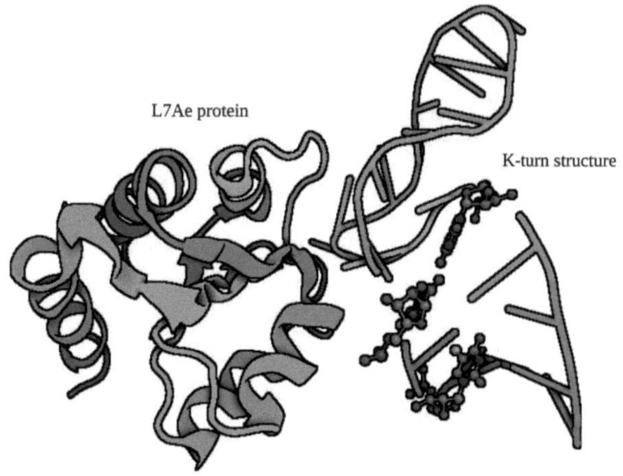

Fig. 1 Crystal structure of the *Archaeoglobus fulgidus* L7Ae protein binding to the box C/D RNA that forms a classical kink-turn (K-turn) structure. Protein structure was determined in[13] (PDB id:1RLG)

(a) To assure protein disruption by TEVp proteolitic cleavage

(b) To have a minimal impact on protein structure and activity

In this configuration, when TEVp is expressed, it cleaves the TCS, disrupting L7Ae structure and inhibiting its function. By disrupting L7Ae-CS repression, the reporter gene is ON (Fig. 2, State 1). On the contrary, in the absence of TEVp, L7Ae-CS retains its repressive activity, and the reporter gene is OFF (Fig. 2, State 2).

MS2-cNOT7 is a chimeric protein. MS2 protein binds specific α-helical repeats located at the 3′ UTR of the target mRNA. cNOT7 is a deadenylase that chops the poly(A) of the target. MS2-cNOT7 was designed to promote target mRNA degradation, via RBP–RNA interaction (13).

Similar to L7Ae, also MS2-cNOT7 was re-engineered to respond to TEVp (Fig. 3). Cleavage site for TEVp protease was inserted in the linker between MS2 and cNOT. Thus, the co-expression of the protease disrupts RBP activity restoring the reporter expression (Fig. 3, State 1), whereas in the absence of TEVp, MS2-TCS-CNOT7 represses the target mRNA inducing its degradation (Fig. 3, State 2).

2 Materials

2.1 In Silico Protein Engineering

1. Personal computer equipped with PyMOL software (10).

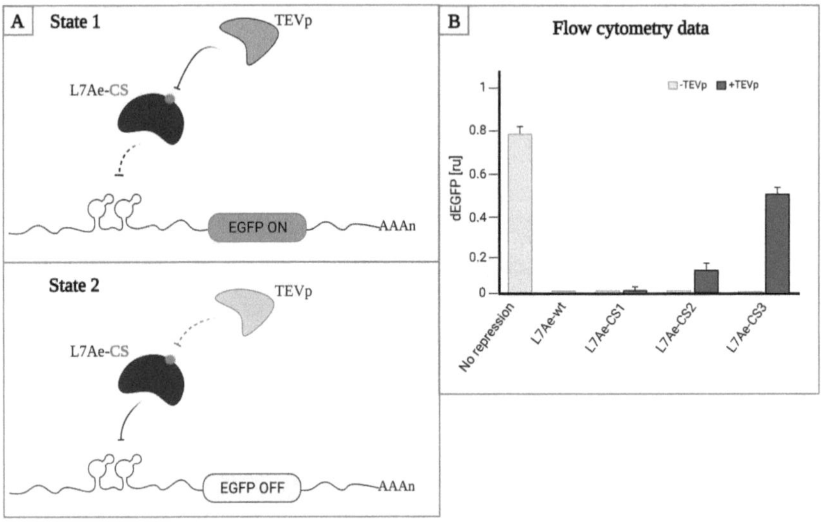

Fig. 2 Translational regulation of engineered TEV protease-responsive L7Ae. (**a**) *State 1*: when TEVp is expressed, it recognizes and cleaves the TCS, disrupting L7Ae and inhibiting its function (EGFP ON). *State 2*: in the absence of TEVp, L7Ae-CS binds and represses the K-turn motifs in the 5'UTR of mRNA target (EGFP OFF). (**b**) Flow cytometry analysis of the three engineered L7Ae-CS (L7Ae-CS1, L7Ae-CS2, L7Ae-CS3) in the presence or absence of TEVp compared to the wild-type L7Ae protein and to the "no repression" condition in which only the reporter gene is present

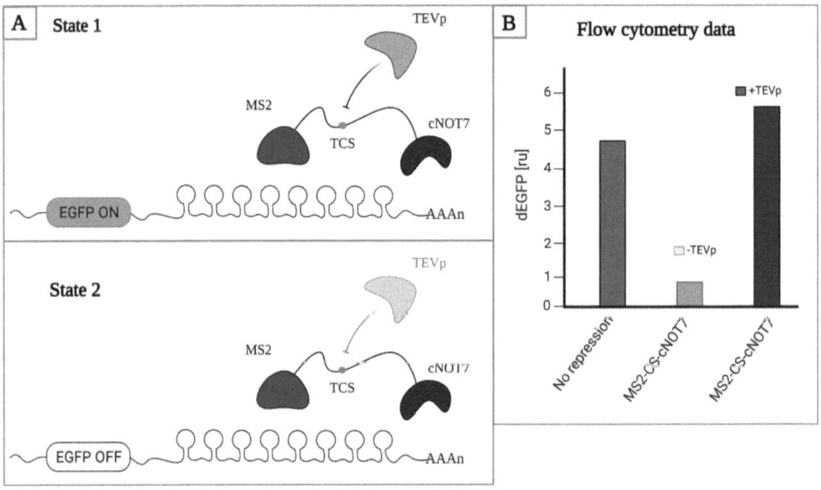

Fig. 3 Schematics of the engineered protease-dependent MS2-cNOT7 protein. (**a**) *State 1:* when TEVp is expressed, MS2-TCS-cNOT7 is cleaved and the two domains are separated. This turns in EGFP translation (EGFP ON). *State 2:* MS2 binds its cognate sequences in the 3'UTR of target EGFP mRNA, resulting in RNA de-adenylation by cNOT7. The mRNA is then degraded (EGFP OFF). (**b**) Flow cytometry analysis of engineered MS2-CS-cNOT7 in the presence or absence of TEVp compared to the "no repression" condition in which only the reporter gene is present

2.2 PCR and Cloning

1. Accuprime PFX Supermix (Thermo Fisher Scientific).
2. BamHI-HF restriction enzyme (NEB).
3. PacI restriction enzymes (NEB).
4. CutSmart® buffer (NEB) to a final concentration of 1× (provided with enzymes).
5. In-Fusion ® HD cloning kit (Clontech) is used to a final concentration of 1×.
6. *E. coli* competent cells.
7. Software to visualize plasmid maps and for in silico primer design (*see* **Note 1**).

2.3 Cell Culture

1. HEK293FT cell line (Invitrogen).
2. Dulbecco's Modified Eagle's Medium (DMEM) phenol red containing 4.5 g/L glucose, L-glutamine, pyruvate.
3. Fetal bovine serum (FBS).
4. Penicillin/streptomycin.
5. L-Glutamine.
6. MEM nonessential amino acids.
7. CO_2-humidified incubator.
8. Trypsin–EDTA 0.25% phenol red.
9. Dulbecco's phosphate-buffered saline (DPBS) no calcium no magnesium.

2.4 Transient Transfection, Cell Imaging, and Flow Cytometry

1. Attractene transfection reagent (Qiagen).
2. Lipofectamine™ 3000 transfection reagent (Invitrogen).
3. Opti-MEM reduced serum medium.
4. 24-well flat bottom plates
5. Countess™ II Automated Cell Counter and Countess® Cell Counting Chamber Slides (Invitrogen).
6. Trypan blue.
7. EVOS® Cell Imaging System (Life Technologies) using 10× objective with EVOS® Light Cubes Texas Red, GFP, and DAPI.
8. LSRFortessa flow cytometer equipped with 405, 488, and 561 nm lasers (BD Biosciences).
9. Round-bottom polystyrene FACS tubes.
10. FlowJo software to perform data analysis (*see* **Note 2**).

3 Methods

The efficiency of the re-engineered devices is tested indirectly using fluorescent reporters responsive to RBPs' activity. After cloning, all plasmids' sequence must be confirmed by sequencing.

3.1 Cell Culture

To maintain the HEK293FT cells, use Dulbecco's Modified Eagle's Medium (DMEM) phenol red supplemented with 10% fetal bovine serum (FBS), 1% penicillin/streptomycin, 1% L-glutamine, and 1% MEM nonessential amino acids. Culture cells at 37 °C in a 5% CO2-humidified incubator.

3.2 Protein Structure Analysis for Re-engineering of RBPs and Plasmid Cloning

3.2.1 L7Ae

1. L7Ae crystal structure is available on Protein Data Bank (PDB) with the id:IRLG (14) (*see* **Note 3**). Visualize the structure in PyMOL to identify possible insertion points for TEV protease cleavage site (TCS): three loci were identified.
2. Synthetize the three L7Ae-CS variants' coding sequence as gBlocks (IDT) for cloning in the destination pL-Al plasmid (GenBank code: MH107777) with BamHI and PacI restriction enzymes.
3. Insert the gBlocks into pL-Al backbone by In-Fusion between BamHI and PacI restriction sites. The ratio of backbone/gBlock used in the ligation step should be 1:2.
4. The reporter plasmid used to test L7Ae activity in the presence or absence of TEV protease is pLS-18 (pBoxCDGC_2xK-Met_DD-EGFP) which encodes for two kink-turn motives upstream an EGFP fluorescent protein with a domain called death domain (DD). The plasmid design is described in Wroblewska et al. (15). pLS-18 plasmid and its sequence are available upon reasonable request from Velia Siciliano (velia.siciliano@iit.it).

3.2.2 MS2-cNOT7

1. Insert the cleavage site for TEVp in the linker between MS2 and NOT7 by PCR (*see* **Note 4**). Note that in this case, the crystal structure is not needed, since this is a fusion protein.
2. Amplify by PCR both MS2 (namely, PCR1) and NOT7 (namely, PCR2) from pL-R1 (GenBank code: MH883359). To do this, use Accuprime PFX DNA Polymerase.
3. Clone the PCR fragments by In-Fusion cloning in pL-Al plasmid previously cut with BamHI and PacI restriction enzymes (*see* **Note 5**). Backbone/PCR1/PCR2 ratio of 1:2:2.
4. pLC-1 reporter plasmid encoding for eight Ms2-binding motives downstream an EGFP fluorescent protein (pBoxCDGC- mut_KMet-EGFP-8xMS2-pA) is used to detect MS2-CS-cNOT7 activity and, indirectly, TEVp function. The

plasmid was published in Wroblewska et al. (15) (GenBank code: MH883358), and it is available upon request from Hirohide Saito (hirohide.saito@cira.kyoto-u.ac.jp) or Ron Weiss (rweiss@mit.edu).

3.3 Test of the Engineered Proteins and Their Interaction with TEV Protease

To test protein–protein interaction devices, measure fluorescent reporters' expression by flow cytometry at steady state 48 h post-transfection. The plasmid coding for TEVp used was p96 (pGTW6-hEF1a-TEV-p, GenBank code: MH883330).

3.3.1 Preparation of Transfection Mixes

1. To test L7Ae-CS, perform transfections with Attractene transfection reagent in HEK293FT cells in 24-well plates format. Aliquot 60 uL of uncomplemented DMEM for each transfection mix. Add a total of 400 ng of DNA per reaction mix (50 ng of fluorescent reporter, 150 ng of L7Ae variant, 60 ng of wild-type protease, 50 ng of transfection marker, empty plasmid to 400 ng). Add 1.5 μL of Attractene transfection reagent for each mix and vortex (*see* **Note 6**). Incubate for 15–20 min.
2. Perform transfections to test MS2-CS-NOT7 and protease–RBP circuit with Lipofectamine™ 3000 transfection reagent in HEK293FT cells in 24-well plates. From now on, each reaction mix volume described is for one well. Scale volumes proportionally for additional wells.

   Prepare two master mixes (i) Opti-MEM + p3000 and (ii) Opti-MEM + Lipofectamine™ 3000 considering the following:

   (a) 25 μL of Opti-MEM and 1 μL of P3000 reagent for each sample

   (b) 25 μL of Opti-MEM and 0.75 μL of Lipofectamine™ 3000 for each sample.

Aliquot 26 μL of master mix (a) per sample in separate Eppendorf tubes (e.g., in a condition with duplicate, 52 μL in total). Add 400 ng of DNA per reaction mix to the master mix (a) (25 ng of fluorescent reporter, 50 ng of Ms2-CS-NOT7 variant, 30 ng of engineered protease, 50 ng of wild-type protease, 50 ng of transfection marker, empty plasmid to 400 ng). Then, add 25.75 μL of master mix (b) to each sample and mix them (*see* **Note 6**). Incubate the reactions for 15 min.

3.3.2 Cell Seeding

1. During the 15 min of incubation time, plate HEK293FT cells in 24-well plates (*see* **Note** 7).
2. To detach cells, first, remove the medium from the flask, and then gently wash the flask with PBS (10 mL for T75 and 5 mL for T25). Add trypsin (1.5 mL for T75 and 0.5 mL for T25) and keep the flasks for 2 min in the incubator (*see* **Note 8**).

Next, add fresh new complemented DMEM to the trypsinized cells (5.5 for T75 and 3.5 mL for T25), mix well, and transfer to a 15 mL tube. Mix 10 μL of resuspended cells with 10 μL of trypan blue and load 10 μL of this mix in a Countess® Cell Counting Chamber Slide and load in the Countess® Cell Counter II for cell counting.

3. Seed a total of 140,000 cells/well in a final volume of 500 μL of complete DMEM. First, prepare the master mix, and then aliquot 500 μL to each well.
4. Add the DNA–lipid complexes to cells.
5. After 24 h, add 1 mL of fresh complemented DMEM to each well.

3.3.3 Flow Cytometry

1. 48 h after transfection, observe the transfected cells at EVOS® Cell Imaging System and acquire images of the transfection in all the fluorescent channels (EGFP and the color of the chosen transfection marker).
2. Analyze the cells with flow cytometer. First, remove DMEM from the wells and detach cells adding 50 μL of trypsin to each well. Keep the plate in an incubator for 2 min. Add 300 μL of DMEM supplemented with 10% of FBS to each well.
3. Transfer the cells into round-bottom FACS tubes and keep them on ice. Vortex each tube for few seconds before loading it into the flow cytometer (*see* **Notes 9–10**).
4. Record 20,000 events in the single cell population (FSC-A vs. FSC-H plot to exclude doublets). Correct laser voltages were set using controls: unstained (wild-type cells), and single color controls.
5. Data were exported from Diva software as FACS files and analyzed with FlowJo software.

4 Results

L7Ae-CS

The L7Ae structure analysis suggested three optimal translation repressors: L7Ae-CS1, L7Ae-CS2, and L7Ae-CS3. These circuits were then built and tested along with the reporter gene (dEGFP) regulated by two repeats of the K-turn motif in its 5'UTR. In Fig. 2b, flow cytometry analysis of engineered L7Ae-CS shows that the presence of TEVp can rescue dEGFP expression by disrupting L7Ae. In particular, L7Ae-CS3 shows more efficient and sustained repression/derepression, meaning that the CS3 insertion caused the least impact on L7Ae structure. It is important to notice that the study of the protein sequence and its three-dimensional structure is of paramount importance for not to have the complete loss of the protein function.

This new protease-dependent translation system can be used to create several circuits; an example is the HCV sensing device able to detect NS3, a viral serine protease (8). The device relies on the use of two single-chain fragment intrabodies for two distinct epitopes of specific NS3 (scFv35 and scFv162). In this configuration, L7Ae-CS and TEVp are fused to scFv35 and scFv162, respectively. The presence of the target protein NS3 and subsequent binding of the two intrabodies results in TEVp cleavage of the L7Ae-CS and derepression of dEGFP reporter.

MS2-CS-cNOT7

The TEV protease-responsive MS2-cNOT7 repressor device showed similar repression to its wild-type counterpart. Figure 3b shows flow cytometry analysis of engineered MS2-CS-cNOT7: in the presence of TEVp, MS2-CS-cNOT7 is disrupted, consequently restoring dEGFP expression levels. Since MS2-cNOT7 is a fusion protein with a linker region, cNOT7 may be hypothetically changed with another protein, depending on the demand. This change could lead to steric restrictions in protein folding and subsequent disruption of protein function.

The platform that was established allows further tailoring of features such as type of proteases that the RNA-binding protein can respond to (*see* **Note 11**). A library of MS2-CS-cNOT7 variants can be built by using other viral proteases such as SuMMVp, TUMVp, and TVMVp proteases (8) enabling novel protein–protein regulation systems.

5 Notes

1. In this study, SnapGene software (www.snapgene.com) or Geneious software (https://www.geneious.com) were used, but any other software for in silico plasmid design would work.
2. Other software for flow data analysis can be used.
3. If the crystal structure of your protein of interest is not resolved, but the structure of homologous proteins is available, you may infer it by homology modeling, e.g., using SWISS-MODEL (11).
4. In the case of a chimeric protein, the crystal structure is not needed as the protease cleavage site is inserted in the linker between the two domains.

5. In this step, make sure to design primers with the right overhangs so that the correct order is maintained: MS2-CS-cNOT7.
6. Mix by light vortexing right after the addition of the liposome to the DNA.
7. This step can be performed before mix preparation since for the beginners, it could require more than 15 min.
8. Trypsin incubation time may vary according to the cell type.
9. To reduce cellular aggregation and increase the number of single cells for the correct analysis, add EDTA to a concentration of 2 mM.
10. Washing cells with PBS before detachment reduces the clumps of cells.
11. The re-engineering of L7Ae and MS2-cNOT7 provides a versatile tool that can be rewired to respond to novel demands. One possibility, for example, is to swap the cleavage site inserted in the protein structure to respond to novel, orthogonal proteases.

Acknowledgments

We thank Daniela Perna for technical support. FC was supported by NextGenerationEU PNRR MUR [M4C2], National Center for Gene Therapy, and Drugsbased on RNA Technology.

References

1. MacDonald JT, Siciliano V (2017) Computational sequence design with R2oDNA designer. Methods Mol Biol 1651:249–262
2. Cella F, de Martino I, Piro F, Siciliano V (2021) Engineering programmable RNA synthetic circuits in mammalian cells. Curr Opin Syst Biol 28:100395
3. Crone MA, MacDonald JT, Freemont PS, Siciliano V (2022) gDesigner: computational design of synthetic gRNAs for Cas12a-based transcriptional repression in mammalian cells. NPJ Syst Biol Appl 8:1–7
4. di Blasi R, Marbiah MM, Siciliano V et al (2021) A call for caution in analysing mammalian co-transfection experiments and implications of resource competition in data misinterpretation. Nat Commun 12:2545
5. Frei T, Cella F, Tedeschi F et al (2020) Characterization and mitigation of gene expression burden in mammalian cells. Nat Commun 11: 4641
6. Jones RD, Qian Y, Siciliano V et al (2020) An endoribonuclease-based feedforward controller for decoupling resource-limited genetic modules in mammalian cells. Nat Commun 11:5690
7. Cella F, Siciliano V (2019) Protein-based parts and devices that respond to intracellular and extracellular signals in mammalian cells. Curr Opin Chem Biol 52:47–53
8. Cella F, Wroblewska L, Weiss R, Siciliano V (2018) Engineering protein-protein devices for multilayered regulation of mRNA translation using orthogonal proteases in mammalian cells. Nat Commun 9:4392
9. Culler SJ, Hoff KG, Smolke CD (2010) Reprogramming cellular behavior with RNA controllers responsive to endogenous proteins. Science 330:1251–1255
10. Schrödinger L, DeLano W (2020) PyMOL. Available at: http://www.pymol.org/pymol

11. Waterhouse A, Bertoni M, Bienert S et al (2018) SWISS-MODEL: homology modelling of protein structures and complexes. Nucleic Acids Res 46:W296–W303
12. Saito H, Kobayashi T, Hara T et al (2010) Synthetic translational regulation by an L7Ae-kink-turn RNP switch. Nat Chem Biol 6:71–78
13. van Etten J, Schagat TL, Hrit J et al (2012) Human Pumilio proteins recruit multiple deadenylases to efficiently repress messenger RNAs. J Biol Chem 287:36370–36383
14. Moore T, Zhang Y, Fenley MO, Li H (2004) Molecular basis of box C/D RNA-protein interactions; cocrystal structure of archaeal L7Ae and a box C/D RNA. Structure 12: 807–818
15. Wroblewska L, Kitada T, Endo K et al (2015) Mammalian synthetic circuits with RNA binding proteins for RNA-only delivery. Nat Biotechnol 33:839–841

Chapter 6

Mechanistic Model-Driven Biodesign in Mammalian Synthetic Biology

Yin Hoon Chew and Lucia Marucci

Abstract

Mathematical modeling plays a vital role in mammalian synthetic biology by providing a framework to design and optimize design circuits and engineered bioprocesses, predict their behavior, and guide experimental design. Here, we review recent models used in the literature, considering mathematical frameworks at the molecular, cellular, and system levels. We report key challenges in the field and discuss opportunities for genome-scale models, machine learning, and cybergenetics to expand the capabilities of model-driven mammalian cell biodesign.

Key words Mathematical modeling, Synthetic biology, Mammalian cells, Gene network, Context-aware design, Metabolic models, Whole-cell models

1 Introduction

Mammalian synthetic biology is a rapidly growing field that aims to engineer mammalian cells to perform specific functions, such as the production of therapeutic proteins or the detection of disease biomarkers. Mathematical modeling is a key tool in this field, allowing researchers to predict the behavior of genetic circuits and parts and to optimize their design before experimental implementation.

Indeed, synthetic biology is similar to other engineering activities where designing and testing models is crucial to optimize the final product [1]. Modeling has been a key element of synthetic biology since the very first works in the field, which combined the derivation and numerical analysis of simple ordinary differential equations (ODEs) models of gene networks to describe and predict the optimal parameter space ensuring the systems to behave as expected [2, 3]. Modeling in biology is usually an iterative process and requires validation by comparing simulation results with experimental measurements. This is particularly true for kinetic

Francesca Ceroni and Karen Polizzi (eds.), *Mammalian Synthetic Systems*, Methods in Molecular Biology, vol. 2774,
https://doi.org/10.1007/978-1-0716-3718-0_6,

modeling, which describes changes in a system over time. Modeling has become more accessible with the availability of software tools and community standards and is used in parallel with experimental approaches. Nonetheless, modeling remains an interdisciplinary task that requires both biological and mathematical knowledge, as well as computational and experimental experience.

We structure this chapter by presenting different types of models which can be used to describe and predict cellular behaviors at different representation scales. The last part of the chapter is devoted to summarize ongoing challenges and to suggest opportunities for the adoption of new types of models and computational approaches in the field.

2 Models at the Molecular and Cellular Level

At the molecular level, mathematical modeling can be used to predict the behavior of individual genetic elements. These elements can be engineered to control the expression of genes of interest, and their behavior can be predicted using mathematical models. For example, various computational frameworks (including machine learning-based ones) have been used to predict the strength of mammalian cell promoters and enhancers [4], represent promoter binding preferences, and predict their genome-wide binding locations [5].

At the cellular level, mathematical models can be used to predict the behavior of genetic circuits in specific cell types and environments. Genetic circuits are composed of elements such as promoters, enhancers, and transcription factors that are combined to produce a desired output [6]. Mathematical models can be used to predict the behavior of these circuits and to optimize their function [7].

A range of gene networks with predictable dynamics have been designed, modelled, and engineered in mammalian cells to date [8]. For example, various circuits have been engineered to enable cells to show multistability (i.e., the presence of multiple steady states, a mechanism often exploited by nature in cell fate-determining processes such as development and the cell cycle [9, 10]). Early circuits demonstrating hysteresis included tunable positive feedback loops [11] (Fig. 1a); these gene networks were relatively simple and so easy to characterize, and combination of modeling and experiments was crucial to establish the right balance of circuit components to enable hysteresis. Other early synthetic biology works which combined modeling and experiments to engineer biological oscillators include gene networks able to behave as a circadian clock in mammalian cells [12] and a low-frequency oscillator based on a synthetic tetracycline transactivator and amplified repression via siRNA-mediated silencing [13] (Fig. 1b). Hybrid

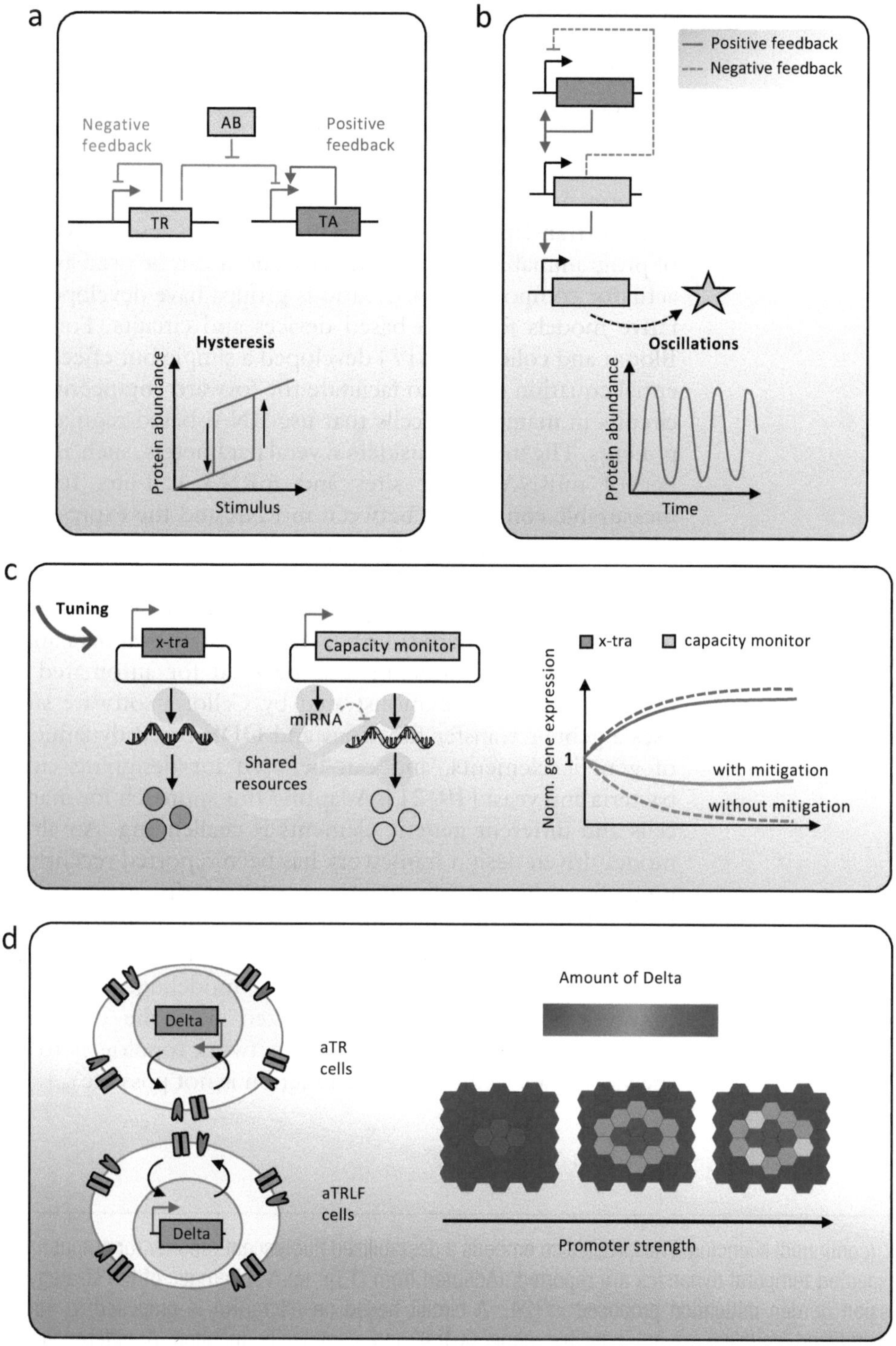

Fig. 1 Mathematical models can predict engineered cellular phenotypes. (**a**, **b**) Examples of gene networks and model-predicted dynamics. (**a**) Topology of a bistable gene network in mammalian cells. A transactivator (TA) activates its own transcription by binding its promoter, which can also be repressed by a transrepressor (TR) whose activity is modulated by an antibiotic (AB). The expected dynamics are reported. (Adapted from [11]). (**b**) Topology of a low-frequency synthetic oscillator in mammalian cells. It consists of an autoregulated gene under the control of an inducible promoter. The negative feedback loop relies on interference-based

systems, composed of combined engineered and natural parts, have also been proposed [14, 15].

RNA molecules are essential components of cellular signalling pathways that modulate gene expression and function by transducing extracellular and intracellular signals. RNAs interact with nucleic acids, proteins, and small molecules, making them versatile components that regulate genes at the transcriptional, translational, and posttranslational levels; this makes RNA a desirable component of programmable synthetic devices, and it can be used as a sensor–actuator component [16]. Various groups have developed quantitative models for RNA-based devices and circuits. For example, Bloom and colleagues [17] developed a simple but effective differential equation model to facilitate the forward engineering of gene circuits in mammalian cells that use RNAi-based regulatory components. The model considers several parameters, such as the number of miRNA target sites and mRNA half-life, to create a measurable connection between miRNA and the expression levels of target genes. Townshend and colleagues [18] employed next-generation sequencing (NGS) technologies to create a screening method capable of evaluating the activity of numerous RNA devices based on the hammerhead ribozyme and statistical data analysis.

Mathematical modeling can be used for automated genetic program design, as demonstrated by Cello: a software suite that uses algebraic transfer functions and ODE-based dynamic models of genetic elements, and can be used for designing circuits in bacteria and yeast [19–21]. Adapting this approach for mammalian cells and different genetic elements is challenging. An alternative model-driven design framework has been reported recently for the predictive design of genetic programs in mammalian cells [22]. This tool combines use of COMET (a toolkit of promoters and transcription factors for orthogonal gene expression control) [23] and differential equation-based modeling. The process is however not completely automatized (i.e., the exploration of large combinatorial spaces of gene network topologies to check if they satisfy a specific objective function is not possible).

Fig. 1 (continued) silencing. The circuit also embeds a destabilized fluorescent reporter for output monitoring. The expected temporal dynamics are reported. (Adapted from [13]). (**c**) A schematic of the strategy for gene expression burden mitigation proposed in [24]. A circuit based on microRNA is proposed to mitigate the burden-induced coupling of an inducible gene (X-tra) and a capacity monitor. A differential equation mathematical model (whose reaction rates capture the reduced resource availability due to competing genes) is used to describe the process. (Adapted from [24]). (**d**) A multicellular engineered system, implemented via positive feedback loops, which enables co-cultured "trigger" cells to propagate induction of Delta to adjacent cells. A model recapitulates the system dynamics depending on the strength of exogenous promoters. (Adapted from [31])

2.1 Models at the System Level

At the system level, mathematical models can be used to predict the behavior of complex biological systems and to design and optimize their function.

One of the key goals of synthetic biology is output reproducibility and robustness; as such, it is important to address possible gene expression burden issues and to develop quantitative models that can predict the effect of exogenous parts on endogenous cellular processes. Frei and colleagues [24] showed in various mammalian cell lines a promoter-independent negative correlation between the transient expression of two otherwise independently expressed genes. They proposed a miRNA-based network to reduce burden and developed a simple but elegant resource-aware kinetic model, where the rates of reactions which involve use of shared resources account for reduced availability due to the presence of competition among different processes (Fig. 1c). Model simulations were able to recapitulate nonintuitive dose responses in different gene network topologies. More recently, an extension of this model (where the reaction rates account for the availability of resource pools given the overall gene expression demand) was used to predict the effect of miRNA regulation on resource availability [25]. In a different work, a combination of modeling results and experimental observations of network-imposed cell burden led to the creation of gene networks that are "context-aware" as they utilize biomolecular controllers such as endoribonucleases [26].

Models at the system level can also be instrumental to design more complex phenotypes in mammalian systems. Synthetic developmental programs have been created for multicellular mammalian systems to control features like patterning and morphogenesis (*see* refs. 27,28] for recent reviews). These programs draw inspiration from natural genetic circuits that pattern and construct embryos, linking cell-to-cell communication channels, multicellular gene regulatory networks for patterning, and effector genes to change cell biophysical parameters for morphogenesis (Fig. 1d) [16, 29–32]. The use of models can reduce the time associated with the engineering of these complex multicellular structures. Sets of differential equations are commonly used to describe multicellular signalling networks and the levels of reactive species/proteins on a lattice that represents the cells [33–35]. ODEs and partial differential equations (PDEs) can be used for describing temporal and spatial patterning, respectively. Diffusible signals and their effect on cellular patterns can also be represented mathematically [33], as well as mechanical aspects (e.g., cell-to-cell forces and cell growth) both within continuum and agent-based models [36–38].

3 Next-Generation Models and Methods for Mammalian Cell Synthetic Biology

Over the next years, we would expect synthetic biology to enable the engineering of more complex functions in mammalian cells and to drive transformative advances in healthcare. From the modeling perspective, models should move from the description of parts to the representation of entire cells and organisms, ultimately enabling a full quantitative prediction of how genomic engineering would be mapped into target phenotypes.

The availability of genome-scale high-throughput data has enabled the derivation of genome-scale models; they are extremely well qualified to support such research, as they promise to quantitatively describe and predict genotype–phenotype relationships in both native and re-engineered living systems.

The most used genome-scale models in synthetic biology are genome-scale metabolic models (GEMs); they are constrained based, use mathematical equations to describe the flow of metabolites through various biochemical pathways, and can be used to simulate the behavior of cells under different conditions, such as changes in nutrient availability or genetic perturbations [39–42]. Constraint-based modeling can integrate various factors such as dynamics and gene expression [43, 44]; machine learning also has enormous potential for the derivation, validation, and application of metabolic models (see ref. 41 for a review). GEMs see a direct application in biomanufacturing. For example, metabolic engineering, which aims at optimizing fluxes and ultimately bioprocesses, can clearly benefit from the combination of metabolic models with optimization algorithms (e.g., linear programming-based methods) to predict genes to remove, overexpress, or silence to design cells with the desired phonotype, such as production of a specific metabolite [45]. The combination of machine learning and metabolic modeling, CRISPR-derived genome engineering tools, and synthetic genetic circuits can be disruptive for designing synthetic pathways for bioproduction [46]. The development of GEMs for mammalian cells has been achieved only quite recently by developing frameworks that generate and maintain coherent GEMs combining multiple data [47, 48]. The use of mammalian systems' GEMs in the engineering biology or biomedical space (e.g., to design a desirable healthy phenotype which reduces a disease-associated one) has a huge and not yet fully explored potential [49, 50].

Going further up in the spectrum of mathematical model complexity, we can find whole-cell models (WCMs) that are computational models of single cells that represent every gene and gene function [51–56]. These models integrate multiple types of data and biological knowledge and mechanistically represent the major biochemical processes in cells. As such, they can simulate and predict emergent behaviors that arise from complex interactions among the processes. This makes them useful for guiding genome design to achieve cells with desired traits and medical intervention to regulate cellular behavior. The potential use of WCMs in synthetic biology has been explored for the bacterium *Mycoplasma genitalium* (*M. genitalium*) [57–60]. Its small genome makes this prokaryotic bacterium an attractive system both for building the first WCM and for constructing a minimal cell. Following on from accomplishing the *M. genitalium* model, other WCMs have been developed including for *Escherichia coli* (*E. coli*) [56], the most used cell factory in the biomanufacturing industry. We anticipate wide adoption of the *E. coli* model for rational design of strains with improved productivity and added capabilities.

Given the utility of WCMs for model-guided design of prokaryotic cells, we believe that similar models for mammalian cells could be as useful and revolutionize design–build–test–learn cycles in mammalian cell synthetic biology (Fig. 2) [61]. Additionally, there is potential for using WCMs of human cells to guide regenerative medicine such as tissue engineering and personalized medical therapies. Work has begun to prototype a WCM of H1 human embryonic stem cells (https://www.wholecell.org/models). Stem cells are suitable choice for pioneering a mammalian WCM because they can self-renew and produce identical daughter cells, much like the clonal reproduction in prokaryotes. Therefore, many assumptions used for modeling prokaryotic cells are applicable.

The challenges to whole-cell modeling of mammalian cells are exemplified in the H1 modeling work. Due to the much larger genome, general challenges such as lack of data and high computational cost are magnified. Additionally, there are challenges specific to eukaryotes and mammalian cells. First, these cells tend to exhibit high redundancy. Many genes have different splice variants and/or form families of genes that may be structurally and functionally similar but also are also context-dependent. Insufficient understanding aside, the bulk of data in the literature and repositories come in a mixed level of granularity. Some measurements lump variants and gene families, and the metadata that document this are not always available. Second, signalling pathways and regulatory networks in mammalian cells are more complex than in bacteria: signals are transmitted via branched cascades of reactions that usually involve interconnected feedback loops. Third, molecules and reactions in mammalian cells are compartmentalized into organelles with intricate structures. These compartments are not just static containers; some are dynamic structures that can fuse and divide. Studies have suggested that changing shape and size of compartments could provide another means of adaptation for eukaryotes [62]. This adds a new dimension that is not currently represented in WCMs.

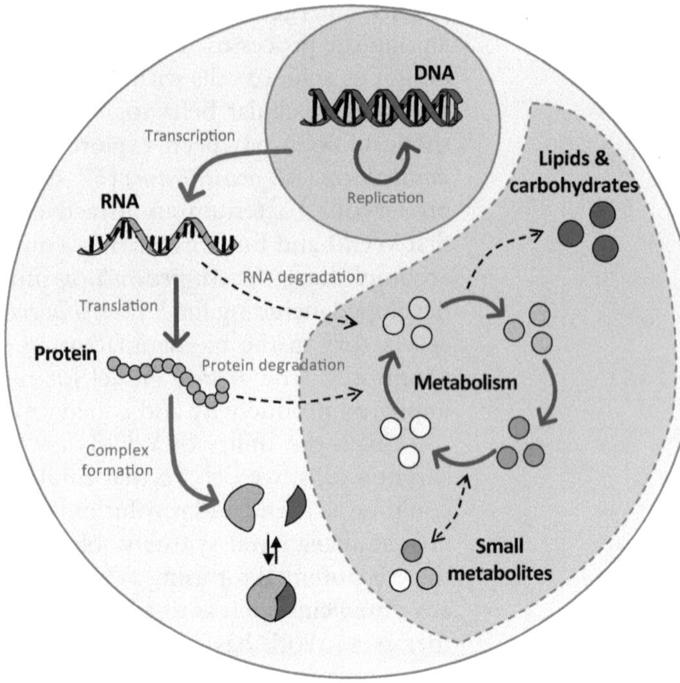

Fig. 2 Possible processes to be included in a WCM for human cells. Simplified diagram illustrating the core molecular species and processes to be represented in whole-cell models for human cells

4 Challenges and Opportunities

Despite its many benefits, mathematical modeling also has several challenges and limitations when applied to mammalian synthetic biology. One major challenge is the need for accurate and comprehensive experimental data to calibrate and validate models; experimental data may be noisy or incomplete, making it difficult to develop accurate models. The development of new experimental platforms will continue to improve the type, quality, and resolution of experimental data. This will also lead to increased understanding of cell biology including gene function and their mechanism. We recommend creating centralized databases to store these data using FAIR principles [63–65].

Another challenge is the need to capture the complexity of biological systems in mathematical models. Biological systems are characterized by nonlinearity, feedback, and stochasticity, which can make them difficult to model using traditional mathematical methods. Many mathematical models in mammalian synthetic biology are based on simplified assumptions and may not fully capture the complexity of biological systems. Looking forward, a collaborative effort is needed to build computational digital twins of living cells to be used to guide mammalian cell re-engineering.

Due to current interest in data science and AI, we are seeing heavy investments that aim to enhance computing power through better algorithms and hardware designs. Advancements in this area will be beneficial to cellular modeling. Machine learning and AI could support both the analysis of genome-scale models outputs (e.g., via surrogate modeling [66, 67]) and the automation of model derivation/calibration.

We also propose building models collaboratively, particularly for mammalian and other eukaryotic cells. This calls for platforms where efforts can be coordinated, interfaces for connecting work from different research groups, and scalable methods that can handle large-scale data and models.

We have an opportunity to extend models of mammalian cells to also include structure function representation. For example, a recent WCM for synthetic minimal bacterial cells represents cell structure to some extent [68]. Beyond that, the explicit functions of compartments and how they dynamically regulate cellular behavior are not considered as research in this area is in its infancy. From the perspective of synthetic biology, adding structure function would enhance the capability of mathematical models' predictions, for example, to help design protocells. Our suggestions are not an exhaustive list. We anticipate that the emerging field of mathematical modeling will continue to evolve, together with progress in other fields.

Parallel to developing models, scalable software tools need to be built to partially automate data curation, data organization, and model generation. Another important aspect to consider is related to model languages. Many existing models are difficult to reuse due to unclear coding, inappropriate languages, or inadequate documentation. Software engineering can address these issues and promote software reuse. Ideally, languages should be modular, human-readable, and open [69].

We foresee opportunities for use of cybergenetics for the derivation and validation of models in synthetic biology. Cybergenetics is an emerging field at the interface of control engineering with synthetic biology that is developing experimental and computational tools for the robust control of cellular processes across different species [70, 71]. It can be implemented via embedded controllers (i.e., gene networks) [72–74], multicellular controllers in cellular consortia [75–77], and external feedback controllers (i.e., via microfluidics or optogenetics-based platforms) [80–84]. The last approach, in particular, has high potential to characterize the dynamics of gene circuits in a model-free way [85], suggest how to revert undesired cellular phenotypes [86], and collect online reach datasets for model derivation and calibration, possibly aided by optimal experimental design approaches [78].

Acknowledgments

LM was funded by the Engineering and Physical Sciences Research Council (EPSRC, EP/S01876X/1) and the Biotechnology and Biological Sciences Research Council (BBSRC, Bristol Centre for Engineering Biology, BB/W013959/1; Breakthrough Award BB/W012235/1).

References

1. Zheng Y, Sriram G (2010) Mathematical modeling: bridging the gap between concept and realization in synthetic biology. J Biomed Biotechnol 2010:541609. https://doi.org/10.1155/2010/541609
2. Elowitz MB, Leibler S (2000) A synthetic oscillatory network of transcriptional regulators. Nature 403(6767):335–338. https://doi.org/10.1038/35002125
3. Gardner TS, Cantor CR, Collins JJ (2000) Construction of a genetic toggle switch in Escherichia coli. Nature 403(6767):339–342. https://doi.org/10.1038/35002131
4. Zehnder T, Benner P, Vingron M (2019) Predicting enhancers in mammalian genomes using supervised hidden Markov models. BMC Bioinf 20(1):157. https://doi.org/10.1186/s12859-019-2708-6
5. Elnitski L, Jin VX, Farnham PJ, Jones SJ (2006) Locating mammalian transcription factor binding sites: a survey of computational and experimental techniques. Genome Res 16(12):1455–1464. https://doi.org/10.1101/gr.4140006
6. Verbic A, Praznik A, Jerala R (2021) A guide to the design of synthetic gene networks in mammalian cells. FEBS J 288(18):5265–5288. https://doi.org/10.1111/febs.15652
7. di Bernardo D, Marucci L, Menolascina F, Siciliano V (2012) Predicting synthetic gene networks. Methods Mol Biol 813:57–81. https://doi.org/10.1007/978-1-61779-412-4_4
8. Mathur M, Xiang JS, Smolke CD (2017) Mammalian synthetic biology for studying the cell. J Cell Biol 216(1):73–82. https://doi.org/10.1083/jcb.201611002
9. Brandman O, Meyer T (2008) Feedback loops shape cellular signals in space and time. Science 322(5900):390–395. https://doi.org/10.1126/science.1160617
10. Marucci L (2017) Nanog dynamics in mouse embryonic stem cells: results from systems biology approaches. Stem Cells Int 2017:7160419. https://doi.org/10.1155/2017/7160419
11. Kramer BP, Fussenegger M (2005) Hysteresis in a synthetic mammalian gene network. Proc Natl Acad Sci U S A 102(27):9517–9522. https://doi.org/10.1073/pnas.0500345102
12. Chilov D, Fussenegger M (2004) Toward construction of a self-sustained clock-like expression system based on the mammalian circadian clock. Biotechnol Bioeng 87(2):234–242. https://doi.org/10.1002/bit.20143
13. Tigges M, Denervaud N, Greber D, Stelling J, Fussenegger M (2010) A synthetic low-frequency mammalian oscillator. Nucleic Acids Res 38(8):2702–2711. https://doi.org/10.1093/nar/gkq121
14. Toettcher JE, Mock C, Batchelor E, Loewer A, Lahav G (2010) A synthetic-natural hybrid oscillator in human cells. Proc Natl Acad Sci U S A 107(39):17047–17052. https://doi.org/10.1073/pnas.1005615107
15. Santorelli M, Perna D, Isomura A, Garzilli I, Annunziata F, Postiglione L et al (2018) Reconstitution of an ultradian oscillator in mammalian cells by a synthetic biology approach. ACS Synth Biol 7(5):1447–1455. https://doi.org/10.1021/acssynbio.8b00083
16. Black JB, Perez-Pinera P, Gersbach CA (2017) Mammalian synthetic biology: engineering biological systems. Annu Rev Biomed Eng 19:249–277. https://doi.org/10.1146/annurev-bioeng-071516-044649
17. Bloom RJ, Winkler SM, Smolke CD (2015) Synthetic feedback control using an RNAi-based gene-regulatory device. J Biol Eng 9:5. https://doi.org/10.1186/s13036-015-0002-3
18. Townshend B, Kennedy AB, Xiang JS, Smolke CD (2015) High-throughput cellular RNA device engineering. Nat Methods 12(10):989–994. https://doi.org/10.1038/nmeth.3486
19. Jones TS, Oliveira SMD, Myers CJ, Voigt CA, Densmore D (2022) Genetic circuit design automation with Cello 2.0. Nat Protoc 17(4):1097–1113. https://doi.org/10.1038/s41596-021-00675-2

20. Chen Y, Zhang S, Young EM, Jones TS, Densmore D, Voigt CA (2020) Genetic circuit design automation for yeast. Nat Microbiol 5(11):1349–1360. https://doi.org/10.1038/s41564-020-0757-2
21. Nielsen AA, Der BS, Shin J, Vaidyanathan P, Paralanov V, Strychalski EA et al (2016) Genetic circuit design automation. Science 352(6281):aac7341. https://doi.org/10.1126/science.aac7341
22. Muldoon JJ, Kandula V, Hong M, Donahue PS, Boucher JD, Bagheri N et al (2021) Model-guided design of mammalian genetic programs. Sci Adv 7(8). https://doi.org/10.1126/sciadv.abe9375
23. Donahue PS, Draut JW, Muldoon JJ, Edelstein HI, Bagheri N, Leonard JN (2020) The COMET toolkit for composing customizable genetic programs in mammalian cells. Nat Commun 11(1):779. https://doi.org/10.1038/s41467-019-14147-5
24. Frei T, Cella F, Tedeschi F, Gutierrez J, Stan GB, Khammash M et al (2020) Characterization and mitigation of gene expression burden in mammalian cells. Nat Commun 11(1):4641. https://doi.org/10.1038/s41467-020-18392-x
25. Cella F, Perrino G, Tedeschi F, Viero G, Bosia C, Stan GB et al (2023) MIRELLA: a mathematical model explains the effect of microRNA-mediated synthetic genes regulation on intracellular resource allocation. Nucleic Acids Res 51(7):3452–3464. https://doi.org/10.1093/nar/gkad151
26. Jones RD, Qian Y, Siciliano V, DiAndreth B, Huh J, Weiss R et al (2020) An endoribonuclease-based feedforward controller for decoupling resource-limited genetic modules in mammalian cells. Nat Commun 11(1):5690. https://doi.org/10.1038/s41467-020-19126-9
27. Santorelli M, Lam C, Morsut L (2019) Synthetic development: building mammalian multicellular structures with artificial genetic programs. Curr Opin Biotechnol 59:130–140. https://doi.org/10.1016/j.copbio.2019.03.016
28. Aydin O, Passaro AP, Raman R, Spellicy SE, Weinberg RP, Kamm RD et al (2022) Principles for the design of multicellular engineered living systems. APL Bioeng 6(1):010903. https://doi.org/10.1063/5.0076635
29. Wieland M, Fussenegger M (2012) Engineering molecular circuits using synthetic biology in mammalian cells. Annu Rev Chem Biomol Eng 3:209–234. https://doi.org/10.1146/annurev-chembioeng-061010-114145
30. Matsuda M, Koga M, Woltjen K, Nishida E, Ebisuya M (2015) Synthetic lateral inhibition governs cell-type bifurcation with robust ratios. Nat Commun 6:6195. https://doi.org/10.1038/ncomms7195
31. Matsuda M, Koga M, Nishida E, Ebisuya M (2012) Synthetic signal propagation through direct cell-cell interaction. Sci Signal 5(220):ra31. https://doi.org/10.1126/scisignal.2002764
32. Morsut L, Roybal KT, Xiong X, Gordley RM, Coyle SM, Thomson M et al (2016) Engineering customized cell sensing and response behaviors using synthetic notch receptors. Cell 164(4):780–791. https://doi.org/10.1016/j.cell.2016.01.012
33. Li P, Markson JS, Wang S, Chen S, Vachharajani V, Elowitz MB (2018) Morphogen gradient reconstitution reveals Hedgehog pathway design principles. Science 360(6388):543–548. https://doi.org/10.1126/science.aao0645
34. Sprinzak D, Lakhanpal A, LeBon L, Garcia-Ojalvo J, Elowitz MB (2011) Mutual inactivation of Notch receptors and ligands facilitates developmental patterning. PLoS Comput Biol 7(6):e1002069. https://doi.org/10.1371/journal.pcbi.1002069
35. Shaya O, Binshtok U, Hersch M, Rivkin D, Weinreb S, Amir-Zilberstein L et al (2017) Cell-cell contact area affects notch signaling and Notch-dependent patterning. Dev Cell 40(5):505–11 e6. https://doi.org/10.1016/j.devcel.2017.02.009
36. Montes-Olivas S, Marucci L, Homer M (2019) Mathematical models of organoid cultures. Front Genet 10:873. https://doi.org/10.3389/fgene.2019.00873
37. Gorochowski TE, Hauert S, Kreft JU, Marucci L, Stillman NR, Tang TD et al (2020) Toward engineering biosystems with emergent collective functions. Front Bioeng Biotechnol 8:705. https://doi.org/10.3389/fbioe.2020.00705
38. Montes-Olivas S, Legge D, Lund A et al (2023) In-silico and in-vitro morphometric analysis of intestinal organoids. PLoS Comput Biol 19(8):e1011386. https://doi.org/10.1371/journal.pcbi.1011386
39. Monk JM, Lloyd CJ, Brunk E, Mih N, Sastry A, King Z et al (2017) iML1515, a knowledgebase that computes Escherichia coli traits. Nat Biotechnol 35(10):904–908. https://doi.org/10.1038/nbt.3956
40. Passi A, Tibocha-Bonilla JD, Kumar M, Tec-Campos D, Zengler K, Zuniga C (2021) Genome-scale metabolic modeling enables

in-depth understanding of big data. Meta 12(1). https://doi.org/10.3390/metabo12010014
41. Fang X, Lloyd CJ, Palsson BO (2020) Reconstructing organisms in silico: genome-scale models and their emerging applications. Nat Rev Microbiol 18(12):731–743. https://doi.org/10.1038/s41579-020-00440-4
42. Gu C, Kim GB, Kim WJ, Kim HU, Lee SY (2019) Current status and applications of genome-scale metabolic models. Genome Biol 20(1):121. https://doi.org/10.1186/s13059-019-1730-3
43. Chen K, Gao Y, Mih N, O'Brien EJ, Yang L, Palsson BO (2017) Thermosensitivity of growth is determined by chaperone-mediated proteome reallocation. Proc Natl Acad Sci U S A 114(43):11548–11553. https://doi.org/10.1073/pnas.1705524114
44. Lloyd CJ, Ebrahim A, Yang L, King ZA, Catoiu E, O'Brien EJ et al (2018) COBRAme: a computational framework for genome-scale models of metabolism and gene expression. PLoS Comput Biol 14(7):e1006302. https://doi.org/10.1371/journal.pcbi.1006302
45. Landon S, Rees-Garbutt J, Marucci L, Grierson C (2019) Genome-driven cell engineering review: in vivo and in silico metabolic and genome engineering. Essays Biochem 63(2): 267–284. https://doi.org/10.1042/EBC20180045
46. Lv X, Hueso-Gil A, Bi X, Wu Y, Liu Y, Liu L et al (2022) New synthetic biology tools for metabolic control. Curr Opin Biotechnol 76: 102724. https://doi.org/10.1016/j.copbio.2022.102724
47. Wang H, Robinson JL, Kocabas P, Gustafsson J, Anton M, Cholley PE et al (2021) Genome-scale metabolic network reconstruction of model animals as a platform for translational research. Proc Natl Acad Sci U S A 118(30). https://doi.org/10.1073/pnas.2102344118
48. Robinson JL, Kocabas P, Wang H, Cholley PE, Cook D, Nilsson A et al (2020) An atlas of human metabolism. Sci Signal 13(624). https://doi.org/10.1126/scisignal.aaz1482
49. Kachhawaha K, Singh S, Joshi K, Nain P, Singh SK (2022) Bioprocessing of recombinant proteins from Escherichia coli inclusion bodies: insights from structure-function relationship for novel applications. Prep Biochem Biotechnol 53:1–25. https://doi.org/10.1080/10826068.2022.2155835
50. Strain B, Morrissey J, Antonakoudis A, Kontoravdi C (2023) Genome-scale models as a vehicle for knowledge transfer from microbial to mammalian cell systems. Comput Struct Biotechnol J 21:1543–1549. https://doi.org/10.1016/j.csbj.2023.02.011
51. Karr JR, Phillips NC, Covert MW (2014) WholeCellSimDB: a hybrid relational/HDF database for whole-cell model predictions. Database (Oxford) 2014. https://doi.org/10.1093/database/bau095
52. Karr JR, Sanghvi JC, Macklin DN, Arora A, Covert MW (2013) WholeCellKB: model organism databases for comprehensive whole-cell models. Nucleic Acids Res 41(Database issue):D787–D792. https://doi.org/10.1093/nar/gks1108
53. Karr JR, Sanghvi JC, Macklin DN, Gutschow MV, Jacobs JM, Bolival B Jr et al (2012) A whole-cell computational model predicts phenotype from genotype. Cell 150(2):389–401. https://doi.org/10.1016/j.cell.2012.05.044
54. Ahn-Horst TA, Mille LS, Sun G, Morrison JH, Covert MW (2022) An expanded whole-cell model of E. coli links cellular physiology with mechanisms of growth rate control. NPJ Syst Biol Appl 8(1):30. https://doi.org/10.1038/s41540-022-00242-9
55. Carrera J, Covert MW (2015) Why build whole-cell models? Trends Cell Biol 25(12): 719–722. https://doi.org/10.1016/j.tcb.2015.09.004
56. Macklin DN, Ahn-Horst TA, Choi H, Ruggero NA, Carrera J, Mason JC et al (2020) Simultaneous cross-evaluation of heterogeneous E. coli datasets via mechanistic simulation. Science 369(6502):eaav3751. https://doi.org/10.1126/science.aav3751
57. Rees-Garbutt J, Chalkley O, Landon S, Purcell O, Marucci L, Grierson C (2020) Designing minimal genomes using whole-cell models. Nat Commun 11(1):836. https://doi.org/10.1038/s41467-020-14545-0
58. Rees-Garbutt J, Rightmyer J, Chalkley O, Marucci L, Grierson C (2021) Testing theoretical minimal genomes using whole-cell models. ACS Synth Biol 10(7):1598–1604. https://doi.org/10.1021/acssynbio.0c00515
59. Marucci L, Barberis M, Karr J, Ray O, Race PR, de Souza AM et al (2020) Computer-aided whole-cell design: taking a holistic approach by integrating synthetic with systems biology. Front Bioeng Biotechnol 8:942. https://doi.org/10.3389/fbioe.2020.00942
60. Landon S, Chalkley O, Breese G, Grierson C, Marucci L (2021) Understanding metabolic flux behaviour in whole-cell model output. Front Mol Biosci 8:732079. https://doi.org/10.3389/fmolb.2021.732079

61. Szigeti B, Roth YD, Sekar JAP, Goldberg AP, Pochiraju SC, Karr JR (2018) A blueprint for human whole-cell modeling. Curr Opin Syst Biol 7:8–15. https://doi.org/10.1016/j.coisb.2017.10.005
62. Elsemman IE, Rodriguez Prado A, Grigaitis P, Garcia Albornoz M, Harman V, Holman SW et al (2022) Whole-cell modeling in yeast predicts compartment-specific proteome constraints that drive metabolic strategies. Nat Commun 13(1):801. https://doi.org/10.1038/s41467-022-28467-6
63. Wilkinson MD, Dumontier M, Jan Aalbersberg I, Appleton G, Axton M, Baak A et al (2019) Addendum: the FAIR guiding principles for scientific data management and stewardship. Sci Data. 6(1):6. https://doi.org/10.1038/s41597-019-0009-6
64. Wilkinson MD, Dumontier M, Aalbersberg IJ, Appleton G, Axton M, Baak A et al (2016) The FAIR guiding principles for scientific data management and stewardship. Sci Data 3:160018. https://doi.org/10.1038/sdata.2016.18
65. Goldberg AP, Szigeti B, Chew YH, Sekar JA, Roth YD, Karr JR (2018) Emerging whole-cell modeling principles and methods. Curr Opin Biotechnol 51:97–102. https://doi.org/10.1016/j.copbio.2017.12.013
66. Gherman IM, Abdallah ZS, Pang W, Gorochowski TE, Grierson CS, Marucci L (2023) Bridging the gap between mechanistic biological models and machine learning surrogates. PLoS Comput Biol 19(4):e1010988. https://doi.org/10.1371/journal.pcbi.1010988
67. Gherman IM, Rees-Garbutt J, Pang W et al (2023) Accelerated design of *Escherichia coli* genomes with reduced size using a whole-cell model and machine learning surrogate. bioRxiv 2023.10.30.564402. https://doi.org/10.1101/2023.10.30.564402
68. Thornburg ZR, Bianchi DM, Brier TA, Gilbert BR, Earnest TM, Melo MCR et al (2022) Fundamental behaviors emerge from simulations of a living minimal cell. Cell 185(2):345–60 e28. https://doi.org/10.1016/j.cell.2021.12.025
69. Scholzel C, Blesius V, Ernst G, Dominik A (2021) Characteristics of mathematical modeling languages that facilitate model reuse in systems biology: a software engineering perspective. NPJ Syst Biol Appl. 7(1):27. https://doi.org/10.1038/s41540-021-00182-w
70. Del Vecchio D, Dy AJ, Qian Y (2016) Control theory meets synthetic biology. J R Soc Interface 13(120):20160380. https://doi.org/10.1098/rsif.2016.0380
71. Filo M, Chang CH, Khammash M (2023) Biomolecular feedback controllers: from theory to applications. Curr Opin Biotechnol 79:102882. https://doi.org/10.1016/j.copbio.2022.102882
72. Briat C, Gupta A, Khammash M (2016) Antithetic integral feedback ensures robust perfect adaptation in noisy biomolecular networks. Cell Syst 2(2):133. https://doi.org/10.1016/j.cels.2016.02.010
73. Anastassov S, Filo M, Chang CH, Khammash M (2023) A cybergenetic framework for engineering intein-mediated integral feedback control systems. Nat Commun 14(1):1337. https://doi.org/10.1038/s41467-023-36863-9
74. Filo M, Kumar S, Khammash M (2022) A hierarchy of biomolecular proportional-integral-derivative feedback controllers for robust perfect adaptation and dynamic performance. Nat Commun 13(1):2119. https://doi.org/10.1038/s41467-022-29640-7
75. Annunziata F, Matyjaszkiewicz A, Fiore G, Grierson CS, Marucci L, di Bernardo M et al (2017) An orthogonal multi-input integration system to control gene expression in Escherichia coli. ACS Synth Biol 6(10):1816–1824. https://doi.org/10.1021/acssynbio.7b00109
76. Fiore G, Matyjaszkiewicz A, Annunziata F, Grierson C, Savery NJ, Marucci L et al (2017) In-silico analysis and implementation of a multicellular feedback control strategy in a synthetic bacterial consortium. ACS Synth Biol 6(3):507–517. https://doi.org/10.1021/acssynbio.6b00220
77. McCarty NS, Ledesma-Amaro R (2019) Synthetic biology tools to engineer microbial communities for biotechnology. Trends Biotechnol 37(2):181–197. https://doi.org/10.1016/j.tibtech.2018.11.002
78. Menolascina F, Fiore G, Orabona E, De Stefano L, Ferry M, Hasty J et al (2014) In-vivo real-time control of protein expression from endogenous and synthetic gene networks. PLoS Comput Biol 10(5):e1003625. https://doi.org/10.1371/journal.pcbi.1003625
79. Lugagne JB, Sosa Carrillo S, Kirch M, Kohler A, Batt G, Hersen P (2017) Balancing a genetic toggle switch by real-time feedback control and periodic forcing. Nat Commun 8(1):1671. https://doi.org/10.1038/s41467-017-01498-0
80. Shannon B, Zamora-Chimal CG, Postiglione L, Salzano D, Grierson CS, Marucci L et al (2020) In vivo feedback control of an antithetic molecular-titration motif in Escherichia coli using microfluidics. ACS

Synth Biol 9(10):2617–2624. https://doi.org/10.1021/acssynbio.0c00105

81. Pedone E, de Cesare I, Zamora-Chimal CG, Haener D, Postiglione L, La Regina A et al (2021) Cheetah: a computational toolkit for cybergenetic control. ACS Synth Biol 10(5): 979–989. https://doi.org/10.1021/acssynbio.0c00463
82. Postiglione L, Napolitano S, Pedone E, Rocca DL, Aulicino F, Santorelli M et al (2018) Regulation of gene expression and signaling pathway activity in mammalian cells by automated microfluidics feedback control. ACS Synth Biol 7(11):2558–2565. https://doi.org/10.1021/acssynbio.8b00235
83. Khazim M, Pedone E, Postiglione L, di Bernardo D, Marucci L (2021) A microfluidic/microscopy-based platform for on-chip controlled gene expression in mammalian cells. Methods Mol Biol 2229:205–219. https://doi.org/10.1007/978-1-0716-1032-9_10
84. de Cesare I, Zamora-Chimal CG, Postiglione L, Khazim M, Pedone E, Shannon B et al (2021) ChipSeg: an automatic tool to segment bacterial and mammalian cells cultured in microfluidic devices. ACS Omega 6(4):2473–2476. https://doi.org/10.1021/acsomega.0c03906
85. de Cesare I, Salzano D, di Bernardo M, Renson L, Marucci L (2022) Control-based continuation: a new approach to prototype synthetic gene networks. ACS Synth Biol 11(7):2300–2313. https://doi.org/10.1021/acssynbio.1c00632
86. Smart B, De Cesare I, Renson L, Marucci L (2022) Model predictive control of cancer cellular dynamics: a new strategy for therapy design. Front Control Eng 3(2022). https://doi.org/10.3389/fcteg.2022.935018

Chapter 7

Realizing Antithetic Integral Feedback Control in Mammalian Cells

Timothy Frei and Mustafa Khammash

Abstract

Genetic circuit engineering has emerged as a powerful methodology to program the behavior of mammalian cells to respond to internal and external cues. This approach is now used to develop new therapeutics and improve production processes. However, genetic interaction networks are complex and hard to engineer rationally. Moreover, a design may fail, and it may not be possible to identify the root cause of its breakdown. Introducing designated regulatory circuitry in the form of integral feedback can introduce performance guarantees by ensuring robust and precise operation.

Key words Mammalian synthetic biology, Biomolecular control system, Genetic circuit design, Integral feedback, Molecular sequestration

1 Introduction

In recent years, genetic circuits have emerged as a promising technology to realize custom behavior programs in mammalian cells. Importantly, scientists now apply these programs to treat diseases in novel ways. However, genetic circuits can rarely be designed, built, and expected to perform as intended. This discrepancy is due to the inherent complexity of biological systems and the associated uncertainty about the interactions that may affect circuit components.

Classical engineering disciplines address uncertainty by introducing feedback from a designated control circuit such that the circuit achieves specific performance objectives. In the setting we consider here, we may think of the control circuit's performance in abstract terms as perceiving the levels of a biomolecule of interest, measuring its deviation from an ideal or a desired level (also called the setpoint of the control circuit) in the form of an error, and feeding back this error into the controlled system such that this error is minimized. Integral feedback remains one of the most popular feedback control paradigms in engineering. Its popularity

Francesca Ceroni and Karen Polizzi (eds.), *Mammalian Synthetic Systems*, Methods in Molecular Biology, vol. 2774,
https://doi.org/10.1007/978-1-0716-3718-0_7,

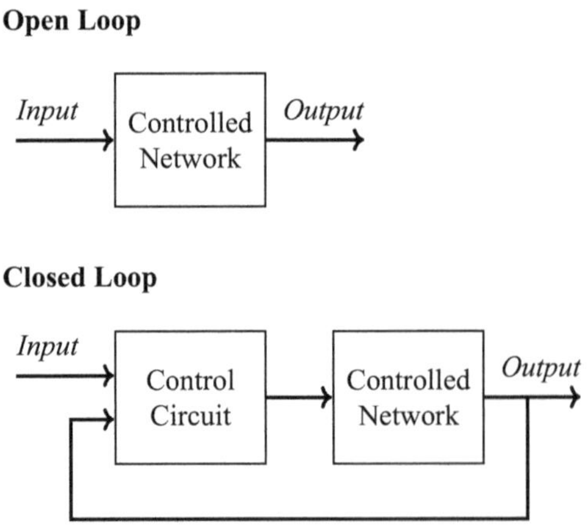

Fig. 1 The controlled network in the open-loop configuration can be thought of as receiving an input and producing an output

can be attributed to its simplicity and effectiveness. In particular, integral feedback control guarantees perfect tracking and disturbance rejection. These properties describe the circuit's ability to follow or track an externally supplied signal or input (tracking) while remaining robust to unwanted changes or disturbances in the parameters or the structure of the controlled network. These two properties can be exploited to protect a network, referred to as the controlled network, from uncertainty and unexpected changes within that network. In particular, this applies to cases where the output is supposed to follow the input robustly. For example, a controlled network may be a circuit component affected by external disturbances or disturbances originating from other circuit components. Alternatively, a controlled network may also be a complex system of interactions that should robustly produce a desired output level. Figure 1 illustrates how a controlled network (open loop) can be connected to a control circuit resulting in a (closed-loop) system such that the network's output will robustly track the input.

Recently, integral feedback was realized in genetic circuits in the form of the antithetic integral feedback motif [1]. Subsequently, the motif was extended to proportional-integral feedback and realized in mammalian cells [2]. At the core of the antithetic integral feedback motif are two biomolecular species that interact by sequestering or annihilating each other's function [3]. This interaction enables the computation of the mathematical integral of the error and bestows the properties of integral feedback control. In this control circuit, the setpoint is determined by the ratio of the production rate of species $\mathbf{Z_1}$ over the production rate of species $\mathbf{Z_2}$ ($h_1(\mu, \mathbf{Z}'_1)/h_2(\theta, \mathbf{Z}'_2, \mathbf{X_n})$ in Fig. 2).

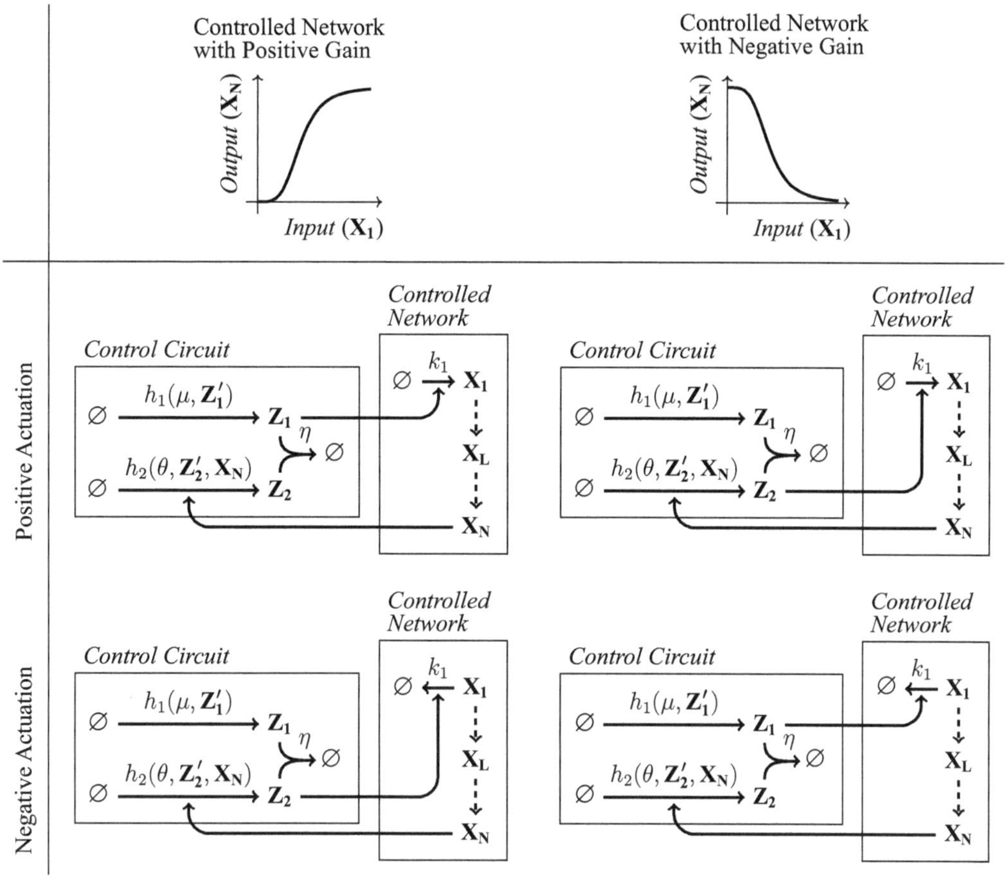

Fig. 2 Different choices of controller configurations given the input/output response of the controlled network. There are two different controller configurations for the two different input/output responses of the controlled network (positive gain and negative gain). One where species $\mathbf{Z_1}$ increases the production of species $\mathbf{X_1}$ (Positive Actuation) and one where it decreases the levels of species $\mathbf{X_1}$ (Negative Actuation). We summarize the dependence of species $\mathbf{Z_1}$ and $\mathbf{Z_2}$ on precursors ($\mathbf{Z_1'}$ and $\mathbf{Z_2'}$, respectively) by depicting their production rates as abstract reactions with rates $h_1(\mu, \mathbf{Z_1'})$ and $h_2(\theta, \mathbf{Z_2'}, \mathbf{X_N})$

In the rest of this chapter, we will go through the design and biological implementation of antithetic integral feedback circuits in detail and address specific concerns that affect circuit performance. We will use the implementation of antithetic integral feedback via the sequestration of sense and antisense mRNA from [2] as a running example (Fig. 3).

2 Materials

Adjust all enzymatic reactions to their final volume with nuclease-free water. All other solutions are prepared in ultrapure water.

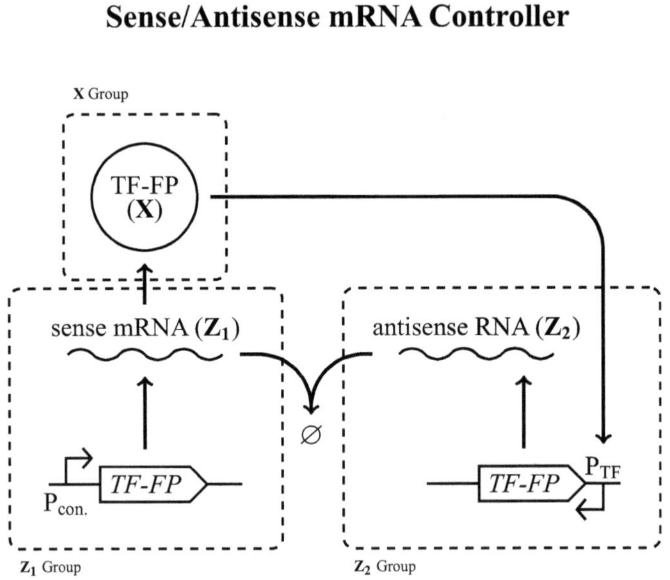

Fig. 3 Closed-loop configuration of a sense/antisense mRNA antithetic integral feedback controller which regulates the expression of a transcription factor (TF). To be able to measure the levels of transcription factor, it is fused to a fluorescent protein (FP). The expression of species $\mathbf{Z_1}$ is driven by a constitutive promoter ($P_{con.}$). In contrast, the expression of species $\mathbf{Z_2}$ is driven by a promoter that responds to the levels of the transcription factor (P_{TF}). The whole circuit can be partitioned into three groups, a group $\mathbf{Z_1}$ for the gene producing the sense mRNA ($\mathbf{Z_1}$), a group $\mathbf{Z_2}$ for the gene producing the antisense RNA ($\mathbf{Z_2}$), and a group **X** for the controlled network (here, only the transcription factor fused to a fluorescent protein; TF-FP)

2.1 Plasmid Construction

We recommend using a plasmid assembly method based on Golden Gate Assembly and the Modular Cloning (MoClo) plasmid system to construct genetic circuits in mammalian cells. Detailed descriptions and protocols can be found in [4, 5].

1. 37 °C shaking incubator for growing bacterial cultures
2. NanoDrop™ 2000/2000c from Thermo Scientific™
3. Chemically competent *Escherichia coli* strain for cloning (e.g., TOP10)
4. Plasmid purification kit (e.g., ZR Plasmid Miniprep™-*Classic* from ZYMO RESEARCH or equivalent)
5. 1.5 mL Eppendorf tubes®
6. 0.2 mL PCR tubes
7. Ampicillin stock solution (1000 ×): Dissolve 300 mg of Ampicillin powder in 30 mL of water at room temperature. Filter sterilize (0.2 μm) and aliquot at 1 mL. Store at − 20 °C.

8. Chloramphenicol stock solution (1000 ×): Dissolve 75 mg of Chloramphenicol powder in 30 mL of reagent-grade EtOH (non-denatured absolute ethanol without additives) at room temperature. Filter sterilize (0.2 μm) and aliquot at 1 mL. Store at −20 °C.
9. Kanamycin stock solution (1000 ×): Dissolve 150 mg of Kanamycin powder in 30 mL of water at room temperature. Filter sterilize (0.2 μm) and aliquot at 1 mL. Store at −20 °C.
10. Lysogeny Broth (LB): Mix 5 mg of NaCl, 5 mg of tryptone, and 2.5 mg of yeast extract in a 500 mL flask and dissolve in 490 mL of water at room temperature. Autoclave and store at room temperature.
11. LB Agar plates: Dissolve 5 mg of NaCl, 5 mg of tryptone, 2.5 mg of yeast extract, and 7.5 mg of agar in 500 mL of water and autoclave. Once the solution has cooled down, add the desired antibiotic solution.
12. Optionally, thermocycler and PCR master mix for creating new Level.

2.2 Mammalian Cell Culture

1. Laminar flow cabinet
2. 37 °C CO_2-humidified incubator
3. 37 °C bead or water bath
4. T25 or T75 flasks
5. Multi-well tissue culture-treated flat-bottom microplates
6. HEK293T cells (ATCC, strain number CRL-3216)
7. Phosphate-buffered saline (PBS) solution
8. Trypsin-EDTA solution (0.25%)
9. Fetal Bovine Serum (FBS)
10. L-Glutamine or GlutaMAX solution (commonly available as 100 × concentration)
11. Dulbecco's Modified Eagle's Medium (DMEM; it may be acquired with GlutaMAX already added)
12. Complete DMEM: Add 50 mL of heat-inactivated (30 min at 56 °C in a water bath) FBS and 5 mL of L-Glutamine or GlutaMAX solution to a bottle of 500 mL DMEM. Omit the L-Glutamine or GlutaMAX if the DMEM already contains it.

2.3 Transfection

1. Polyethyleneimine (PEI). Prepare solutions according to the vendor's instructions.
2. Opti-MEM I

2.4 Measurement

1. Accutase solution (which can be substituted by Trypsin-EDTA solution)
2. Flow cytometer

2.5 Data Analysis

1. Personal computer
2. An integrated development environment (IDE) for R (such as RStudio) with the expressalyzer package (https://github.com/freitim/expressalyzr) installed or any other preferred data processing and visualization method

3 Methods

We assign roles to specific biomolecular species in the controlled network and the control circuit to simplify the design process and give them names to refer to. We will refer to the input to the controlled network through which the network is manipulated as $\mathbf{X_1}$. The actual output of the controlled network may not be directly observable because it could be an mRNA species. Therefore, we assign the name $\mathbf{X_L}$ to the observable molecular species, which will robustly track the input upstream. We call $\mathbf{X_N}$ the actual unobservable output, which acts as the input to the control circuit. Note that $\mathbf{X_L}$ will only be robust to an upstream parameter or structural disturbances. We call $\mathbf{Z_1}$ and $\mathbf{Z_2}$ the two species that comprise the control circuit, respectively. One of these two species then interacts as the input to the controlled network $\mathbf{X_1}$.

3.1 Genetic Circuit Assembly

The following subsection describes the assembly of the DNA sequences and plasmids required to realize the genetic controller design obtained via the procedure described in Subheading 4. A detailed protocol for genetic circuit assembly according to the Modular Cloning (MoClo) plasmid system can be found in [5]. The protocol may be simplified by replacing the thermocycling steps with incubation at 37 °C overnight in an incubator, as suggested in [6].

1. Obtain the necessary DNA sequences through synthesis services or from AddGene.
2. Build each genetic circuit component on a separate plasmid using your preferred molecular cloning technique. As stated above, we recommend Golden Gate Assembly and the Modular Cloning (MoClo) plasmid system. For example, construct separate plasmids for both control circuit species, $\mathbf{Z_1}$ and $\mathbf{Z_2}$, and every additional species that is required to interface the control circuit and the controlled network.

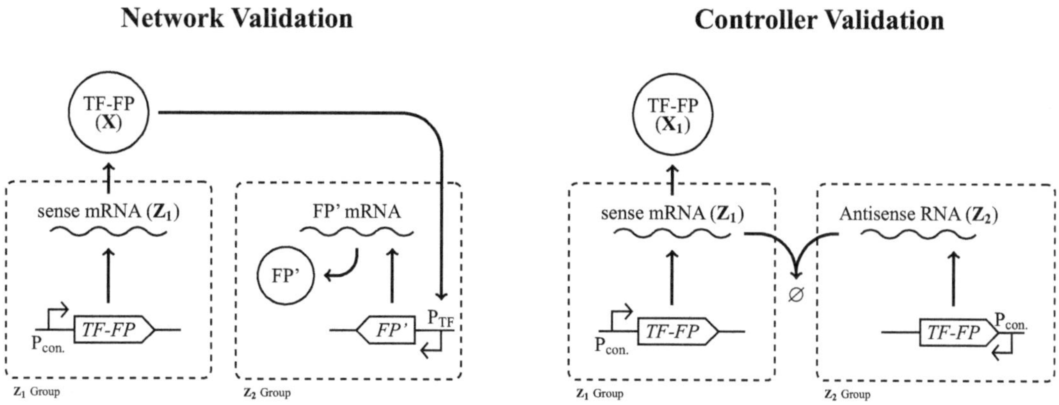

Fig. 4 Shown are two validation configurations of the circuit. On the left is a configuration for characterizing the steady-state input/output response of the controlled network. The main idea is to remove the control circuit while retaining the sensing and actuating interface to the controlled network. Here, FP' represents a fluorescent protein other than the one fused to the transcription factor. This configuration may also be used to determine the gain of the controlled network. Shown on the right is a configuration for validating the sequestration of the sense mRNA ($\mathbf{Z_1}$) by the antisense RNA ($\mathbf{Z_2}$). The main idea is to remove the controlled network to determine the input/output mapping of the control circuit in isolation

3. Build additional plasmids required for validation as depicted in Fig. 4. Among others, these include a reporter plasmid expressing a fluorescent protein driven by the input from the controlled network and a constitutively expressing version of the $\mathbf{Z_2}$ plasmid.
4. If not yet available, assemble plasmids constitutively expressing the fluorescent proteins that are used as readouts, as well as an inert plasmid consisting only of an *E. coli* replication origin and an antibiotic resistance cassette. The latter will be used to ensure that all transfection reactions contain the same amount of DNA.

3.2 Experimental Validation

Once the individual circuit components are assembled, their function in isolation and combination should be validated to ensure the proper function of the feedback system. This is commonly done via transient co-transfection of different combinations of the circuit components. Transient transfections can be either performed on adhered cells or adherent cells in suspension just before they are seeded. In our experience, the latter method can lead to an increase in transfection efficiency. However, roughly double the amount of cells is required at the time of transfection/seeding of the cells.

To evaluate the function of the antithetic integral controller, it is best to divide its components into three groups, as is discussed in Step 8 in Subheading 4. By having separate groups for species $\mathbf{Z_1}$ together with its precursors and $\mathbf{Z_2}$ together with its precursors, it is possible to manipulate their production rates by changing the amount of plasmid transfected. The amount of plasmid will directly affect the production of precursors in each group and will, therefore, also affect the production levels of species $\mathbf{Z_1}$ or $\mathbf{Z_2}$. This property can be used to design plasmid titration experiments, where different amounts of the plasmids driving the expression of species $\mathbf{Z_1}$ are co-transfected with different amounts of the plasmids driving the expression of species $\mathbf{Z_2}$. Titration experiments can be used to determine the steady-state input/output response of genetic circuits and therefore provide valuable information for determining optimal working ranges and troubleshooting circuit designs.

The following will describe a general recipe for designing and executing titration experiments. These experiments can be applied to the validation experiments depicted in Fig. 4.

3.2.1 Preparation

1. If transfections are preferably performed on already adherent cells, then ensure 2 days before transfection that the HEK293T cell culture has reached a confluency of approximately 85 % to 90 %. The next day, seed the cells in the desired multi-well plate format such that they will reach 85 % to 95 % confluency approximately 72 h after. Otherwise, the same needs to be checked only 1 day before the planned transfection, and the cells will be seeded during the transfection process.
2. If transfection of already adhered cells is preferred, then a day before the planned transfection, seed the HEK293T cells.
3. At least a day before the transfection of the circuit components, inoculate liquid cultures of the *E. coli* strains carrying the circuit component plasmids in 5 mL of LB media supplemented with the appropriate antibiotic (add 5 μL of a 1000 × stock) and let them grow overnight.
4. Extract the plasmids from the liquid cultures the following day using your preferred plasmid extraction method. We recommend using a plasmid miniprep kit that removes endotoxins (e.g., the ZYMO RESEARCH ZR Plasmid Miniprep—Classic kit). Use nuclease-free water to elute the plasmid DNA.
5. Ensure that the extracted plasmid DNA has a concentration of at least 100ng μL^{-1} and is of high purity using a spectrophotometer such as a NanoDrop™ 2000/2000c or equivalent.

3.2.2 Transfections Plan

1. Choose the plasmids that will be co-transfected, the necessary single-color control plasmids if more than one fluorescent color is used, and an inert plasmid to balance total transfected DNA amounts.
2. Determine the total number of conditions based on the desired number of titration steps.
3. Set the number of replicates.

4. Based on the total number of conditions and controls, choose a multi-well plate format that fits all of them together, with transfection and single-color controls. Alternatively, use more than one plate.
5. Based on the chosen plate format, determine the total amount of DNA to be transfected per condition.
6. Based on the total number of conditions and controls, work out how much PEI solution to dilute in how much Opti-MEM I. In our experience, a ratio of 3:1 μg PEI to μg plasmid DNA works well.

3.2.3 Transfections

1. Distribute the plasmid DNA according to the planned co-transfection. Each condition should be prepared in a separate tube. Depending on the chosen multi-well plate format, use either 0.2 mL PCR tubes or 1.5 mL Eppendorf tubes®.
2. If the transfection will be performed on suspended cells, prepare enough complete DMEM to seed the cells in the desired multi-well plate format and to propagate the cells. Place the prepared DMEM to warm up in the 37 °C bead or water bath.
3. Move tubes with co-transfection conditions into the laminar flow cabinet.
4. Move Opti-MEM I and PEI solution to laminar flow cabinet.
5. In an empty tube, dilute PEI solution in Opti-MEM I as determined previously.
6. Adjust the volume in each condition and control to half the total transfection reaction volume.
7. Mix each condition or control with half the total reaction volume of PEI Opti-MEM I solution to achieve the total reaction volume in each.
8. Let the transfection reactions incubate at room temperature for 25 min.
9. In the meantime, if suspended cells are going to be transfected, passage and count the HEK293T cells. Seed the cells in the chosen multi-well plate format such that they will reach 85 % to 95 % confluency in 48 h.
10. After 25 min distribute the transfection reactions to the wells of the seeded multi-well plate. If suspended cells are transfected, do so while the cells are still in suspension. Light pipetting up and down of suspended cells may help distribute the transfection reaction and increase transfection efficiency. Do not pipette up and down if cells were transfected while adhered.
11. Place the transfected HEK293T cells into a 37 °C CO_2-humidified incubator.
12. Let the cells incubate for approximately 48 h.

3.2.4 Measurement

1. After 48 h aspirate the cell culture media in all wells of transfected cells.
2. Carefully wash cells with PBS solution.
3. Add Accutase solution to cells.
4. Incubate cells with Accutase in a 37 °C CO_2-humidified incubator for at least 5 min.
5. When the cells have fully detached, transfer them into an appropriate vessel for use with the specific flow cytometer available.
6. Proceed to the flow cytometer and set it up according to the manufacturer's or the operator's instructions.
7. Acquire the samples.
8. Once all samples have been acquired, transfer the data for storage and analysis.
9. Shut down the flow cytometer according to the manufacturer's or the operator's instructions.

3.2.5 Analysis

1. Load data into your preferred analysis environment.
2. Gate the data for live cells and singlets.
3. If more than one fluorescent protein color was acquired, perform fluorescence compensation to remove the spillover signal.
4. Compute the mean fluorescence for all fluorescence channels for each sample.
5. Compute the mean and standard error across the replicates of each condition.
6. Plot the mean and standard error along the titration steps.

3.3 Characterization of the Steady-State Input/Output Response of the Controlled Network

Characterizing the input/output response of the controlled network at steady state is important: first to confirm the gain of the controlled network, which influences the choice of control circuit configuration (Subheading 4 and Fig. 2), and second to determine the range of admissible setpoints. The second point follows from the fact that the control circuit cannot push the controlled network to points that lay below its constitutive or leaky production level and above its saturation level.

To obtain the steady-state input/output response of the controlled network of the example circuit shown in Fig. 3, one can adopt a circuit configuration as in Fig. 4 left and titrate the amounts of transfected $\mathbf{Z_1}$-group (which is only the TF-FP plasmid in this case). Ideally, this titration is also performed for different amounts of $\mathbf{Z_2}$-group plasmid (in this case, only the readout plasmid FP'). Titrating these two groups should identify ratios of the $\mathbf{Z_1}$-group versus the $\mathbf{Z_2}$-group that yield good starting points for finding admissible setpoints in the closed-loop circuit.

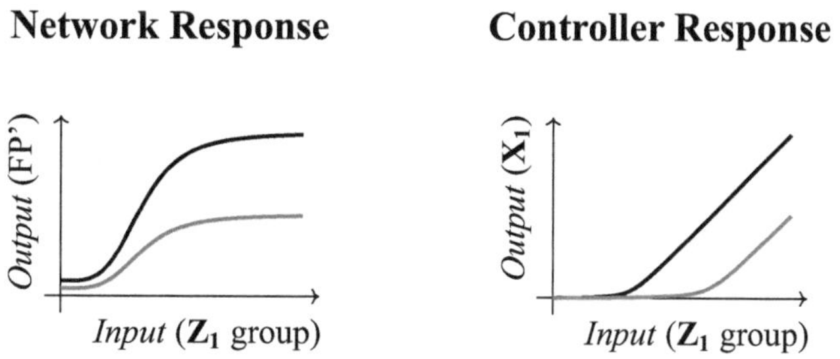

Fig. 5 Expected responses for the two validation approaches. The graphs depict the input/output response of the circuit if the components are interacting in the desired way

To obtain this data, apply the procedure described in Subheading 3.2 and plot the titration of the $\mathbf{Z_1}$-group against the FP' output for different amounts of the $\mathbf{Z_2}$-group. For a positive-gain controlled-network, the response should look similar to the response depicted in Fig. 5 left. Based on the response curve, select a combination of $\mathbf{Z_1}$-group amounts and $\mathbf{Z_2}$-group amounts such that the output is large and lies somewhere in between the leaky production and saturation levels. This will serve as a starting point for determining the range of admissible setpoint in the closed-loop circuit.

3.4 Validation of the Sequestration Reaction

Similarly to the previous section, a titration experiment can be performed to assess the sequestration of species $\mathbf{Z_1}$ by species $\mathbf{Z_2}$. The sequestration produces a characteristic threshold response, where the expression of species $\mathbf{X_1}$ only occurs once all the free species $\mathbf{Z_2}$ is sequestered by $\mathbf{Z_1}$ and the excess free species $\mathbf{Z_1}$ can produce species $\mathbf{X_1}$. Since unintended interactions with other species or excessive degradation will interfere with this threshold response, it is a good way to troubleshoot a controller design.

The experimental procedure follows the approach described in Subheading 3.2. Here, the $\mathbf{Z_1}$-group should be titrated against different amounts of the $\mathbf{Z_2}$-group. Importantly, a high-enough titration resolution should be used such that the threshold is clearly distinguishable. Finding an optimal titration range may require several iterations of the experiment. Figure 4 right shows an example circuit configuration. The measured output should produce responses similar to the ones shown in Fig. 5 right. A different threshold should be obtained for different amounts of the $\mathbf{Z_2}$-group.

3.5 Validation of the Integral Feedback

Once the sequestration has been verified (Subheading 3.4) and suitable starting conditions for setpoint titrations are obtained (Subheading 3.3), disturbance rejection in the full close-loop circuit can be tested. To assess this property, we compare the closed-loop circuit to an open-loop counterpart in the presence of a disturbance to the controlled network (adding the ability to disturb the controlled network is described in Step 9 in Subheading 4). The validation configuration for the control circuit can be used as an open-loop configuration. However, here, the whole minimal or full controlled network should be connected so that species $\mathbf{X_L}$ can be measured as output. The design of a minimal controlled network is described in Step 10 in Subheading 4. Since disturbance rejection will only succeed for setpoints that lay within the admissible region, a simultaneous titration of the setpoint will ensure that disturbance rejection may be observed. We recommend titrating the setpoint by fixing the level of the $\mathbf{Z_2}$-group to the levels determined in Subheading 3.3 and decreasing the amounts of the $\mathbf{Z_1}$-group. If a minimal controlled network has been constructed, it is advisable to first test the control circuit on this minimal implementation before advancing to the full controlled network.

Again, these experiments can be performed according to the procedure in Subheading 3.2. The titrations are prepared for both the open- and closed-loop circuits, with and without a disturbance. If the control circuit is working correctly, the expression levels of species $\mathbf{X_L}$ in the closed-loop circuit should remain within a 10% error bound relative to the undisturbed condition. In the open-loop configuration, the expression levels of species $\mathbf{X_N}$ should change significantly when the circuit is disturbed. Disturbance rejection may break down for high setpoints that lie close to the saturation level, and a relative error larger than 10% may be observed. If a minimal controlled network was used to verify disturbance rejection, it should now also be verified on the full controlled network.

4 Notes

1. The following notes outline the conceptual and genetic design of biomolecular integral feedback controllers. They should act as guides to help find circuit implementations given a specific controlled network.
2. Determine which species in the controlled network will be measured and should therefore act as $\mathbf{X_L}$. Ideally, choose the species as far downstream or close to the output $\mathbf{X_N}$ as possible. Optimally, the output $\mathbf{X_N}$ will be $\mathbf{X_L}$, so it will be maximally robust to all possible disturbances.
3. Based on what is known about the controlled network from the literature, prior experiments, and intuition, estimate if the mapping from input $\mathbf{X_1}$ to output $\mathbf{X_N}$ in the open-loop controlled network has a positive or a negative gain as shown in the top of Fig. 2. Ideally, this is determined experimentally. We describe how to do so in Subheading 3.2. Specifically, the mapping has a positive gain if an increase in the input $\mathbf{X_1}$

increases the output $\mathbf{X_N}$. Conversely, when an increase in the input $\mathbf{X_1}$ decreases the output $\mathbf{X_N}$, the mapping has a negative gain.

4. Determine if the input $\mathbf{X_1}$ can be actuated positively (e.g., by inducing production through other species) or if it can be actuated negatively (e.g., by inducing degradation through other species). Negative actuation leads to the realization of proportional-integral feedback, as has been recently discovered in [7].
5. Keep in mind that species $\mathbf{X_1}$ needs to interact with the control circuit output (either $\mathbf{Z_1}$ or $\mathbf{Z_2}$). Therefore, choose it so there can be a direct interaction between the two. By direct interaction, we mean interactions of the form protein to mRNA (here, with protein, we mean a transcription factor) or mRNA to protein.
6. Based on Steps 3, 4, and 5 in Subheading 4, identify biomolecules that can implement the corresponding control circuit motif shown in Fig. 2. Two adjustable implementation strategies are presented in [2] and [8] based on mRNA hybridization and split intein trans-splicing, respectively.
7. Determine how the output of the controlled network $\mathbf{X_N}$ will connect to the control circuit and if you may need to introduce extra species to accomplish this. For example, if your control circuit implementation is realized as RNA species and your output species $\mathbf{X_N}$ is an RNA as well, you may need to introduce a protein species, such as a transcription factor, to connect the two.
8. For implementing the circuit on plasmids, it is helpful to partition it into groups. Where the groups are formed by $\mathbf{Z_1}$ and its precursors $\mathbf{Z}'_1$, $\mathbf{Z_2}$ and its precursors $\mathbf{Z}'_2$, and everything relating to the controlled network ($\mathbf{X} = \{\mathbf{X_1}, \ldots, \mathbf{X_L}, \ldots, \mathbf{X_N}\}$). The groups of species $\mathbf{Z_1}$ and $\mathbf{Z_2}$ are ideally each realized on a separate plasmid to allow for independent tuning and titration experiments (Subheading 3.2).
9. To test the robustness of the circuit, come up with a method of introducing disturbances into the controlled network. Depending on the controlled network, it might be possible to modulate the activity or stability of species upstream of species $\mathbf{X_L}$ using a small-molecule drug. Alternatively, activity or stability modifying protein tags may be fused to species $\mathbf{X_1}$ (e.g., the SMASh-tag from [9]). Be aware that the disturbance must translate into species' expression levels $\mathbf{X_L}$. For example, if species $\mathbf{X_L}$ is the synthetic transcription factor tTA fused to a fluorescent protein as in [2], then using its inhibitor Doxycycline will affect the activity of the transcription factor but will not affect its abundance.

10. If possible, a minimal controlled network should be designed so that the function of the control circuit can be evaluated without having to deal with the complexities of the full controlled network. The requirements for a minimal controlled network are that it is measurable (there is a fluorescent signal in the network) and that the connection from its input to its output consists of as few species as possible. For example, in the circuit depicted in Fig. 3, the minimal controlled network is a transcription factor fused to a fluorescent protein.

References

1. Aoki SK, Lillacci G, Gupta A, Baumschlager A, Schweingruber D, Khammash M (2019) A universal biomolecular integral feedback controller for robust perfect adaptation. Nature 570:533–537
2. Frei T, Chang CH, Filo M, Arampatzis A, Khammash M (2022) A genetic mammalian proportional–integral feedback control circuit for robust and precise gene regulation. Proc Nat Acad Sci U S A 119:e2122132119
3. Briat C, Gupta A, Khammash M (2016) Antithetic integral feedback ensures robust perfect adaptation in noisy biomolecular networks. Cell Syst 2:15–26
4. Fonseca JP, Bonny AR, Kumar GR, Ng AH, Town J, Wu QC, et al (2019) A toolkit for rapid modular construction of biological circuits in mammalian cells. ACS Synth Biol 8:2593–2606
5. Fonseca JP, Bonny AR, Town J, El-Samad H (2020) Assembly of genetic circuits with the mammalian toolkit. Bio-protocol 10:e3547–e3547
6. Pryor JM, Potapov V, Bilotti K, Pokhrel N, Lohman GJ (2022) Rapid 40 kb genome construction from 52 parts through data-optimized assembly design. ACS Synth Biol 11:2036–2042
7. Filo M, Hou M, Khammash M (2023) A hidden proportional feedback mechanism underlies enhanced dynamic performance and noise rejection in sensor-based antithetic integral control. bioRxiv
8. Anastassov S, Filo M, Chang CH, Khammash M (2023) A cybergenetic framework for engineering intein-mediated integral feedback control systems. Nat Commun 14:1337
9. Chung HK, Jacobs CL, Huo Y, Yang J, Krumm SA, Plemper RK, et al (2015) Tunable and reversible drug control of protein production via a self-excising degron. Nat Chem Biol 11: 713–720

Chapter 8

A Computational Modeling Approach for the Design of Genetic Control Systems that Respond to Transcriptional Activity

Carlos D. Llanos, Tianyi Xie, Ha Eun Lim, and Laura Segatori

Abstract

Recent progress in synthetic biology has enabled the design of complex genetic circuits that interface with innate cellular functions, such as gene transcription, and control user-defined outputs. Implementing these genetic networks in mammalian cells, however, is a cumbersome process that requires several steps of optimization and benefits from the use of predictive modeling. Combining deterministic mathematical models with software-based numerical computing platforms allows researchers to quickly design, evaluate, and optimize multiple circuit topologies to establish experimental constraints that generate the desired control systems. In this chapter, we present a systematic approach based on predictive mathematical modeling to guide the design and construction of gene activity-based sensors. This approach enables user-driven circuit optimization through iterations of sensitivity analyses and parameter scans, providing a universal method to engineer sense and respond cells for diverse applications.

Key words Genetic control systems, Sense-and-respond, Genetic circuits, Mammalian synthetic biology, Cell therapies

1 Introduction

Engineering mammalian cells to respond to environmental stimuli in a user-controlled fashion is a powerful approach that is revolutionizing the way we think about medical science. Sense-and-respond cellular devices typically comprise multiple units of operation designed to sense a specific biological input, process the signal, and respond with a user-defined biomolecular output (Fig. 1a). The resulting designer cells can be used for a variety of applications requiring input-induced programmable responses, including the production of reporter molecules for diagnostics purposes or the delivery of therapeutic molecules for feedback-controlled treatment approaches [1, 2]. Current strategies to build cellular sensors capitalize mainly on recent progress in the field of receptor

Francesca Ceroni and Karen Polizzi (eds.), *Mammalian Synthetic Systems*, Methods in Molecular Biology, vol. 2774,
https://doi.org/10.1007/978-1-0716-3718-0_8,

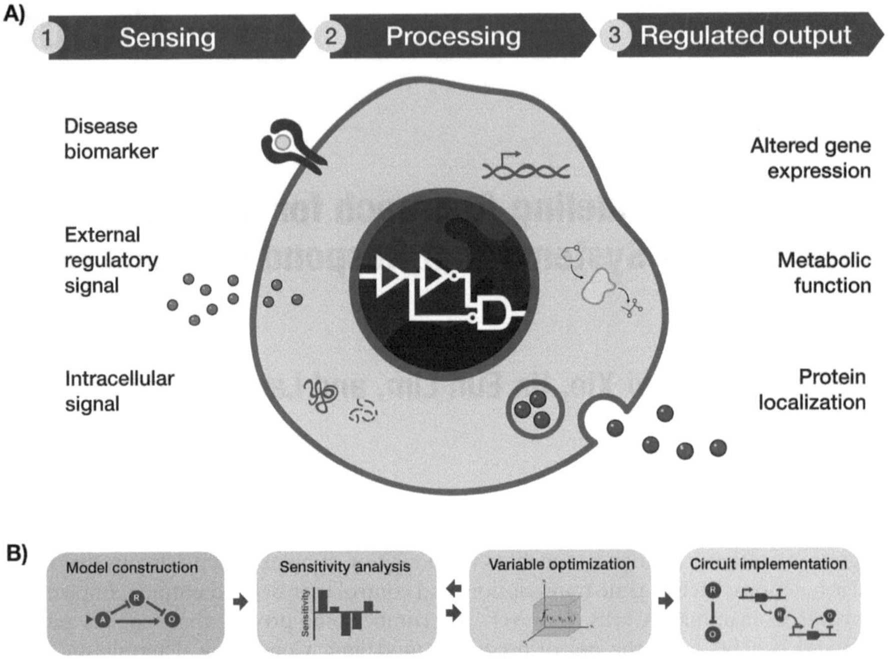

Fig. 1 Design and optimization of sense-and-respond cells with genetic control systems. (**a**) Schematic representation of cellular sensors as comprising modules for (1) sensing inputs including molecular signatures of diseases such as inflammatory or toxic molecules, environmental features such as changes in pH or oxidative stress, and intracellular cues such as proteotoxic stress; (2) processing inputs through logic operations for conversion into a programmable output; and (3) regulating user-defined outputs including gene expression, metabolic functions, and protein localization. (**b**) Simplified workflow of the computational approach for the design of genetic control systems including the construction of a model for profiling the concentration of the circuits' components, a sensitivity analysis for determining parameters for optimization, optimization of parameters based on user-defined design goals, and generation of design rules for experimental implementation

engineering, which provides innovative avenues to endow cells with seemingly endless sensing capabilities [3]. The design of cellular sensors based on membrane receptors typically provides signal amplification for enhanced sensitivity and modular components for interchangeable inputs and output responses [4]. The system's dynamic properties, however, are dictated by the signal transduction mechanism that converts input detection into output control and may not accurately recapitulate the dynamic behavior of the input signal. Cellular diagnostics and therapeutics that operate over physiologically relevant timescales require control systems with an appropriate dynamic resolution of the biomarker concentration profile [5].

Gene regulation determines cell functionality during physiological and pathological processes and is dynamically modulated to respond to external and internal stimuli. Engineering cellular sensors based on the detection of gene activity from the chromosomal context might thus provide a more efficient strategy to create designer cells that respond to desired extracellular cues with superior dynamic resolutions. In addition, genetic sensors for the detection of relevant transcriptional signatures will expand the capabilities of sense-and-respond systems to the detection of extracellular as well as intracellular cues that may not be detected through ligand–receptor interactions, such as acidic and hypoxic environments, or proteotoxic stress [5].

Since very early efforts in the design of biological systems with novel functionalities, predictive modeling has provided a powerful tool to establish the design rules of synthetic gene networks that generate desired outputs, such as oscillatory or hysteretic behaviors [6, 7]. Specifically, modeling synthetic gene networks allows identifying key parameters that affect the network's behavior and specific values that satisfy user-defined objectives (*See* **Note 1**). In this chapter, we present a computational modeling approach designed to support the construction and optimization of genetic control systems for engineering designer cells (Fig. 1b). This approach will be generally useful to build synthetic cells that sense and respond to environmental as well as intracellular stimuli associated with characteristic transcriptional signatures. As such, this strategy will enable the design of cellular devices for real-time monitoring of disease biomarkers and therapeutic approaches that translate the detection of a biomarker into the execution of a therapeutic program.

2 Computational Modeling Approach for the Design of Genetic Control Systems

We present a series of genetic control systems designed to link the expression of a target gene from the chromosomal context to a user-designed output. Specifically, we investigated four circuit topologies expected to result in the expression of the output gene with different dynamic behaviors. In all the topologies, the expression of the target gene is linked to that of a transcription factor (an activator [A] or a repressor [R]). This transcription factor functions as a master regulator of the genetic circuit that controls the output's expression. Such an approach can be implemented experimentally by genome editing to insert a cassette encoding an internal ribosome entry site (IRES) or a 2A self-cleaving peptide and the gene encoding the master regulator downstream of the target gene (*See* **Note 2**) [8, 9]. This strategy ensures a constant ratio of expression of the target gene and the master regulator. The following topologies are presented: (A) a direct amplification loop in which the target is linked to the output's activator (Fig. 2a), (B) an inverter configuration in which the target is linked to the output's repressor (Fig. 2b), (C) an amplification loop (as in A) with negative output's regulation expected to lower the output's basal expression (Fig. 2c), and (D) an inverter configuration (as in

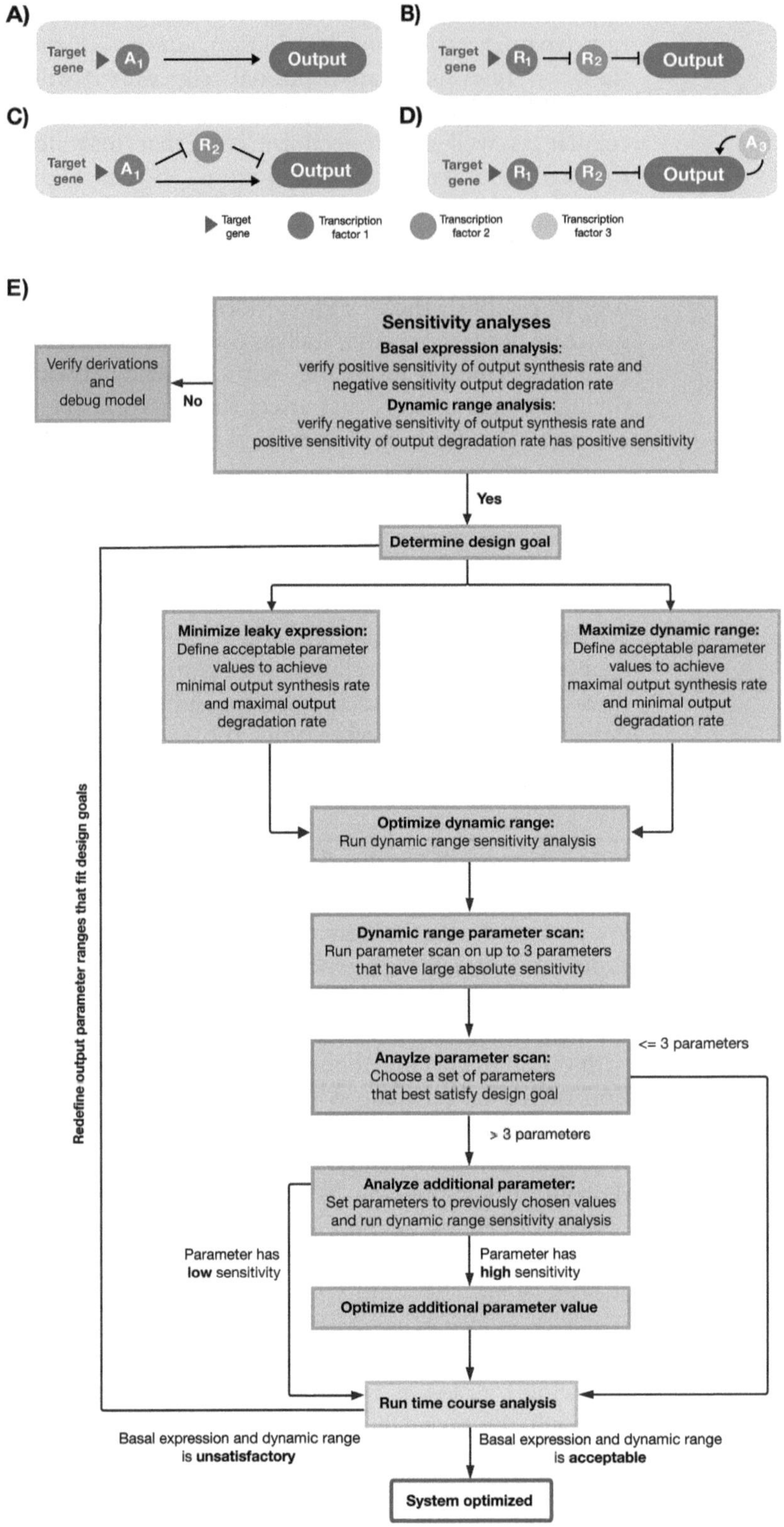

Fig. 2 A modeling approach for the design of genetic control systems. (**a**–**d**) Circuit topologies explored in this study to build genetic control systems that link the expression of a chromosomal gene to a user-defined

B) with output's self-amplification expected to maximize output expression (Fig. 2d).

We developed a systematic optimization workflow that employs mathematical simulations, including sensitivity analysis and parameter scans, to identify the parameter values that generate desired output behaviors. The first step of this workflow consists of building a deterministic mathematical model that simulates the behavior of the genetic circuits. The next step involves performing a sensitivity analysis on all parameters with respect to the output to identify the parameters for optimization of the circuit design. We illustrate that user-defined goals, such as minimizing leaky expression or maximizing output dynamic range, are likely to affect the selection of relevant parameters for circuit optimization. Following parameter selection, a parameter scan is performed to determine the parameter values that satisfy the desired performance goals. If more than three parameters are selected for optimization, additional iterations of sensitivity analysis and parameter scan are likely required to facilitate visualization and analysis of the parameter scan results. Lastly, a simulation is performed using the optimized parameters to determine whether the established goals of dynamic range and basal output expression are satisfied or if further modifications are needed.

3 Methods

3.1 Predictive Mathematical Model

The mathematical model describing the four circuit topologies (Fig. 2a–d) is built by deriving a set of ODEs to profile the concentration of each circuit's component. The concentration profiles are derived under the assumption that transcription is fast compared to translation and protein dimerization is fast compared to dissociation and degradation [10, 11]. The states of promoters in which transcription factors are bound to single operator sequences are neglected. The degradation of all proteins and the dynamic of translation are considered linear. The change in concentration of each protein component depends on the component's rate of production (α), degradation (γ), and dilution due to cell growth (μ) and can be expressed using the following differential equation:

Fig. 2 (continued) output. (**a**) Amplifier linking the target gene to the output's activator. (**b**) Genetic inverter linking the target gene to the output's repressor. (**c**) Amplifier linking the target gene to the output's activator that also functions as a repressor of the output's repressor. (**d**) Genetic inverter linking the target gene to the output's repressor with self-amplification of the output. (**e**) Workflow of the modeling approach to the design of genetic control systems including sensitivity analyses (pink), parameter scans (green), and time course analysis (yellow)

$$\frac{d[X]}{dt} = \alpha_X - (\gamma_X + \mu) \cdot [X] \quad (1)$$

The rate of production of a circuit component depends on whether the component's expression is linked to that of a target gene or regulated by a transcription factor.

1. For a component linked to the target gene, the rate of production depends on the synthesis rate of the target gene at basal condition (β_0) and the fold change of expression of the target gene upon detection of the input (f_c) and is affected by the mechanism linking the expression of the target gene to the circuit component (i.e., IRES or 2A peptide) as accounted by the phi factor (ϕ). The rate of production of the master regulator is expressed as:

$$\alpha_X = \beta_0 \cdot \phi \cdot f_c \quad (2)$$

2. The rates of production of all other circuit components depend on the component's synthesis rate and are described using Hill functions and modified to account for basal expression due to leakage [12–14].
3. The rate of production of a component that is regulated by a repressor, such as the transcription factor R_2 in topology B (Fig. 2b), which is regulated by the transcription factor R_1, depends on the synthesis rate β_{R2} and the repression factor f_{R1}. The unbound state of the operator leads to the basal expression of R_2 with synthesis rate β_{R2}. The binding of R_1 to the operator leads to the repression of R_2 synthesis by a factor f_{R1}. The rate of production of R_2 is thus:

$$\alpha_{R2} = \left[\frac{\beta_{R2}}{1 + \frac{[R1]^{n_{R1}}}{K_{DR1}}}\right] \cdot \left[1 + f_{R1} \cdot \frac{[R1]^{n_{R1}}}{K_{DR1}}\right] \quad (3)$$

where K_{DR1} is the equilibrium dissociation constant of R_1 binding to the operator, $[R_1]$ is the concentration of free R_1, and n_{R1} is the degree of cooperativity of R_1 binding to the operator.
4. The rate of production of a component that is regulated by both a repressor and an activator is defined by a Hill function that accounts for activation and repression. For instance, the rate of production of the output (α_O) in topology C (Fig. 2c) depends on the relative contribution of the unbound state of the promoter, the state of the promoter bound only to the activator A_1, and the state of the promoter bound to the repressor R_2. The unbound state of the promoter leads to the basal expression of the output with the synthesis rate β_O. Binding of only transcription factor A_1 to the operator leads to the activation of output synthesis by a factor f_{A1}, while binding of transcription factor R_2 to the operator, regardless

of whether A_1 is also bound, leads to repression by a factor f_{R2}. The rate of production of the output is thus:

$$\alpha_O = \frac{\beta_O}{\left[1+\left(\frac{[R2]^{n_{R2}}}{K_{DR2}}\right)\right]\cdot\left[1+\left(\frac{[A1]^{n_{A1}}}{K_{DA1}}\right)\right]}\cdot[1+(f_{A1})\cdot\left(\frac{[A1]^{n_{A1}}}{K_{DA1}}\right) + \left[(f_{R2})\cdot\left(\frac{[R2]^{n_{R2}}}{K_{DR2}}\right)\right]\cdot\left[1+\left(\frac{[A1]^{n_{A1}}}{K_{DA1}}\right)\right]] \quad (4)$$

where K_{DA1} is the equilibrium dissociation constant of A_1 binding to the operator, $[A_1]$ is the concentration of free A_1, n_{A1} is the degree of cooperativity of A_1 binding to the operator, K_{DR2} is the equilibrium dissociation constant of R_2 binding to the operator, $[R_2]$ is the concentration of free R_2, and n_{R2} is the degree of cooperativity of R_2 binding to the operator.

3.1.1 Modeling Results

The complete set of ODEs (Box 1) was used to profile the concentration of each component in the four topologies (Fig. 2a–d) using the parameters reported in Table 1. The components' concentrations were simulated over 500 h in response to a transient stimulus (from 200 to 400 h) that induces a twofold change in the target gene expression. The simulations were conducted using MATLAB (MathWorks) using the built-in function *ode45* (Fig. 3).

Box 1. Mathematical Model

Mathematical expressions for each component in topologies A, B, C, and D:

Topology A:

$$\frac{d[A_1]}{dt} = \alpha_{A_1} - (\gamma_{A_1} + \mu)\cdot[A_1]$$

$$\alpha_{A_1} = \alpha_T \cdot \phi \cdot f_c$$

$$\frac{d[O]}{dt} = \alpha_O - (\gamma_O + \mu)\cdot[O]$$

$$\alpha_O = \left[\frac{\beta_O}{1+\frac{[A1]^{n_{A1}}}{K_{DA1}}}\right]\cdot\left[1 + f_{A1}\cdot\frac{[A1]^{n_{A1}}}{K_{DA1}}\right]$$

Topology B:

$$\frac{d[R_1]}{dt} = \alpha_{R_1} - (\gamma_{R_1} + \mu)\cdot[R_1]$$

$$\frac{d[O]}{dt} = \alpha_O - (\gamma_O + \mu)\cdot[O]$$

$$\alpha_{R_2} = \left[\frac{\beta_{R_2}}{1+\frac{[R_1]^{n_{R_1}}}{K_{D_{R_1}}}}\right]\cdot\left[1 + f_{R_1}\cdot\frac{[R_1]^{n_{R_1}}}{K_{D_{R_1}}}\right]$$

$$\frac{d[R_2]}{dt} = \alpha_{R_2} - (\gamma_{R_2} + \mu)\cdot[R_2]$$

$$\alpha_{R_1} = \alpha_T \cdot \phi \cdot f_c$$

$$\alpha_O = \left[\frac{\beta_O}{1+\frac{[R_2]^{n_{R_2}}}{K_{D_{R_2}}}}\right]\cdot\left[1 + f_{R_2}\cdot\frac{[R_2]^{n_{R_2}}}{K_{D_{R_2}}}\right]$$

Topology C:

$$\frac{d[A_1]}{dt} = \alpha_{A_1} - (\gamma_{A_1} + \mu)\cdot[A_1]$$

$$\frac{d[R_2]}{dt} = \alpha_{R_2} - (\gamma_{R_2} + \mu)\cdot[R_2]$$

$$\alpha_{R_2} = \left[\frac{\beta_{R_2}}{1+\frac{[A_1]^{n_{A_1}}}{K_{D_{A_1}}}}\right]\cdot\left[1 + f_{A_1}\cdot\frac{[A_1]^{n_{A_1}}}{K_{D_{A_1}}}\right]$$

$$\frac{d[O]}{dt} = \alpha_O - (\gamma_O + \mu)\cdot[O]$$

$$\alpha_{A_1} = \alpha_T \cdot \phi \cdot f_c$$

$$\alpha_O = \frac{\beta_O}{\left[1+\left(\frac{[R2]^{n_{R2}}}{K_{DR2}}\right)\right]\cdot\left[1+\left(\frac{[A1]^{n_{A1}}}{K_{DA1}}\right)\right]}\cdot[1+(f_{A1})\cdot\left(\frac{[A1]^{n_{A1}}}{K_{DA1}}\right) + \left[(f_{R2})\cdot\left(\frac{[R2]^{n_{R2}}}{K_{DR2}}\right)\right]\cdot\left[1+\left(\frac{[A1]^{n_{A1}}}{K_{DA1}}\right)\right]]$$

Topology D:

(continued)

Box 1 (continued)

$$\frac{d[R_1]}{dt} = \alpha_{R_1} - (\gamma_{R_1} + \mu) \cdot [R_1] \qquad \frac{d[O]}{dt} = \alpha_O - (\gamma_O + \mu) \cdot [O]$$
$$\frac{d[R_2]}{dt} = \alpha_{R_2} - (\gamma_{R_2} + \mu) \cdot [R_2] \qquad \alpha_{R_1} = \alpha_T \cdot \phi \cdot f_c$$
$$\frac{d[A_3]}{dt} = \alpha_{A_3} - (\gamma_{A_3} + \mu) \cdot [A_3] \qquad \alpha_{A_3} = \alpha_O \cdot \phi_{A3}$$
$$\alpha_{R_2} = \left[\frac{\beta_{R_2}}{1 + \frac{[R_1]^{n_{R_1}}}{K_{D_{R_1}}}}\right] \cdot \left[1 + f_{R_1} \cdot \frac{[R_1]^{n_{R_1}}}{K_{D_{R_1}}}\right]$$

(continued)

Table 1
List of model parameters

Parameter	Description	Value [units]	Optimized	Values for sensitivity analysis	Source
f_{Ax}	Activation factor	100	No	–	[1]
f_{Rx}	Repression factor	0.1	No	–	[2]
ϕ_x	Linker factor	1	No	0.3–1([a])	[1]
k_{Dx}	Equilibrium dissociation constant of transcription factor–DNA	1 [nM]	No	–	[3]
n_x	Cooperativity coefficient	2	No	–	[4]
γ_x	Degradation rate	0.3 [h^{-1}]	Yes	0.0267–1.347	[1]
μ	Cell dilution rate	0.023 [h^{-1}]	No	–	[5]
β_x	Synthesis rate	2 [nM h^{-1}]	Yes	0.25–10	[2]
β_O	Synthesis rate of target gene at basal conditions	1 [nM h^{-1}]	Yes	0.25–10	[2]
f_C	Fold change of expression of target gene	2×	No	–	[1]

[a]This parameter was only analyzed for topology D ($\emptyset_{A3}$)

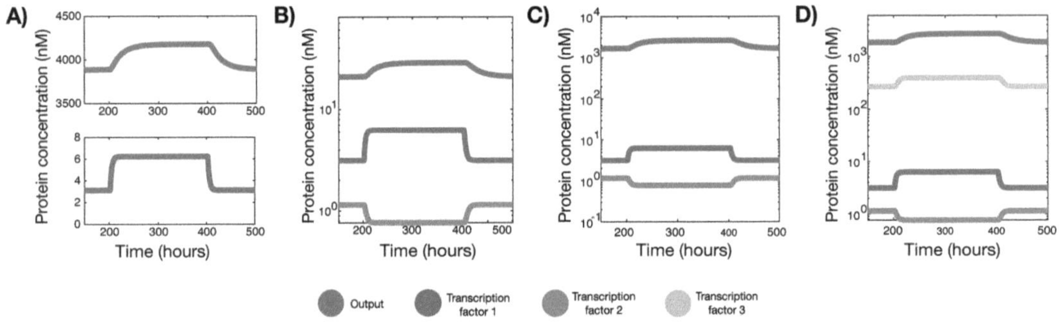

Fig. 3 Simulations of the genetic control systems. (**a**–**d**) Concentration of output (gray) and transcription factors (pink, blue, yellow) of topologies **A**, **B**, **C**, and **D** reported in Fig. 2a–d simulated for 500 h

Box 1 (continued)

$$\alpha_O = \frac{\beta_O}{\left[1+\left(\frac{[R2]^{n_{R2}}}{K_{DR2}}\right)\right]\cdot\left[1+\left(\frac{[A3]^{n_{A3}}}{K_{DA3}}\right)\right]}\cdot\left[1+(f_{A3})\cdot\left(\frac{[A3]^{n_{A3}}}{K_{DA3}}\right)+\left[(f_{R2})\cdot\left(\frac{[R2]^{n_{R2}}}{K_{DR2}}\right)\right]\cdot\left[1+\left(\frac{[A3]^{n_{A3}}}{K_{DA3}}\right)\right]\right]$$

3.1.2 Sensitivity Analysis

A sensitivity analysis is typically conducted to identify parameters for system optimization. Sensitivity analyses allow evaluating the output's response to changes in the system's parameters and identifying parameters that have the most dramatic effect on the system's performance. This analysis can be performed using MATLAB and focuses on parameters that can be altered experimentally, such as the synthesis rates, which can be modulated through the choice of appropriate promoters, or the degradation rates, which can be modulated using degron tags (*See* **Notes 3–5**) [15]. The effect of a given parameter on the output is evaluated by calculating the parameter's sensitivity coefficient, which is defined as the partial derivative of the change in output in response to a change in input. Sensitivity coefficients are typically normalized by multiplying the sensitivity coefficient by the ratio between the input (x_i) and output obtained from the specific change in input ($y(x_i)$), generating scaled sensitivity coefficients (ε_{x_i}) that allow comparing the effect of variations in different parameters. Scaled sensitivity coefficients are thus calculated as follows:

$$\varepsilon_{x_i} = \frac{\partial y}{\partial x_i}\cdot\frac{x_i}{y(x_i)} \tag{5}$$

where y is the model simulation output (i.e., the output concentration or the output dynamic range) and x_i is the design parameter. A positive sensitivity coefficient indicates a positive correlation between the output and parameter perturbation, while a negative coefficient indicates a negative correlation. The absolute value of the sensitivity coefficient correlates with the degree of change in output in response to the parameter perturbation.

1. A five-point stencil approximation is used here to numerically evaluate the partial derivative of the output with respect to the input parameter (i.e., the term $\frac{\partial y}{\partial x_i}$ in Eq. 5). A five-point stencil approximation evaluates the derivative by calculating a weighted average of outputs calculated from four adjacent inputs. The derivatives are calculated using the following general equation for calculating the derivative of a function ($f(x)$) based on the weighted average of the function at multiple input values separated by a perturbation value h:

$$f'(x) \approx \frac{-8f(x+2h)+8f(x+h)-8f(x-h)+f(x-2h)}{12h} \tag{6}$$

2. A sensitivity analysis is first conducted using values from the initial simulation. Such an analysis provides a preliminary evaluation of the simulated output response to changes in the

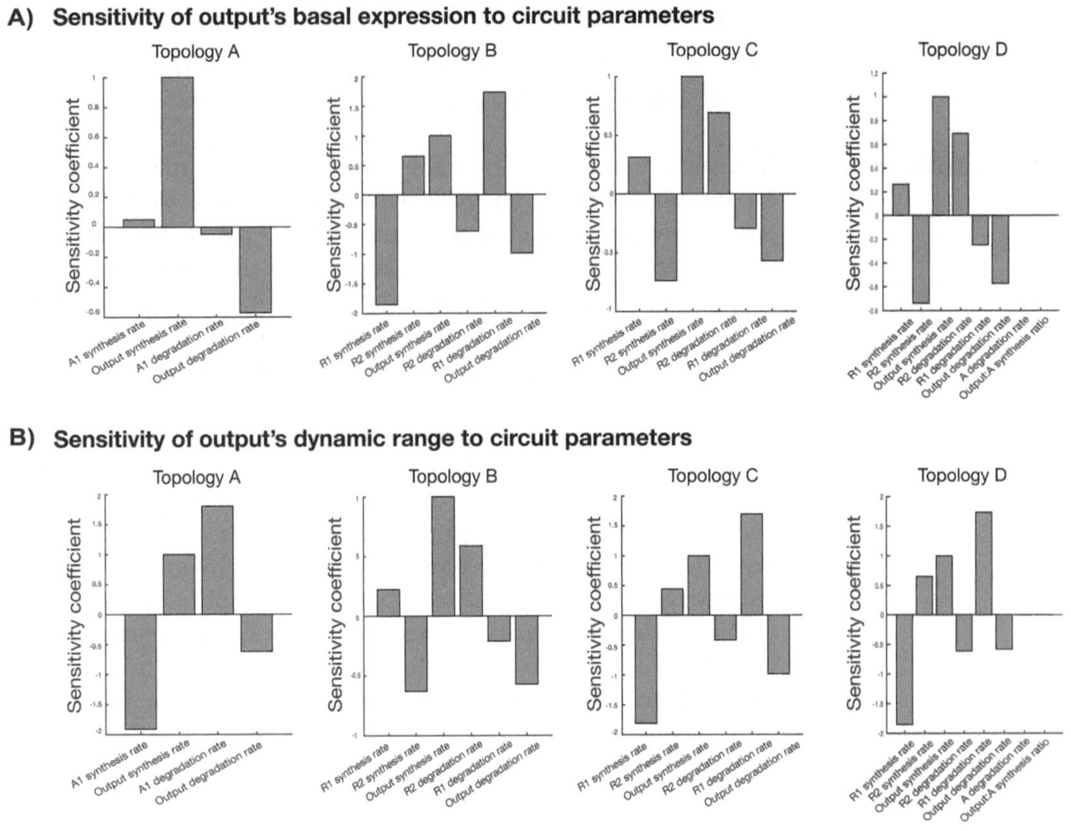

Fig. 4 Sensitivity analysis. (**a**) Sensitivity coefficients of circuit parameters with respect to the output's basal expression for topologies A–D. (**b**) Sensitivity coefficients of circuit parameters with respect to the output's dynamic range for topologies A–D

design parameter around initial values. Since the output's response to parameters' perturbations depends on the initial value of each parameter, a more extensive sensitivity scan was performed to examine parameters across a range of biologically relevant values.

3. As a sensitivity coefficient at a given parameter value approaches zero, parameter perturbations at that value are deemed not to affect the output.

3.1.3 Sensitivity Analysis Results

The sensitivity analysis revealed that the sensitivity coefficients of most of the parameters analyzed depend on the circuit topology. For instance, the sensitivity analysis conducted to evaluate the sensitivity of the output's basal expression to parameters' perturbation revealed all of the design parameters of topology B are highly relevant (i.e., high absolute values of the sensitivity coefficients), while the design parameters of topology A have a more diverse range of effects, with the synthesis and degradation rates of the transcriptional factor A_1 having a minimal effect (Fig. 4).

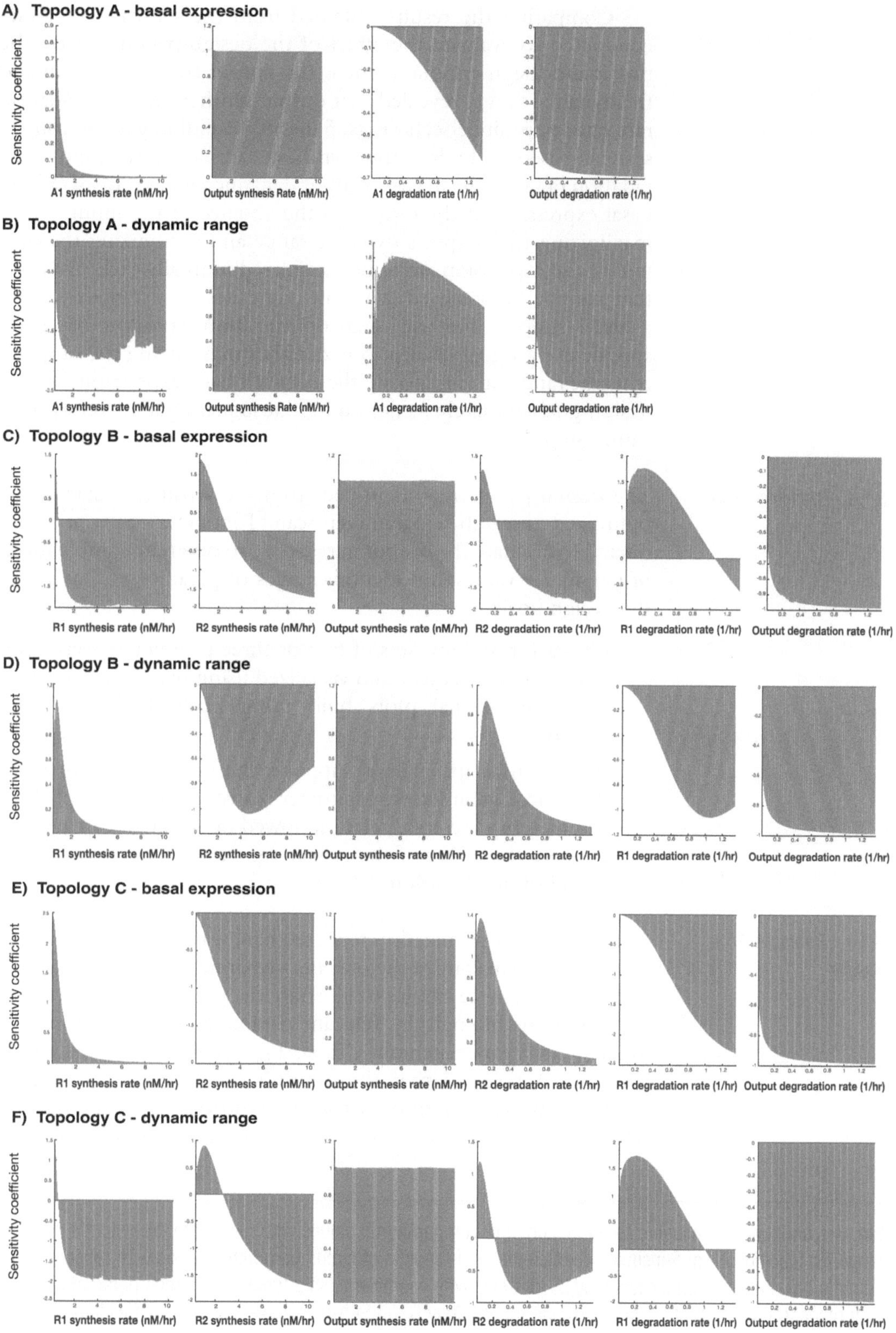

Fig. 5 Results of the sensitivity analysis of topologies A, B, and C. (**a**) Sensitivity coefficients as a function of circuit parameter values with respect to the output's basal expression for topology A. (**b**) Sensitivity

Comparing the results obtained from the sensitivity analyses conducted to evaluate the effect of the design parameters on the two main design criteria, namely, the output basal expression and the dynamic range, revealed that output synthesis and degradation rate have opposite effects (Figs. 5 and 6). Building genetic control systems based on high output synthesis rates and low output degradation rates results in large output dynamic ranges but also a high basal expression of the output in the absence of the input. Thus, maximizing the output's dynamic range and minimizing the output's basal expression are to be considered mutually exclusive goals that cannot be achieved with the same design parameters. Our analyses support the use of an optimization workflow based on specific design goals that prioritize either maximizing the output's dynamic range or minimizing the output's basal expression, as such design goals are likely to determine subsequent parameter optimization steps.

3.1.4 Parameter Scan

The design parameters identified from the sensitivity analysis are optimized through a parameter scan. Parameter scans are conducted to evaluate the output in response to different combinations of parameter values and identify ranges of parameters that satisfy the design goals.

1. For each topology, sets of two or three parameters were optimized simultaneously and visualized using two-dimensional or three-dimensional plots built using pcolor3 (Figs. 7 and 8) [16].
2. If the sensitivity analysis supports the optimization of more than three parameters, parameter scans based on sets of three parameters were conducted, followed by sensitivity analyses to evaluate the effect of the remaining parameters after the first round of optimization.

3.1.5 Parameter Scan Results

The parameter optimization reported here was conducted with the design goal to maximize the outputs' dynamic range. Topologies A and B are used as examples to illustrate the parameter optimization process using biologically relevant parameter values that can be implemented experimentally (Table 1). The sensitivity analysis of topology A revealed that the output's dynamic range is highly sensitive to transcription factor A_1 synthesis and degradation rates. The parameter scan indicated that maximal output's dynamic

Fig. 5 (continued) coefficients as a function of circuit parameter values with respect to the output's dynamic range for topology A. (**c**) Sensitivity coefficients as a function of circuit parameter values with respect to the output's basal expression for topology B. (**d**) Sensitivity coefficients as a function of circuit parameter values with respect to the output's dynamic range for topology B. (**e**) Sensitivity coefficients as a function of circuit parameter values with respect to the output's basal expression for topology C. (**f**) Sensitivity coefficients as a function of circuit parameter values with respect to the output's dynamic range for topology C

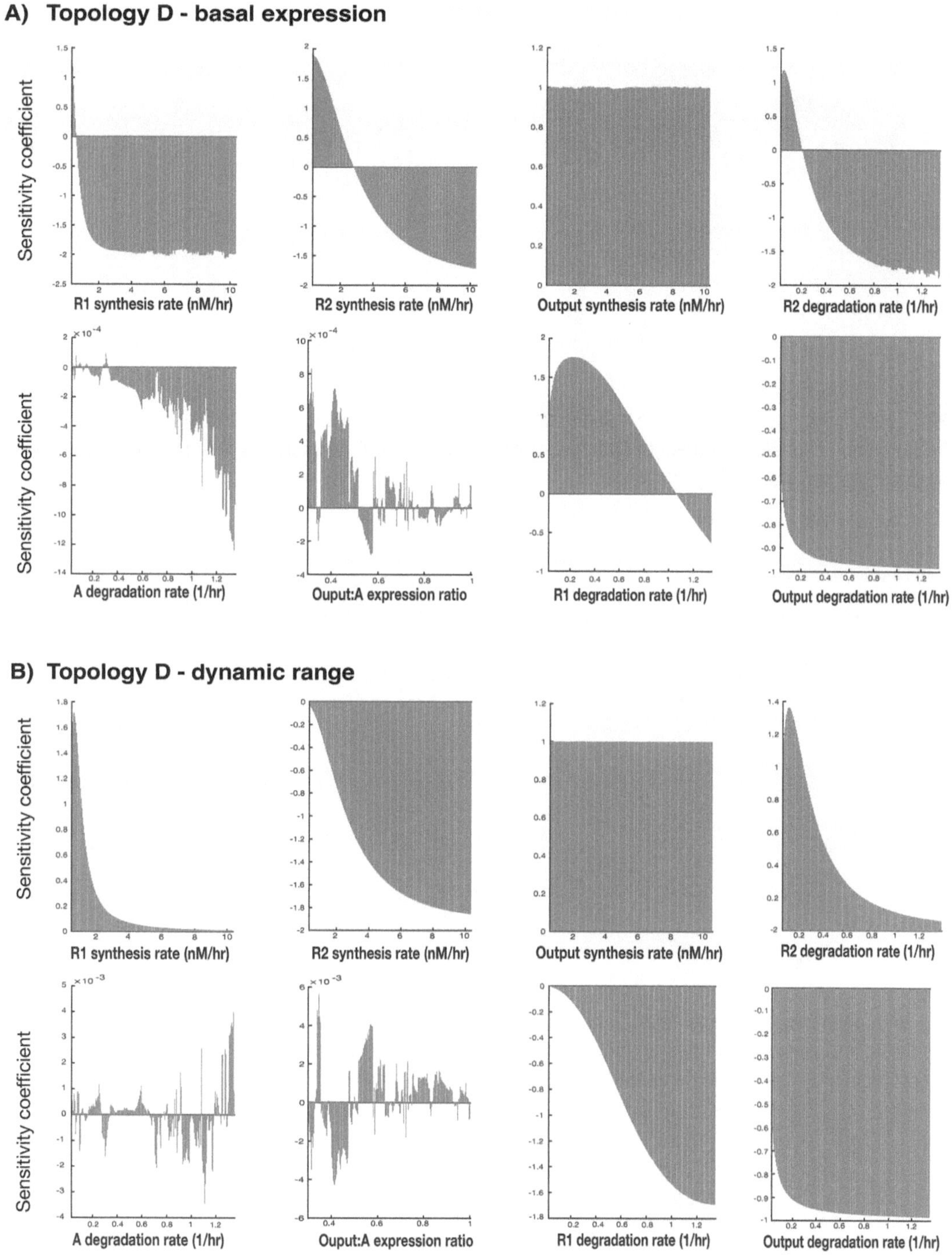

Fig. 6 Results of the sensitivity analysis of topology D. (**a**) Sensitivity coefficients as a function of circuit parameter values with respect to the output's basal expression for topology D. (**b**) Sensitivity coefficients as a function of circuit parameter values with respect to the output's dynamic range for topology D

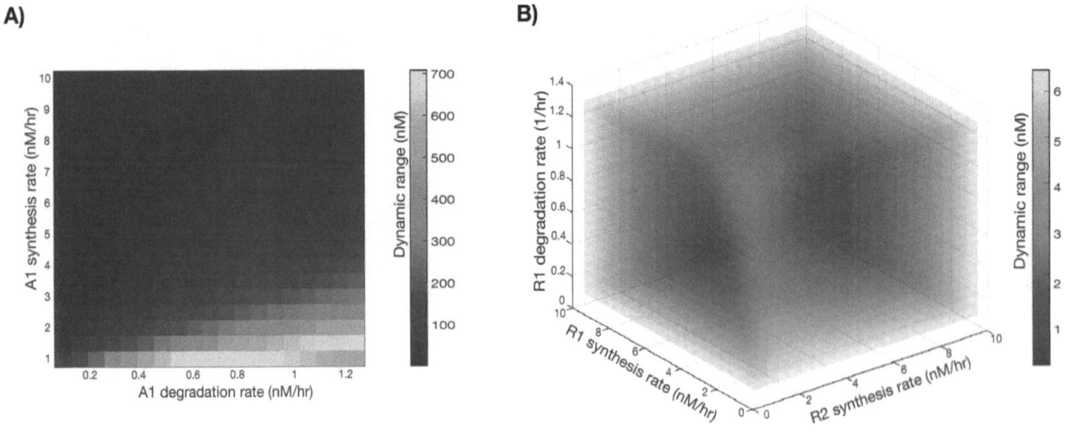

Fig. 7 Parameter scan for topologies A and B. (**a**) Dynamic range of the output of topology A as a function of A_1 synthesis and degradation rates. (**b**) Dynamic range of the output of topology B as a function of R_1 synthesis and degradation rates and R_2 synthesis rate

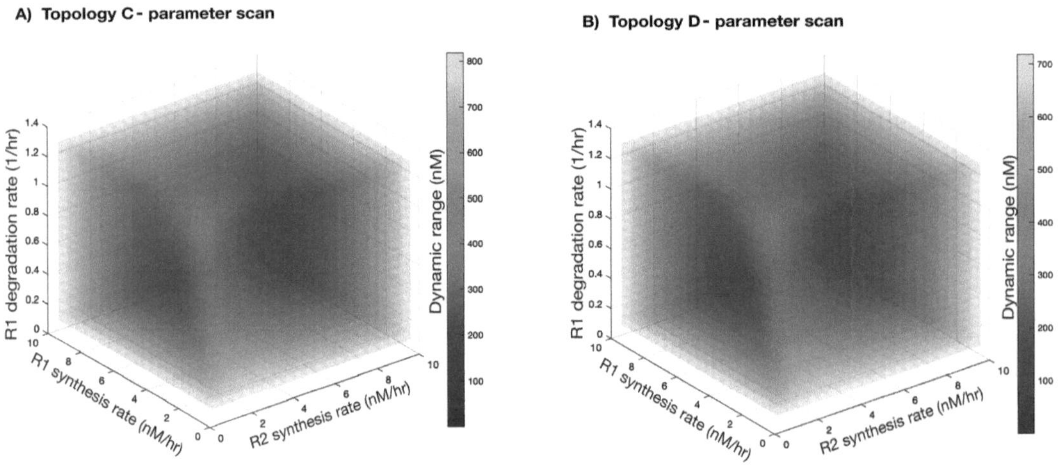

Fig. 8 Parameter scan for topologies C and D. (**a**) Dynamic range of the output of topology C as a function of R_1 synthesis and degradation rates and R_2 synthesis rate. (**b**) Dynamic range of the output of topology D as a function of R_1 synthesis and degradation rates and R_2 synthesis rate

range is achieved using biological parts that result in low synthesis and high degradation rates of the transcription factor A_1 (Fig. 7a). Combinations of parameters that can be implemented experimentally can be selected within the range of values of synthesis and degradation rates that lead to maximal output's dynamic range.

The sensitivity analysis of topology B revealed that the output's dynamic range depends on the synthesis and degradation rates of the transcription factors R_1 and R_2. A scan of the values of R_1 synthesis rate, R_2 synthesis rate, and R_1 degradation rate revealed that maximal output's dynamic range is achieved using low R_1

and R_2 synthesis rates regardless of R_1 degradation rate (Fig. 7b). Constraining the model with optimal values for the three parameters scanned (i.e., R_1 synthesis rate, R_2 synthesis rate, and R_1 degradation rate) results in maximal output's dynamic range that is no longer affected by the remaining design parameter (i.e., R_2 degradation rate) as can be observed from additional sensitivity analyses. The parameter scan analysis of topologies C and D (Fig. 8) generated similar results and conclusions for parameter optimization.

The final step in the optimization workflow consists in repeating the model simulations using the parameter values selected from the parameter scan to verify whether the design goals are satisfied (*See* **Notes 6–9**). Further optimization of the genetic control systems is otherwise achieved by redefining the design goals. For instance, increasing the threshold for acceptable output's basal expression (in an optimization process aimed at maximizing the output's dynamic range and minimizing the output's basal expression) by increasing the output's synthesis rates and decreasing the output's degradation rate would shift the range of parameters that optimize the system. Such parameters are compatible with increased levels of basal expression but also larger dynamic ranges. It is important to note that the range of parameters that can be translated experimentally is limited by the availability of biological parts. For instance, the synthesis rate depends on the promoter sequence, which restricts the selection of parameters to a discrete, limited number of values. If design goals cannot be achieved with parameters' values corresponding to available biological parts, the user may need to select an alternative circuit topology to satisfy the design goals.

4 Notes

1. The mammalian synthetic biology toolbox offers a diverse but limited set of standard biological parts for building genetic networks. Such parts can be used to recapitulate the design rules obtained from predictive modeling analyses and generate cell sensors with optimized behaviors (Fig. 9).
2. The first requirement for building cells that sense and respond to the activity of specific genes is to link the expression of a master regulator of the genetic control system to that of the target gene such that the master regulator's expression reflects that of the target gene from the chromosomal context. This feature can be achieved experimentally by genome editing to insert a cassette for the expression of the master regulator linked to that of the target gene through a 2A peptide sequence or an IRES. The use of 2A peptides results in monocistronic

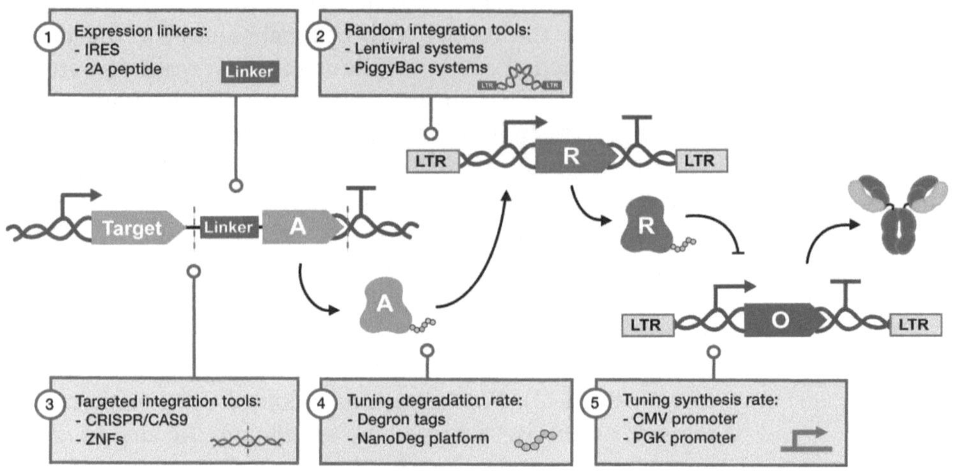

Fig. 9 The synthetic biology toolbox for the construction of genetic control systems. (1) Tools for the integration of exogenous cassettes and specific chromosomal loci; (2) tools for random integration of exogenous cassettes; (3) sequences linking the transcription or translation of multiple genes to maintain constant ratios of protein expression; (4) tuning tools for modulating protein degradation rates; and (5) tools for modulating the gene synthesis rates. *LTR* long terminal repeat, *A* activator, R repressor, *O* output

transcription units and posttranslational processing of the fusion protein, which leads to the same expression level of the two proteins [9]. The use of IRES results in the expression of a bicistronic transcription unit, which leads to a constant ratio of expression of the two genes [8]. The ratio of expression levels of the master regulator relative to the target gene depends on the specific IRES sequence and is accounted for by the phi factor (ϕ) in the mathematical model [17]. As a low expression level of the master regulator leads to high output's dynamic range, IRES sequences resulting in low ratios of expression are recommended for maximizing the output's dynamic range. The integration of the master regulator requires site-directed insertion integration methods and can be achieved using programmable nucleases (e.g., ZNFs and CRISPR-CAS9) [18].

3. Other components of the genetic control systems can be expressed independently under the control of user-defined regulatory sequences, enabling control of the synthesis rates. Specifically, the desired synthesis rates can be achieved using promoters resulting in different degrees of expression, such as the UBC, SV40, and CMV promoters, which are associated with minimal, medium, and high expression, respectively, in HEK293T cells [19]. Furthermore, as these components do not require insertion at specific chromosomal loci, stable cell lines can be generated using random integration methods such as virus-based integration systems (i.e., lentiviral vectors)

or transposon-based integration systems (i.e., PiggyBac system) [20].

4. The degradation rate of the circuit components is another important parameter for the optimization of the cellular sensors and can be modulated through different protein engineering approaches. Degron tags are peptide sequences that confer susceptibility to proteasomal degradation to the fusion protein with rates that depend specifically on the nature of the tag's amino acid sequence [21, 22]. Appropriate design of the degron tag thus allows modulating the degradation rate of the circuit's component to achieve the design goals. The NanoDeg is a modular system that consists of a nanobody for target recognition and a degron tag for degradation of the nanobody–target complex. Such an approach allows controlling a target's degradation rate without the need to genetically engineer the target [15, 23]. Incorporating the NanoDeg in the genetic circuits analyzed here would allow tuning the degradation rate of various circuit components such as the transcription factors or the output to achieve user-defined design goals.
5. Similar to the transcription factors controlling intermediate nodes, the output can be expressed independently under the control of user-defined regulatory sequences, enabling control of the synthesis rates. The main output parameters affecting the genetic control system's behavior are the synthesis and degradation rates.
6. Selecting design goals and appropriate biological parts depends on the application of the cellular sense-and-respond system. Cell therapies for the delivery of proteins with narrow therapeutic windows are likely to benefit from a circuit designed to minimize the output's basal expression using biological parts mediating low output synthesis rate and high degradation rates. The design of diagnostics, on the other hand, depends on the detection limit and sensitivity of the instrument, which provide constraints for selecting acceptable output's basal expression and dynamic range, respectively.
7. While the effect of each parameter varies depending on the circuit topology, some parameters are generally associated with large sensitivity coefficients and should be prioritized regardless of the design goal. For instance, all four topologies analyzed depend on the output's synthesis and degradation rates. The parameter scan revealed that designing cellular sensors with high output dynamic range, which is a desirable feature of diagnostics requiring signal amplification, requires master regulators with low synthesis rates and high degradation rates. While the degradation rate of the master regulator can be modulated through protein engineering approaches, the

synthesis rate depends strictly on the target gene. Such optimization criteria can thus be achieved by selecting the appropriate target genes if possible. Alternatively, the use of linkers that lower the ratio of expression between the target gene and the master regulator can also allow maximizing the output signal amplification.

8. The design of cellular sensors with minimal basal output signal, which is often a desirable feature of cellular therapies for drug delivery only in response to the detection of disease biomarkers, is optimally achieved using topology B. In such circuit topology, the output is directly regulated by a transcriptional repressor, allowing precise control of the output basal expression.
9. The design of cellular sensors with maximal output dynamic range, which is often a desirable feature of diagnostics, is optimally achieved using topologies C and D. Both topologies are based on a hybrid promoter that responds to both a repressor and an activator to control the output production, which results in amplification of the output signal upon detection of the input. Topology D provides a more versatile approach for maximizing the output signal amplification as it relies on a positive feedback loop for self-activation of the output, which could be tuned by selecting a transcriptional activator with the appropriate features.

Acknowledgments

This work was supported by the National Institutes of Health grant EB030030 and by the National Science Foundation grants 2128370 and 2036109.

References

1. Tastanova A, Folcher M, Müller M et al (2018) Synthetic biology-based cellular biomedical tattoo for detection of hypercalcemia associated with cancer. Sci Transl Med 10: eaap8562. https://doi.org/10.1126/scitranslmed.aap8562
2. Scheller L, Strittmatter T, Fuchs D et al (2018) Generalized extracellular molecule sensor platform for programming cellular behavior. Nat Chem Biol 14:723–729. https://doi.org/10.1038/s41589-018-0046-z
3. Stefanov B-A, Fussenegger M (2022) Biomarker-driven feedback control of synthetic biology systems for next-generation personalized medicine. Front Bioeng Biotechnol 10: 986210. https://doi.org/10.3389/fbioe.2022.986210
4. Manhas J, Edelstein HI, Leonard JN, Morsut L (2022) The evolution of synthetic receptor systems. Nat Chem Biol 18:244–255. https://doi.org/10.1038/s41589-021-00926-z
5. de Rossi J, Arefeayne Y, Robinson A, Segatori L (2022) Emerging technologies for genetic control systems in cellular therapies. Curr Opin Biotechnol 78:102833. https://doi.org/10.1016/j.copbio.2022.102833
6. Elowitz MB, Leibler S (2000) A synthetic oscillatory network of transcriptional regulators. Nature 403:335–338. https://doi.org/10.1038/35002125

7. Gardner TS, Cantor CR, Collins JJ (2000) Construction of a genetic toggle switch in Escherichia coli. Nature 403:339–342. https://doi.org/10.1038/35002131
8. Mizuguchi H, Xu Z, Ishii-Watabe A et al (2000) IRES-dependent second gene expression is significantly lower than cap-dependent first gene expression in a bicistronic vector. Mol Ther 1:376–382. https://doi.org/10.1006/mthe.2000.0050
9. Shaimardanova AA, Kitaeva KV, Abdrakhmanova II et al (2019) Production and application of multicistronic constructs for various human disease therapies. Pharmaceutics 11:580. https://doi.org/10.3390/pharmaceutics11110580
10. Nevozhay D, Adams RM, Murphy KF et al (2009) Negative autoregulation linearizes the dose–response and suppresses the heterogeneity of gene expression. Proc Natl Acad Sci USA 106:5123–5128. https://doi.org/10.1073/pnas.0809901106
11. Zhao W, Bonem M, McWhite C et al (2014) Sensitive detection of proteasomal activation using the Deg-On mammalian synthetic gene circuit. Nat Commun 5:3612. https://doi.org/10.1038/ncomms4612
12. Kim H, Gelenbe E (2012) Stochastic gene expression modeling with Hill function for switch-like gene responses. IEEE/ACM Trans Comput Biol and Bioinf 9:973–979. https://doi.org/10.1109/TCBB.2011.153
13. Rosenfeld N, Young JW, Alon U et al (2005) Gene regulation at the single-cell level. Science 307:1962–1965. https://doi.org/10.1126/science.1106914
14. Nevozhay D, Zal T, Balázsi G (2013) Transferring a synthetic gene circuit from yeast to mammalian cells. Nat Commun 4:1451. https://doi.org/10.1038/ncomms2471
15. Zhao W, Pferdehirt L, Segatori L (2018) Quantitatively predictable control of cellular protein levels through proteasomal degradation. ACS Synth Biol 7:540–552. https://doi.org/10.1021/acssynbio.7b00325
16. Sanz-Leon P, Stuart-knock (2021) Brain-modelling-group/neural-flows: NIMH-DiDViz-2021
17. Origel Marmolejo CA, Bachhav B, Patibandla SD et al (2020) A gene signal amplifier platform for monitoring the unfolded protein response. Nat Chem Biol 16:520–528. https://doi.org/10.1038/s41589-020-0497-x
18. Yamamoto Y, Gerbi SA (2018) Making ends meet: targeted integration of DNA fragments by genome editing. Chromosoma 127: 405–420. https://doi.org/10.1007/s00412-018-0677-6
19. Qin JY, Zhang L, Clift KL et al (2010) Systematic comparison of constitutive promoters and the doxycycline-inducible promoter. PLoS ONE 5:e10611. https://doi.org/10.1371/journal.pone.0010611
20. Bachhav B, de Rossi J, Llanos CD, Segatori L (2023) Cell factory engineering: challenges and opportunities for synthetic biology applications. Biotech Bioeng bit.28365. https://doi.org/10.1002/bit.28365
21. Matsuzawa S, Cuddy M, Fukushima T, Reed JC (2005) Method for targeting protein destruction by using a ubiquitin-independent, proteasome-mediated degradation pathway. Proc Natl Acad Sci USA 102:14982–14987. https://doi.org/10.1073/pnas.0507512102
22. Gilon T, Chomsky O, Kulka RG (2000) Degradation signals recognized by the Ubc6p-Ubc7p ubiquitin-conjugating enzyme pair. Mol Cell Biol 20:7214–7219. https://doi.org/10.1128/MCB.20.19.7214-7219.2000
23. Jayanthi BEK, Zhao W, Segatori L (2019) Input-dependent post-translational control of the reporter output enhances dynamic resolution of mammalian signaling systems. In: Methods in enzymology. Elsevier, pp 1–27

Chapter 9

Flux Balance Analysis of Mammalian Cell Systems

James Morrissey, Benjamin Strain, and Cleo Kontoravdi

Abstract

Flux balance analysis (FBA) is a computational methodology to model and analyze the metabolic behavior of cells. In this chapter, we break down the key steps for formulating an FBA model and other FBA-derived methodologies in the context of mammalian cell biology, including strain design, developing cell line-specific models, and conducting flux sampling. We provide annotated COBRApy code for each step to show how it would work in practice.

Key words Flux balance analysis, Genome-scale modeling, Stoichiometric modeling, Metabolism, Cell culture

1 Introduction

Flux balance analysis (FBA) is a solution methodology for metabolic models. Its simplicity to implement, fast computational time, and the lack of parameterization have led FBA to become one of the most popular tools in a systems biologist's arsenal. FBA aims to solve for unknown reaction rates, also known as fluxes, using a constrained optimization problem.

FBA has been widely applied in microbial systems, but its applications to mammalian cell systems are still relatively new. With the emergence of genome-scale models (GEMs) for mammalian systems [1–3], FBA is increasing in importance as a method to analyze mammalian cell metabolism. FBA can solve large models, quickly, with little experimental data.

These mammalian GEMs have allowed for the application of FBA in areas such as biopharmaceutical production and human health. FBA can be used to model biopharmaceutical-producing cells, such as Chinese hamster ovary (CHO) or human embryonic kidney (HEK) cells. FBA has been used in the biopharmaceutical industry for media design [4], process control [5], and genetic engineering targeting [6]. In human health, FBA has been used

Francesca Ceroni and Karen Polizzi (eds.), *Mammalian Synthetic Systems*, Methods in Molecular Biology, vol. 2774,
https://doi.org/10.1007/978-1-0716-3718-0_9,

to understand metabolic changes that occur in diseased cells. This information help identify potential therapeutic targets and develop new treatments [7, 8]. Here, we provide a step-by-step guide to formulating an FBA problem, as well as a selection of other FBA-derived methodologies in the context of mammalian synthetic biology. We provide annotated code in a popular FBA software, COBRApy [9].

2 Materials

2.1 Stoichiometric Network Model

The first requirement for running an FBA model is a stoichiometric network that describes the metabolic reactions of the biological system to be modelled. A stoichiometric network is a mathematical representation of the reactions in the system, which can be represented as a set of linear equations that describe the flow of metabolites through the network.

Stoichiometric models can vary from small-scale models, either simplified or focusing on a particular area of metabolism, to genome-scale model (GEM) reconstructions, which aim to encompass all areas of metabolism. Stoichiometric models can be downloaded from several online databases, including:

1. BiGG Models (http://bigg.ucsd.edu/) [10]
2. ModelSEED (https://modelseed.org/) [11]
3. MetaNetX (https://www.metanetx.org/) [12]
4. BioModels (https://www.ebi.ac.uk/biomodels/) [13]

2.2 Input Constraints

FBA, as a constraint-based methodology, relies on system knowledge or experimental data to constrain the reaction fluxes in the model. Constraints guide reaction fluxes toward a more accurate representation of cell metabolism. The types of data that can be applied to FBA models are shown in Table 1.

As FBA is a predictive methodology, constraints do not necessarily need to be derived from measured data, but instead can be simulated to assess cell behavior in a theoretical situation. For example, the impact of media design on cell growth and productivity can be modeled by changing the constraints to represent the new media design.

2.3 Software Implementation

FBA can also be implemented in any software capable of optimization. There are also several specialized tools for implementing FBA either as a standalone software or a package within a major programming language. Table 2 shows some popular software capable of FBA in mammalian cell systems.

Table 1
Data types and utilization in FBA models

Data type	How can it be used in FBA?
Time course viable cell density	Growth rate of FBA model can be constrained
Metabolomic data	Concentration change of metabolites, intra-/extracellularly can be constrained
Transcriptomic/ proteomic data	Expression of genes and proteins informs the model of which reactions are (in)active at a given time point
Thermodynamic	Directionality of reactions constrained to prevent thermodynamically infeasible solutions
Fluxomic data	Radioactively labeled metabolites can be used to estimate reaction rates
Kinetic data	Mechanistic reaction kinetics can estimate and constrain fluxes, e.g., Michaelis–Menten kinetics

Table 2
Popular FBA software packages

Name	Platform	Link	References
COBRA	MATLAB	https://opencobra.github.io/cobratoolbox/	[14]
COBRApy	Python	https://opencobra.github.io/cobrapy/	[9]
OptFlux	Standalone software	http://www.optflux.org/	[15]
Escher-FBA	Web-based	https://sbrg.github.io/escher-fba	[16]
CellNetAnalyzer	MATLAB	http://www2.mpi-magdeburg.mpg.de/projects/cna/cna.html	[17]
PSAMM	Python	https://zhanglab.github.io/psamm/	[18]
CNApy	Python with GUI	https://github.com/cnapy-org	[19]

3 Methodology

3.1 Flux Balance Analysis

FBA is a computational methodology to solve stoichiometric metabolic models. Its primary objective is to determine the unknown reaction fluxes (reaction rates) in a metabolic network by optimizing for an assumed metabolic objective. To accomplish this, FBA involves a series of six key steps that are illustrated in Fig. 1. By following these steps, an FBA optimization problem can be formulated, enabling the determination of the reaction fluxes.

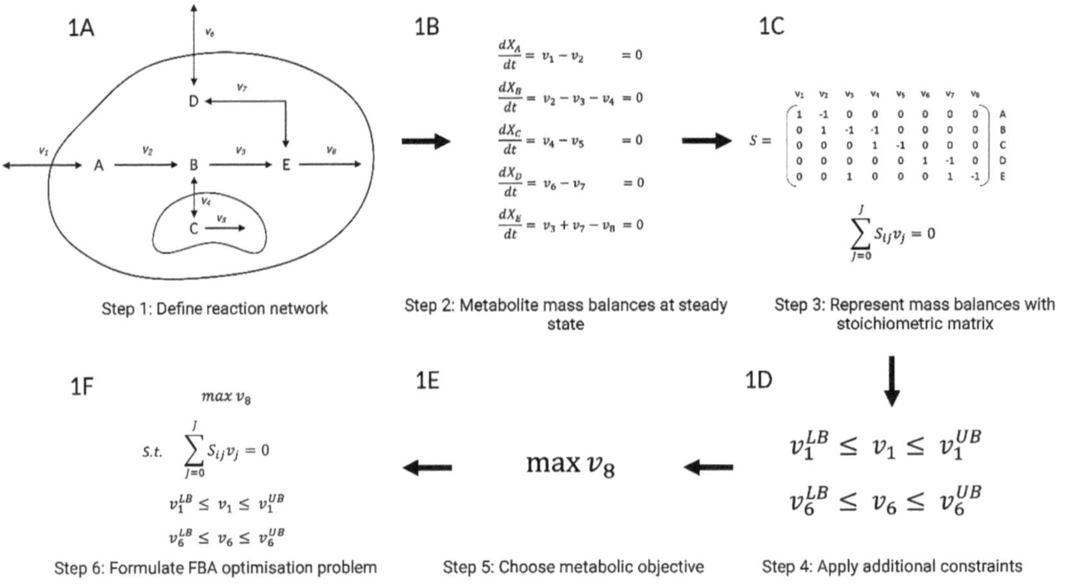

Fig. 1 Overview of the key stages involved in formulating an FBA problem, utilizing a "toy model" as an example. In the first step (**a**), it is necessary to define the area of metabolism to be modeled, which involves identifying a set of metabolites and the reactions that connect them. In this example, there are three enzymatically catalyzed internal reactions (v_2, v_3, v_7), two exchange reactions (v_1, v_6), one transport reaction (v_4), and two demand/sink reactions (v_5, v_8). The second step (**b**) requires writing a mass balance on each metabolite, resulting in a system of linear equations by applying the steady-state assumption. The third step (**c**) involves storing the metabolic network's stoichiometric information in a matrix, *S*. In the fourth step (**d**), the known reaction fluxes, obtained from experimental data, are constrained within measured bounds. For instance, in this example, we have information on the exchange of metabolite A between the cell and the extracellular media. In the fifth step (**e**), a metabolic objective is selected to be maximized (or minimized). In this example, the demand flux of metabolite E (v_8) is the objective. Finally, in the sixth step (**f**), the FBA optimization problem is formulated, using the objective function and constrained by the steady-state mass balance and the bounds on the measured fluxes. The solution to this problem provides estimates for the unknown metabolic fluxes

Step One: Reaction Network The initial step in FBA involves describing the metabolic network. This can be achieved by utilizing a genome-scale metabolic reconstruction or a smaller-scale metabolic models that depict either a subsystem or a simplified version of metabolism. The metabolic network should consist of nodes, which represent metabolites, and edges, which represent reactions connecting the metabolites.

A typical cell metabolism network consists of various types of reactions such as enzymatically catalyzed metabolic reactions, exchange reactions representing the uptake and secretion of metabolites by the cell, transport reactions transferring metabolites between intracellular compartments, and demand/sink reactions for metabolic products that accumulate inside the cell without secretion. Additionally, the pseudo-biomass reaction, which involves the metabolites necessary for the cell to grow, is another critical reaction in the network. In Fig. 1a, a toy model is described to provide an example of a metabolic network.

Step Two: Mass Balance Formulation The second step in FBA involves transforming the given reaction network into a series of metabolite mass balances. For each metabolite in the network, a mass balance equation is written, which represents the change in its intracellular concentration over time. This change in concentration is equal to the sum of all reaction fluxes producing the metabolite, minus the sum of all reaction fluxes consuming the metabolite, including exchange reactions that transport the metabolite inside and outside the cell.

The aim of FBA is to solve this system of mass balances for the reaction fluxes (v_j). The first key assumption made in FBA is the pseudo-steady-state assumption, also known as the balanced growth condition. This assumption implies that the intracellular accumulation of metabolites is negligibly small compared to the reaction fluxes, within the time frame considered by FBA. As a result, the system of mass balances becomes a system of linear equations, which can be solved for the reaction fluxes.

Step Three: Stoichiometric Matrix The stoichiometric coefficient of each metabolite in every reaction from system of mass balances is stored in the stoichiometric matrix, S. Each row represents a unique metabolite, and each column represents a reaction. The S matrix is used to impose the mass balance constraint during FBA, ensuring that the total amount of any metabolite being produced is equal to the total amount being consumed at steady state. This is given by the following equation, where S_{ij} is the stoichiometric coefficient of metabolite i in reaction j, and v_j are the unknown fluxes.

$$\sum_{J=0}^{J} S_{ij} v_j = 0 \text{ for } i \in I$$

Step Four: Additional Constraints In Step Four of the FBA process, additional constraints are added to the metabolic network to bound reaction fluxes. These constraints are derived from biological knowledge of the cell system and measured data, and they are crucial for generating accurate FBA predictions. A common example of such constraints is the exchange fluxes of metabolites between the media and the cell, which can be inferred from concentration changes of metabolites in the media. For instance, in the toy example (Fig. 1d), it is assumed that measured data is available for the exchange of metabolite A with the media. Table 3 provides a non-exhaustive list of constraints that can be

Table 3
Examples of additional constraints that can be imposed on FBA models

Constraint	General mathematical formulation example	References
Steady-state stoichiometric mass balance	$\sum_{J=0}^{J} S_{ij} v_j = 0$	[20]
Thermodynamic feasibility	$v_j \geq 0$	[21–26]
Measured fluxes	$v_j^{m,LB} \leq v_j \leq v_j^{m,UB}$	–
Enzyme capacity	$\sum_{J=0}^{J} \frac{MW_j v_j}{k_j^{cat}} \leq C$	[27–29]
Gene/protein expression	$-e_j \leq v_j \leq e_j$	[30–33]
Regulatory constraints	$v_1 = 0$ if $(v_2 \neq 0)$	[34–36]

applied to mammalian cell metabolic models. By incorporating these additional constraints, the feasible solution space is restricted, leading to more accurate and biologically relevant FBA predictions.

Step Five: Metabolic Objective The system of mass balances contains I linear equations, where I is the total number of metabolites, and J variables, where J is the total number of reactions. Since metabolites are involved in multiple reactions, $J > I$, so we have more unknown variables than equations. The system is therefore *underdetermined*, meaning there is not a unique solution. FBA is used to overcome this by optimizing for a metabolic objective. The distribution of fluxes that maximizes (or minimizes) this objective function is the solution to FBA.

The choice of objective function is key. The objective is typically something that represents assumed cell behavior, for example, in early exponential phase of cell culture, biomass maximization is a popular objective function. The theory behind this is that the cell is allocating its flux toward a high growth rate phenotype. Other objectives include maximizing energy production [37], minimizing uptake of nutrients [38], minimizing total sum of flux [39], and minimizing redox potential [40]. In the toy model, we have chosen the demand flux of metabolite E, v_8, as the objective. Metabolite E could represent an essential component for cell growth or a desired product.

Step Six: Optimization Problem Formulation The metabolic objective, steady-state mass balance, and additional constraints are packaged together into an optimization formulation.

```
import cobra

# Load the metabolic model
model = cobra.io.load_json_model('toy_model.json')

# Set the bounds of exchange reaction v1 between 1.0 and 1.2 mmol/gDCW/hr
model.reactions.get_by_id('v1').bounds = (1.0,1.2)

# Set the bounds of exchange reaction v6 between 0.3 and 0.5 mmol/gDCW/hr
model.reactions.get_by_id('v6').bounds = (0.3,0.5)

# Get the reaction object for v8
v8_reaction = model.reactions.get_by_id('v8')

# Set the objective coefficient for v8 to 1
model.objective = v8_reaction

# Set the optimization solver to maximize the objective function
solution = model.optimize(objective_sense='maximize')

# Print the flux through v8
print('Optimal v8 Flux: '+ str(solution.fluxes['v8']) + 'mmol/gDCW/hr')

-----------------------------------------------------------------------------
---
OUTPUT:
-----------------------------------------------------------------------------
---

Optimal v8 Flux: 1.7mmol/gDCW/hr
```

Fig. 2 Python code demonstrating how to run the optimization of a toy metabolic model using exchange reaction constraints and objective function. The exchange reaction bounds for v1 and v6 are set to a range of 1.0–1.2 mmol gDCW^{-1} h^{-1} and 0.3–0.5 mmol gDCW^{-1} hr^{-1}, respectively. The reaction object for v8 is obtained and set as the objective function with a coefficient of 1. The optimization solver is set to maximize the objective function. The optimal flux through v8 is found to be 1.7 mmol gDCW^{-1} hr^{-1}. These results demonstrate the successful optimization of the metabolic model using exchange reaction constraints and objective function

The FBA formulation in Fig. 1 is written in COBRApy [9] as shown in Fig. 2. In COBRApy, the steady-state mass balances are embedded within the "model" class, just the constraints and objective function must be defined. The output of this code is displayed below, with the optimum flux through the objective, v_8, being 1.7 mmol per gram dry cell weight per hour (mmol gDCW^{-1} h^{-1}).

3.2 FBA-Based Methodologies in Mammalian Synthetic Biology

3.2.1 Strain Design

Strain design algorithms, such as those summarized in Table 4, are powerful tools for designing optimal genetic engineering targets in biological organisms using genome-scale models. These algorithms work by identifying the key metabolic pathways in the model that can be manipulated to enhance the production of desired compounds or to optimize growth rates in specific environments.

The OptKnock algorithm was one of the first published strain design approaches and is based on a bi-level optimization problem whereby the upper-level problem is the maximization of the production of a specific metabolite, while the lower-level problem is the maximization of the growth rate. The algorithm identifies the

Table 4
Summary of key algorithms for optimization-led strain design

Algorithm	Description	References
OptKnock	Suggests gene knockouts based on a bi-level optimization framework that couples the desired overproduction target (i.e., the biopharmaceutical) to growth	Burgard et al. [41]
OptReg	Extension of OptKnock that suggests reactions for regulation (up, down, etc.) based on a bi-level optimization framework that couples the fluxes of two different reactions	Pharkya et al. [42]
OptForce	Extension of OptReg that contrasts the metabolic flux patterns observed in an initial strain and a strain overproducing the chemical at the target yield	Ranganathan et al. [43]
GDLS	Uses logical search to look for multiple paths to improve the production of metabolites	Lun et al. [44]
OptORF	Identifies optimal number of metabolic and regulatory gene knockouts/overexpressions to couple production with growth	Kim et al. [45]
k-OptForce	Identifies regulators in transcription factors and metabolic genes by accounting for gene expression in its objective function	Chowdhury et al. [46]
OptRam	Extension of OptForce that uses kinetic equations to better calculate steady-state fluxes of metabolic network	Shen et al. [47]
gsOpt	Predicts coupling strength of two metabolic reactions	Alter and Ebert [48]
OptCouple	Analyzes both genetic engineering strategies with process engineering solutions for the overproduction of a metabolite	Jensen et al. [49]
OptStrain	Assesses knock-ins of non-native functionalities from a comprehensive database of reactions	Pharkya et al. [50]

optimal set of gene knockouts that can enhance the production of the target metabolite while maintaining a high growth rate of the organism. The rationale behind this approach is that the optimal set of gene knockouts should not compromise the growth rate of the organism, which is essential for its survival and function. The knockouts should also be designed to enhance the production of the target metabolite by removing the competing pathways and reducing the waste of resources.

Many of these algorithms, including OptKnock, can be simply implemented using the StrainDesign package in Python by executing the code shown in Fig. 3.

It is important to note that for larger models containing several thousand reactions, such as those common with mammalian cell systems, these strain design algorithms can be incredibly computationally intensive. As such, if attempting to use these approaches on larger models, using advanced solvers such as Gurobi [51] is advised. Increasing the number of allowed solutions and altering the solution approach to "any" can also speed up computation

```
#import the required packages
import straindesign as sd
import cobra


#load you model of interest
MODEL = cobra.io.load_json_model('model.json')



#You could now constrain your model to experimental uptakes if desired.
Make sure you unconstraint the production reaction of interest

#set up the strain design module for computing OptKnock. The module
consists of an outer and inner objective and additional constraints. The
additional constraints are used to enforce the minimal desired growth rate.

module_optknock = sd.SDModule(MODEL,sd.names.OPTKNOCK,
                              inner_objective='BIOMASS_REACTION,

outer_objective='PRODUCT_OF_INTEREST_REACTION',
                              constraints='BIOMASS_REACTION >= 0.05')

#Compute strain designs with arbitrary parameters for maximum solutions,
maximum strategies and time limit

sols= sd.compute_strain_designs(MODEL,
                                sd_modules = module_optknock,
                                max_solutions = 100,
                                max_cost = 150, time_limit= 6000,
solution_approach = 'BEST')


# Print solution and save
num_sols= sols.get_num_sols()
print(num_sols)
reaction_sol= sols.get_reaction_sd()
print(reaction_sol)
sols.save('STRAIN_DESIGN_RESULTS')
```

Fig. 3 Python code demonstrates the use of the StrainDesign and COBRApy packages to perform strain design analysis in a metabolic model. First, the model of interest is loaded using the cobra.io.load_json_model function. Experimental uptake constraints can be applied if desired, and the production reaction of interest is unconstrained. Next, a strain design module is set up to compute the OptKnock algorithm. The module consists of an outer and inner objective, and additional constraints are used to enforce a minimal desired growth rate. The inner objective is set as the biomass reaction, and the outer objective is set as the product of interest reaction. The minimal desired growth rate constraint is set at 0.05. Strain designs are computed using the OptKnock algorithm with arbitrary parameters for maximum solutions, maximum strategies, and a time limit of 6000 s. A maximum of 100 solutions are considered, and the cost limit is set at 150. The "BEST" solution approach is used. The number of solutions and reaction strain design is printed and saved. The results are saved in a file named "STRAIN_DESIGN_RESULTS"

time. Moreover, reducing the number of reactions in a model by generating cell line-specific model can help reduce computational burden. Methods to achieve this are discussed below.

3.2.2 Cell Line-Specific Models

Mammalian cell genome-scale models are often generic representations of all of metabolism for a given species. For instance, the human genome-scale model Recon3D contains all known reactions in human cells but is not specific to any individual cell type. To take into account the differences that exist between different cell lines within the same organism, generic models may be pruned to remove any non-present reactions. This can be achieved using "omics" data, the GPR (gene–protein–reaction) associations that exist within a model, and model extraction algorithms.

A GPR association is a Boolean expression that links a set of genes and proteins to a reaction in the model. The expression is typically represented as follows:

Rxn1 = (GeneA AND GeneB) OR (GeneC AND GeneD)

This expression indicates that reaction 1 requires either GeneA AND GeneB OR GeneC AND GeneD for its function. Each gene in the expression is associated with a protein, and the proteins work together to catalyze the reaction. This Boolean association allows "omics" data to be mapped to the reactions in the model, which can then be used to decide if the reaction could be catalyzed or not. In this case, if using transcriptomics, reaction 1 would require expression of mRNA above a certain threshold either GeneA AND GeneB OR GeneC AND GeneD. To map a single gene value to each reaction, the minimum expression value of any OR association and the maximum of any AND association is taken forward and mapped to the reaction.

Once expression values are mapped to each reaction, it becomes possible to use model extraction algorithms, such as those summarized in Table 5, to prune reactions associated with low expression that are likely non-present in the cell line of interest.

For example, the Cost Optimization Reaction Dependency Assessment (CORDA) algorithm [[54] takes user-defined high, medium, and negative confidence reactions, which are defined using the mapped transcriptomics data, to produce a viable while maximizing high and medium confidence reactions and minimizing the number of low and negative confidence reactions. This is achieved using a dependency assessment, where negative confidence reactions are assigned an arbitrarily high cost. This cost is then minimized while enforcing a small flux through medium or high confidence reactions to distinguish which negative confidence reactions are beneficial for high confidence reactions to carry flux that should be included in the final reconstruction. This can be simply implemented using the code in Fig. 4.

3.3 Flux Sampling

Flux sampling is another solution methodology for stoichiometric models. Flux sampling uses the same formulation as FBA (*see* Fig. 1a–d); however, it does not require the assumption of a metabolic objective. Instead, flux sampling generates a large number of flux distributions that satisfy the model constraints. Flux sampling can provide valuable insights into the variability of the model predictions and can help identify alternative solutions that are not apparent from the optimal solution alone. Figure 5 shows how flux sampling methodology differs from FBA.

Table 5
Summary of cell line-specific generation methods

Method	Description
Adaptation of Metabolism (AdaM) [52]	Method for integration of temporal gene expression data. For each time point, it finds a minimal functional network consistent with the differential expression pattern
Åkesson-04 [53]	Constrains reaction fluxes to zero if their associated enzyme genes are expressed at low levels
Cost Optimization Reaction Dependency Assessment (CORDA) [54]	Takes user-defined high, medium, and negative confidence reactions to produce a model that is consistent (i.e., all reactions can carry flux) while maximizing high and medium confidence reactions and minimizing the number of negative confidence reactions
Gene Inactivity Moderated by Metabolism and Expression (GIMME) [33]	Takes "omics" expression inputs mapped to reactions, a metabolic reconstruction, and required metabolic functionalities. A reconstruction is mapped through an omics dataset, removing reactions that are below a specified threshold, creating a reduced model
Integrative Metabolic Analysis Tool (iMAT) [31]	Gene expression is used to divide reactions into two groups: highly and lowly expressed, finding a flux distribution that maximizes the consistency with this classification
Integrative Network Inference for Tissues (INIT) [55]	Maximizes the activation of selected reactions based on a qualitative confidence score while minimizing the utilization of reactions associated with absent proteins/transcripts
Metabolic Context-specificity Assessed by Deterministic Reaction Evaluation (mCADRE) [56]	Utilizes network topology data combined with gene expression levels to calculate connectivity-based evidence scores for all reactions within a model. These scores are used to determine which reactions should be removed from the generic model to create a specific model context/cell line-specific model

Flux sampling is valuable for the analysis of mammalian cell GEMs for two key reasons. Firstly, mammalian cell models are much larger than their prokaryotic counterparts. For example, a popular *E. coli* GEM contains 2583 reactions [57], while a recent human GEM contains over 10,000 [3]. This discrepancy means mammalian models have many more degrees of freedom, and so there are a large combination of feasible fluxes that can satisfy a given metabolic objective. Flux sampling is a good method to analyze model outputs with this uncertainty. By generating many

```
import pandas as pd
import cobra

# Load the metabolic model
model = cobra.io.load_json_model('your model here')

#You could now constrain your model to experimental uptakes if required.


# Read CSV file with expression values mapped to model reactions. One
column should be model reactions IDs named 'Reaction' the other should be
the mapped expression value for that reaction named 'Expression'
df = pd.read_csv('expression_data.csv')

# Loop through each row of the data frame and assign confidence scores -
this is using arbitrary expression thresholds which should be manually set
by the user
confidence_scores = {}
for index, row in df.iterrows():
    expression_value = row['Expression']
# assign 0 (unknown) if expression nan
    if pd.isna(expression_value):
        confidence_scores[row['Reaction']] = 0
# assign -1 (not expressed) if expression = 0
    elif expression_value == 0:
        confidence_scores[row['Reaction']] = -1
# assign 1 (low confidence) if expression between 0 and 1
    elif 0 < expression_value < 1:
        confidence_scores[row['Reaction']] = 1
# assign 2 (medium confidence) if expression between 1 and 2
    elif 1 <= expression_value < 2:
        confidence_scores[row['Reaction']] = 2
# assign 3 (high confidence) if expression greater than 3
    else:
        confidence_scores[row['Reaction']] = 3

print(confidence_scores)


#Now make reduced model using the CORDA package
from corda import CORDA

Reduced cell line specific model = CORDA(model, confidence scores)
```

Fig. 4 Python workflow for creating a cell line-specific metabolic model using gene expression data. A GEM is loaded, and gene expression data is obtained for the cell line of interest. The expression data is mapped onto the GEM and confidence scores are assigned to each reaction based on expression levels. Using the CORDA package, a reduced model is generated from the GEM and the confidence scores. The resulting cell line-specific model helps captures the unique metabolic behavior of the cell line

possible flux distributions, it can provide a more realistic and comprehensive understanding of the behavior of the biological system.

Secondly, it is difficult to define a meaningful objective function for mammalian systems. Unlike prokaryotes, mammalian cells do not try to maximize a single metabolic objective. Instead, they have multiple conflicting and context-specific objectives, such as growth, energy production, maintenance, and efficient allocation of cellular resources [58]. It can be difficult to balance these competing objectives and select an objective function that accurately represents cellular physiology. By assuming optimality toward a particular objective, mammalian cell models may find unrealistic flux distributions. Flux sampling, on the other hand, does not require

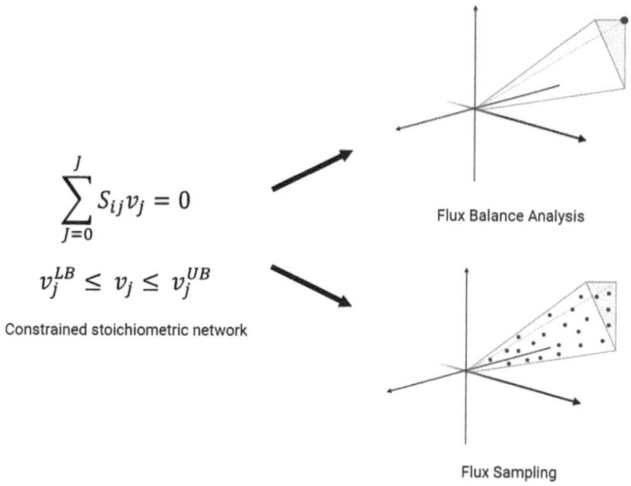

Fig. 5 Illustration of flux sampling and FBA approaches to modeling the solution space. Both methods start with the same set of constraints. FBA then finds the flux distribution that maximizes for a metabolic objective. Flux sampling, on the other hand, creates many flux distributions within the feasible solution space

```
import cobra
from cobra.sampling import sample

# Load the metabolic model
model = cobra.io.load_json_model('toy_model.json')

# Set the bounds of exchange reaction v1 between 1.0 and 1.2 mmol/gDCW/hr
model.reactions.get_by_id('v1').bounds = (1.0, 1.2)

# Set the bounds of exchange reaction v6 between 0.3 and 0.5 mmol/gDCW/hr
model.reactions.get_by_id('v6').bounds = (0.3, 0.5)

# Use flux sampling to explore the solution space of the model
solutions = sample(model, 1000)

# Display the sampled fluxes
print(solutions)

# Print the average flux through v8 in the sampled solutions
print('Average v8 Flux: ' + str(solutions['v8'].mean()) + ' mmol/gDCW/hr')

-----------------------------------------------------------------------------
---
OUTPUT
-----------------------------------------------------------------------------
---

Average v8 Flux: 0.9436 mmol/gDCW/h
```

Fig. 6 Python code demonstrating how to perform flux sampling on a toy metabolic model. Using the COBRApy package, the metabolic model is loaded from a JSON file, and the exchange reaction bounds for v1 and v6 are set to explore the solution space. Flux sampling is used to obtain solutions, and the average flux through v8 in the sampled solutions is calculated and displayed. The average flux through v8 is 0.9436 mmol $gDCW^{-1}$ h^{-1}

the assumption of a metabolic objective; instead, all flux distributions are assessed, subject to the model constraints.

Flux sampling can be implemented in the COBRApy package. The code in Fig. 6 shows how it can be implemented for the toy model. We use the same constraints on v_1 and v_6; however, instead of maximizing for v_8, we sample the solution space 1000 times to find the average flux through v_8. The output of this code is given below. Note how the average flux through v_8 is 0.9436 mmol $gDCW^{-1}$ h^{-1}, which is less than the optimum flux of 1.7 mmol $gDCW^{-1}$ hr^{-1}predicted in FBA. This is because the model will allocate flux toward all feasible distributions, for example, reaction v_5 will have zero flux in FBA but carries flux in sampling.

Acknowledgments

Benjamin Strain would like to thank the UK Biotechnology and Biological Sciences Research Council (BBSRC) and GlaxoSmithKline for their funding and support. James Morrissey thanks the BBSRC and AstraZeneca for their funding and support.

References

1. Hefzi H, Ang KS, Hanscho M et al (2016) A consensus genome-scale reconstruction of Chinese hamster ovary cell metabolism. Cell Syst 3:434–443
2. Yeo HC, Hong J, Lakshmanan M et al (2020) Enzyme capacity-based genome scale modelling of CHO cells. Metab Eng 60:138–147
3. Brunk E, Sahoo S, Zielinski DC et al (2018) Recon3D enables a three-dimensional view of gene variation in human metabolism. Nat Biotechnol 36:272–281
4. Fouladiha H, Marashi SA, Torkashvand F et al (2020) A metabolic network-based approach for developing feeding strategies for CHO cells to increase monoclonal antibody production. Bioprocess Biosyst Eng 43:1381–1389
5. Schinn SM, Morrison C, Wei W et al (2021) A genome scale metabolic network model and machine learning predict amino acid concentrations in Chinese Hamster Ovary cell cultures. Biotechnol Bioeng 118:2118–2123
6. Kol S, Ley D, Wulff T et al (2020) Multiplex secretome engineering enhances recombinant protein production and purity. Nat Commun 11:1908
7. Gatto F, Ferreira R, Nielsen J (2020) Pan-cancer analysis of the metabolic reaction network. Metab Eng 57:51–62
8. Argmann CA, Houten SM, Zhu J et al (2016) A next generation multiscale view of inborn errors of metabolism. Cell Metab 23:13–26
9. Ebrahim A, Lerman JA, Palsson BO et al (2013) COBRApy: COnstraints-based reconstruction and analysis for python. BMC Syst Biol 7:74
10. King ZA, Lu J, Dräger A et al (2016) BiGG models: a platform for integrating, standardizing and sharing genome-scale models. Nucleic Acids Res 44:D515–D522
11. Henry CS, Dejongh M, Best AA et al (2010) High-throughput generation, optimization and analysis of genome-scale metabolic models. Nat Biotechnol 28:977–982
12. Ganter M, Bernard T, Moretti S et al (2013) MetaNetX.org: a website and repository for accessing, analysing and manipulating metabolic networks. Bioinformatics 29:815–816
13. Malik-Sheriff RS, Glont M, Nguyen TVN et al (2020) BioModels-15 years of sharing computational models in life science. Nucleic Acids Res 48:D407–D415
14. Heirendt L, Arreckx S, Pfau T et al (2019) Creation and analysis of biochemical constraint-based models using the COBRA Toolbox v.3.0. Nat Protoc 14:639–702
15. Rocha I, Maia P, Evangelista P et al (2010) OptFlux: an open-source software platform for in silico metabolic engineering. BMC Syst Biol 4:45
16. Rowe E, Palsson BO, King ZA (2018) Escher-FBA: a web application for interactive flux balance analysis. BMC Syst Biol 12:84
17. Klamt S, Saez-Rodriguez J, Gilles ED (2007) Structural and functional analysis of cellular networks with CellNetAnalyzer. BMC Syst Biol 1:2
18. Dufault-Thompson K, Steffensen JL, Zhang Y (2018) Using PSAMM for the curation and analysis of genome-scale metabolic models.

Methods in Molecular Biology, In, pp 131–150

19. Thiele S, Kamp AV, Bekiaris PS et al (2022) CNApy: a CellNetAnalyzer GUI in Python for analyzing and designing metabolic networks. Bioinformatics 38:1467–1469
20. Papoutsakis ET (1984) Equations and calculations for fermentations of butyric acid bacteria. Biotechnol Bioeng 26:174–187
21. Henry CS, Broadbelt LJ, Hatzimanikatis V (2007) Thermodynamics-based metabolic flux analysis. Biophys J 92:1792–1805
22. Qian H, Beard DA (2005) Thermodynamics of stoichiometric biochemical networks in living systems far from equilibrium. Biophys Chem 114:213–220
23. Fleming RMT, Thiele I, Provan G et al (2010) Integrated stoichiometric, thermodynamic and kinetic modelling of steady state metabolism. J Theor Biol 264:683–692
24. Desouki AA, Jarre F, Gelius-Dietrich G et al (2015) CycleFreeFlux: efficient removal of thermodynamically infeasible loops from flux distributions. Bioinformatics 31:2159–2165
25. Schellenberger J, Lewis NE, Palsson B (2011) Elimination of thermodynamically infeasible loops in steady-state metabolic models. Biophys J 100:544–553
26. Zhu Y, Song J, Xu Z et al (2013) Development of thermodynamic optimum searching (TOS) to improve the prediction accuracy of flux balance analysis. Biotechnol Bioeng 110:914–923
27. Beg QK, Vazquez A, Ernst J et al (2007) Intracellular crowding defines the mode and sequence of substrate uptake by Escherichia coli and constrains its metabolic activity. Proc Natl Acad Sci U S A 104:12663–12668
28. Adadi R, Volkmer B, Milo R et al (2012) Prediction of microbial growth rate versus biomass yield by a metabolic network with kinetic parameters. PLoS Comput Biol 8:e1002575
29. Sánchez BJ, Zhang C, Nilsson A et al (2017) Improving the phenotype predictions of a yeast genome-scale metabolic model by incorporating enzymatic constraints. Mol Syst Biol 13: 935
30. Kim MK, Lane A, Kelley JJ et al (2016) E-Flux2 and sPOT: validated methods for inferring intracellular metabolic flux distributions from transcriptomic data. PLoS One 11: e0157101
31. Zur H, Ruppin E, Shlomi T (2010) iMAT: an integrative metabolic analysis tool. Bioinformatics 26:3140–3142
32. Tian M, Reed JL (2018) Integrating proteomic or transcriptomic data into metabolic models using linear bound flux balance analysis. Bioinformatics 34:3882–3888
33. Becker SA, Palsson BO (2008) Context-specific metabolic networks are consistent with experiments. PLoS Comput Biol 4: e1000082
34. Stelling J, Klamt S, Bettenbrock K et al (2002) Metabolic network structure determines key aspects of functionality and regulation. Nature 420:190–193
35. Schilling CH, Covert MW, Famili I et al (2002) Genome-scale metabolic model of Helicobacter pylori 26695. J Bacteriol 184:4582–4593
36. Covert MW, Schilling CH, Palsson B (2001) Regulation of gene expression in flux balance models of metabolism. J Theor Biol 213:73–88
37. Ramakrishna R, Edwards JS, McCulloch A et al (2001) Flux-balance analysis of mitochondrial energy metabolism: consequences of systemic stoichiometric constraints. Am J Phys Regul Integr Comp Phys 280:R695–R704
38. Chen Y, McConnell BO, Gayatri Dhara V et al (2019) An unconventional uptake rate objective function approach enhances applicability of genome-scale models for mammalian cells. NPJ Syst Biol Appl 5:25
39. Lewis NE, Hixson KK, Conrad TM et al (2010) Omic data from evolved E. coli are consistent with computed optimal growth from genome-scale models. Mol Syst Biol 6: 390
40. García Sánchez CE, Vargas García CA, Torres Sáez RG (2012) Predictive potential of flux balance analysis of Saccharomyces cerevisiae using as optimization function combinations of cell compartmental objectives. PLoS One 7:e43006
41. Burgard AP, Pharkya P, Maranas CD (2003) OptKnock: a bilevel programming framework for identifying gene knockout strategies for microbial strain optimization. Biotechnol Bioeng 84:647–657
42. Pharkya P, Maranas CD (2006) An optimization framework for identifying reaction activation/inhibition or elimination candidates for overproduction in microbial systems. Metab Eng 8:1–13
43. Ranganathan S, Suthers PF, Maranas CD (2010) OptForce: an optimization procedure for identifying all genetic manipulations leading to targeted overproductions. PLoS Comput Biol 6:e1000744
44. Lun DS, Rockwell G, Guido NJ et al (2009) Large-scale identification of genetic design strategies using local search. Mol Syst Biol 5: 296

45. Kim J, Reed JL (2010) OptORF: optimal metabolic and regulatory perturbations for metabolic engineering of microbial strains. BMC Syst Biol 4:53
46. Chowdhury A, Zomorrodi AR, Maranas CD (2014) k-OptForce: integrating kinetics with flux balance analysis for strain design. PLoS Comput Biol 10:e1003487
47. Shen F, Sun R, Yao J et al (2019) Optram: in-silico strain design via integrative regulatory-metabolic network modeling. PLoS Comput Biol 15:e1006835
48. Alter TB, Ebert BE (2019) Determination of growth-coupling strategies and their underlying principles. BMC Bioinf 20:447
49. Jensen K, Broeken V, Hansen ASL et al (2019) OptCouple: joint simulation of gene knockouts, insertions and medium modifications for prediction of growth-coupled strain designs. Metab Eng Commun 8:e00087
50. Pharkya P, Burgard AP, Maranas CD (2004) OptStrain: a computational framework for redesign of microbial production systems. Genome Res 14:2367–2376
51. Gurobi Optimization L (2023) Gurobi optimizer reference manual
52. Töpfer N, Jozefczuk S, Nikoloski Z (2012) Integration of time-resolved transcriptomics data with flux-based methods reveals stress-induced metabolic adaptation in Escherichia coli. BMC Syst Biol 6:148
53. Åkesson M, Förster J, Nielsen J (2004) Integration of gene expression data into genome-scale metabolic models. Metab Eng 6:285–293
54. Schultz A, Qutub AA (2016) Reconstruction of tissue-specific metabolic networks using CORDA. PLoS Comput Biol 12:1–33
55. Agren R, Bordel S, Mardinoglu A et al (2012) Reconstruction of genome-scale active metabolic networks for 69 human cell types and 16 cancer types using INIT. PLoS Comput Biol 8:e1002518
56. Wang Y, Eddy JA, Price ND (2012) Reconstruction of genome-scale metabolic models for 126 human tissues using mCADRE. BMC Syst Biol 6:153
57. Orth JD, Conrad TM, Na J et al (2011) A comprehensive genome-scale reconstruction of Escherichia coli metabolism-2011. Mol Syst Biol 7:535
58. Orman MA, Berthiaume F, Androulakis IP et al (2011) Advanced stoichiometric analysis of metabolic networks of mammalian systems. Crit Rev Biomed Eng 39:511–534

Chapter 10

Multiplex Functional Characterization of Protein Variant Libraries in Mammalian Cells with Single-Copy Genomic Integration and High-Throughput DNA Sequencing

Nisha D. Kamath and Kenneth A. Matreyek

Abstract

Sequencing-based, massively parallel genetic assays have enabled simultaneous characterization of the genotype–phenotype relationships for libraries encoding thousands of unique protein variants. Since plasmid transfection and lentiviral transduction have characteristics that limit multiplexing with pooled libraries, we developed a mammalian synthetic biology platform that harnesses the Bxb1 bacteriophage DNA recombinase to insert single promoterless plasmids encoding a transgene of interest into a pre-engineered "landing pad" site within the cell genome. The transgene is expressed behind a genomically integrated promoter, ensuring only one transgene is expressed per cell, preserving a strict genotype–phenotype link. Upon selecting cells based on a desired phenotype, the transgene can be sequenced to ascribe each variant a phenotypic score. We describe how to create and utilize landing pad cells for large-scale, library-based genetic experiments. Using the provided examples, the experimental template can be adapted to explore protein variants in diverse biological problems within mammalian cells.

Key words DNA recombinase, Bxb1 integrase, Landing pad, Functional genetic experiments, Deep mutational scanning, Multiplex assays of variant effect, High-throughput DNA sequencing

1 Introduction

Genomic tools like genome-wide siRNA or lentiCRISPR libraries enable the phenotypic characterization of the impacts of down-regulation, and in some cases upregulation, of proteins encoded by the cell genome. In contrast, similar high-throughput techniques do not exist for studying libraries of transgenic sequences not already encoded by the cell, such as large cDNA libraries encoding missense variants of a particular protein of interest. Traditional experimental formats for studying protein variants are low throughput, with each sample separated into its own tube or well. The characterization of a large number of protein variants through these methods is prohibitively laborious. To overcome this, we have

Francesca Ceroni and Karen Polizzi (eds.), *Mammalian Synthetic Systems*, Methods in Molecular Biology, vol. 2774,
https://doi.org/10.1007/978-1-0716-3718-0_10,

developed a cell engineering platform called "the landing pad" platform, allowing for the stable expression of a large library of up to hundreds of thousands of protein variants with site-specific integration [1, 2].

The key component to the landing pad platform is a genomically integrated Bxb1 attP recombinase site. This attP site is bound by a dimer of the Bxb1 recombinase enzyme, a serine DNA recombinase that is more efficient than the traditionally used Cre, Flp, and PhiC31 recombinases [3], while having no known off-target sites already present in the human genome [4]. Transfection with plasmids encoding the complementary attB recombinase site also becomes bound by a dimer of Bxb1 recombinase enzymes. Once the attP- and attB-bound Bxb1 recombinase dimers come into contact, the recombinase tetramer enacts a DNA cleavage, rotation, and ligation reaction [5], integrating the plasmid molecule into the pre-engineered genomic site.

This approach can be used to stably express circuits of transgenic products [6], but there are additional design requirements to express a single transgenic protein variant per cell (Fig. 1). First, there must only be one genomically integrated attP site. While we initially achieved this through careful screening of cells modified through CRISPR-mediated homology-directed repair [1], we have since adopted low multiplicity of infection DNA transduction with lenti-landing pad lentiviral vectors to more efficiently generate these modified cells [2, 7]. Second, expression of the transgenes must only occur upon recombination since thousands of plasmid molecules are typically transfected into each cell and are only slowly degraded or diluted through cell division [8]. To avoid a complicated mixture of transgenic variants expressed from these plasmids, we generate promoterless attB plasmids encoding our transgenic library. Within the landing pad, we preceded the genomic attP recombination site with a Tet-inducible promoter and provided the reverse tet transactivator rtTA3G in a separate transgenic cassette, so that addition of a tetracycline analog like doxycycline during culture enables the expression of the transgene proceeding the attP recombinase site. Thus, the landing pad promoter is originally provided in trans (Fig. 1a) and only becomes in cis with the transgene upon DNA recombination (Fig. 1b). Since there is only one attP landing pad locus per cell, this means that only one transgene becomes expressed per cell.

With only one protein variant expressed in a given cell, there is a strict genotype–phenotype link allowing for compatibility with high-throughput single-cell assays and the characterization of many protein variants at once by multiplex assays of variant effect, like deep mutational scanning (DMS). Unlike lentiviral transduction, which results in varied expression levels from transgenes integrated randomly across the cell genome, the site-specific integration of the landing pad platform permits uniform transgene

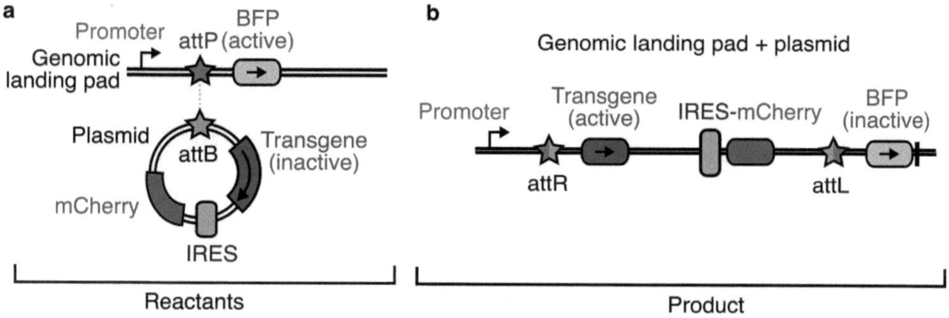

Fig. 1 Schematic representation of the reactants and products of DNA recombination. (**a**) The landing pad consists of an attP site preceded by a Tet-inducible promoter and proceeded by DNA encoding blue fluorescent protein (BFP). The promoterless recombination plasmid consists of an attB site, the transgene, and mCherry fluorescent protein. (**b**) After recombination, the plasmid integrates into the attP site of the landing pad locus through the Bxb1 serine recombinase enzymatic activity. The increased distance between the promoter and BFP turns off its expression. The resulting recombined cells will express the transgene and mCherry only

expression, reducing heterogeneity in protein expression confounding variant characterization [7]. When paired with a phenotypic selection and quantitation before and after selection using high-throughput DNA sequencing, this method generates large mutational datasets and sequence-function maps that provide increased resolution of protein biology unattainable by low-throughput methods.

This protocol details the wet-lab procedures needed to execute a typical DMS workflow using the landing pad platform (Fig. 2). This begins with generating the initial recombinant DNA and cellular materials [1, 2]. The landing pad cells are created through lentiviral transduction of the landing pad locus into the genomes of HEK293T cells or a different cell line of interest. AttB plasmids containing the transgene of interest are generated through PCR and Gibson Assembly. The attB plasmids are recombined into the landing pad cells, and nearly pure populations of recombined cells are obtained through an antibiotic treatment process. While we omit the library generation and phenotypic selection steps from our protocol, as both of these will be very application- and project-specific, we describe the molecular biology steps and important considerations needed to amplify and sequence the transgenic DNA from the landing pad.

2 Materials

2.1 Equipment

1. Thermocycler for polymerase chain reactions (PCRs).
2. Bacterial incubator (shaking and static).

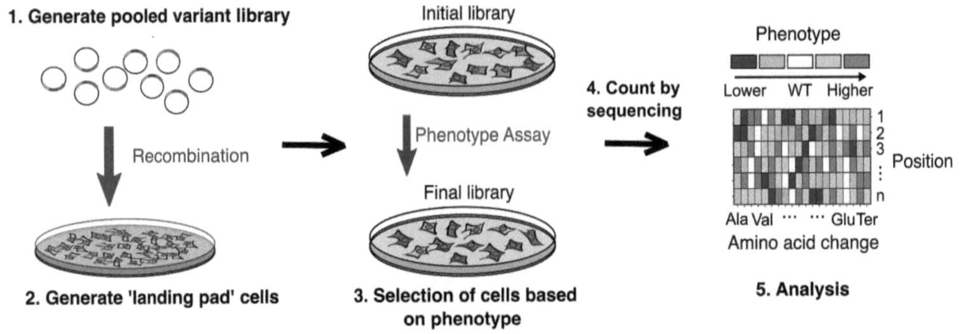

Fig. 2 Overview of processes required for a successful deep mutational scan: Our protocol details each step required for conducting a basic deep mutational scan for a protein of interest. The first step involves generating an attB plasmid with the transgene of interest through our outlined molecular cloning process and making the necessary control constructs or library derivatives using the wild-type construct as a template. Concomitantly, if landing pad cells do not already exist for the desired cell type, create them using lentiviral transduction with the lenti-landing pad construct. These landing pad cells are then recombined with the pooled variant library. The cells, each encoding and expressing a single variant, can be sorted or selected with a phenotypic selection; in this example, we demonstrate having two time points separated by a selection step. Finally, the variant frequencies at each time point are counted by high-throughput sequencing. The resulting counts can be analyzed to determine the phenotypic effect of each variant

3. Benchtop centrifuges compatible with 1.8 mL microcentrifuge tubes, 15 mL, and 50 mL conical tubes.
4. Spectrophotometer to check optical density/absorbance at 600 nm (OD600) of bacterial culture.
5. Qubit™ fluorometer for DNA quantification.
6. Biosafety cabinet for cell culture work.
7. Cell culture incubator at 37 °C and 5% carbon dioxide.
8. Fluorescence-activated cell sorting (FACS) equipment to sort cells based on expression of fluorophores encoded in recombinant DNA constructs.
9. Agarose gel electrophoresis setup with gel holder, gel box, voltage source, and blue light for analysis.

2.2 Plasmids and Cell Lines

1. HEK293T cell line.
2. Lentiviral transfer vector for landing pad platform: pLenti-Tet-coBxb1-2A-BFP_IRES-iCasp9-2A-Blast_rtTA3, Addgene plasmid #171588.
3. Lentiviral packaging and VSV-G envelope vectors: psPAX2, Addgene plasmid #12260 and pMD2.G, Addgene plasmid #12259.
4. Bxb1 attB recombination plasmid backbone to modify to allow expression of the transgene of interest: AttB_ACE2_IRES-mCherry-H2A-P2A-PuroR, Addgene plasmid #171594.
5. attB-EGFP plasmid, Addgene #171597.

6. attB-mCherry plasmid, Addgene #171598.
7. DNA sequence encoding the transgene of interest.

2.3 Reagents for Plasmid Generation

1. TE buffer: MQ H_2O with 10 mM Tris and 1 mM EDTA at pH 8.
2. Tris buffer: MQ H_2O with 10 mM Tris at pH 8.
3. 2× KAPA HiFi Polymerase HotStart ReadyMix.
4. DpnI restriction enzyme.
5. DNA Clean & Concentrator-5 kit.
6. GeneArt™ Gibson Assembly 2× Enzyme.
7. Luria broth (LB).
8. 100 mM calcium chloride.
9. Ampicillin.
10. DNA extraction kit, e.g., GeneJET Plasmid Miniprep Kit.

2.4 Reagents for Landing Pad Generation and Plasmid Recombination

1. Dulbecco's Modified Eagle's Medium (DMEM) with high glucose and GlutaMAX™ Supplement.
2. Complete DMEM: Dulbecco's Modified Eagle's Medium (DMEM) with 10% fetal bovine serum and 100 U/mL penicillin and 0.1 mg/mL streptomycin.
3. PEI solution: 1 μg/μL PEI MAX – Transfection Grade Linear Polyethylenimine Hydrochloride, MW 40,000 dissolved in MQ H_2O. Adjust pH to 7 and filter-sterilize. Store at −20 °C with a working stock at 4 °C.
4. PEI diluent solution: 10 mM HEPES, 150 mM NaCl, dH_2O. Adjust pH to 7, filter, and surround in foil to protect from light. Store at room temperature.
5. Doxycycline hydrochloride.
6. Blasticidin S hydrochloride.
7. Puromycin dihydrochloride.
8. AP1903/rimiducid.
9. FuGENE 6 transfection reagent .

2.5 Reagents for Genomic DNA Extraction and High-Throughput Sequencing Sample Prep

1. Genomic DNA extraction kit, e.g., Qiagen DNeasy Blood and Tissue Kit.
2. Phusion plus high-fidelity PCR mix.
3. Tris-acetate-EDTA (TAE) buffer: 40 mM Tris, 20 mM acetic acid, and 1 mM EDTA.
4. 1% agarose gel: 1 g agarose in 100 mL 1× TAE buffer.
5. SYBR™ Safe DNA Gel Stain.
6. 100 bp DNA Ladder.

7. Freeze 'N Squeeze™ DNA Gel Extraction Spin Columns.
8. DNA Clean & Concentrator-5 kit.
9. Qubit™ dsDNA Quantitation, high sensitivity detection kit.

3 Methods

3.1 Production of Lenti-Landing Pad Lentiviral Vector Particles

1. Start with healthy HEK293T cells that have been cultured in complete DMEM for at least 3 days after thawing.
2. Prepare PEI solution and diluent solution as detailed in the Subheading 2.
3. Plate 10 million cells in a 10 cm plate in a total volume of 10 mL of complete DMEM (*see* **Note 1**).
4. Add 10.5 μg of DNA in 5:5:1 ratio of packaging plasmid psPAX2, transfer vector pLenti-Tet-coBxb1-2A-BFP_IRES-iCasp9-2A-Blast_rtTA3, and pMD2.G. Add diluent to the DNA for a total volume of 500 μL.
5. Add 42 μL PEI to the diluted DNA mixture. Vortex to mix the solution and incubate for 15 min at room temperature.
6. Add DNA + PEI mixture dropwise to cells and place cells in an incubator at 37 °C and 5% CO_2.
7. Replace media in plate with 6 mL complete DMEM between 12 and 24 h post transfection (*see* **Notes 2** and **3**).
8. Collect media over the next 4 days and store pooled supernatants at 4 °C.
9. Centrifuge pooled supernatant at 300× *g* for 3 min at 4 °C.
10. Filter through a 0.45 μm filter and use fresh without freezing to achieve maximal transduction efficiencies.

3.2 Lentiviral Transduction to Generate Landing Pad Cells

1. Start with healthy HEK293T cells (or another cell line to be modified with the landing pad) that have been cultured in complete DMEM for at least 3 days after thawing.
2. Incubate cells with various dilutions of lentiviral vector supernatant (100 μL to 4 mL) to transduce cultures at a range of multiplicities of infection (MOI). Choose the culture with an MOI less than one, confirmed through the expression of blue fluorescent protein (BFP) observed by fluorescent microscopy or flow cytometry (*see* **Note 4**). Ensure that the media contains doxycycline to a final concentration of 2 μg/mL.
3. After 48 h, incubate cells with 10 μg/mL of blasticidin for 5 days to select for cells that have been successfully transduced. While a pure population of transduced cells is unnecessary, having a population of cells at least partially selected for transductants will make the subsequent steps faster and more cost-effective.

4. Sort individual cells on positive BFP expression using fluorescence-activated cell sorting (FACS) into separate wells of 96-well plates. Transfer clones that grow out to appropriate size plates to allow for further propagation. Additional considerations may be needed depending on the cell line (*see* **Note 5**).
5. Maintain the cells in complete DMEM with 2 μg/mL doxycycline and 10 μg/mL blasticidin during routine propagation to ensure a roughly pure population of unsilenced landing pad cells.

3.3 Validation of Landing Pad Clonal Lines

1. Culture HEK293T landing pad cells in complete DMEM with 2 μg/mL doxycycline prior to the transfection. These cells will be used to perform the plasmid transfection with attB-EGFP and attB-mCherry plasmids in a 24-well plate as described below (*see* **Note 6**).
2. Make two transfection mixtures in microcentrifuge tubes labeled "A" and "B" for each sample to be recombined. In "A," add 23 μL of Opti-MEM and 1 μL of Fugene 6. In "B," add volume of plasmid DNA corresponding to 240 ng and Opti-MEM to a final volume of 24 μL. Mix "A" and "B" and incubate for 15–30 min.
3. Trypsinize and count HEK293T landing pad cells. Plate 120,000 cells into each well with DMEM + dox in a final volume of 300 μL. Once incubation is complete, add plasmid mixture dropwise to cells.
4. The following day, add 500 μL of fresh DMEM + dox media to the cells.
5. Perform flow cytometry after 5 days following transfection. Analyze the flow cytometry data to ensure that the cells have taken a fluorescence profile consistent with the presence of a single landing pad (Fig. 3).

3.4 Recombination Plasmid Generation

1. Use Benchling or another plasmid editor program to create primers flanking the transgene to incorporate the desired coding sequence into the recombination plasmid (Fig. 4). The primer design consists of creating ~17 nucleotide overhangs at the 3′ and 5′ ends of the DNA, such that overlapping DNA fragments can be joined by one isothermal reaction through Gibson Assembly [9]. Make sure the primer has a melting temperature of around 65 °C to ensure proper ligation.
2. Start an overnight culture of *E. coli* bacteria, such as NEB 10-beta, in 2 mL of LB media for bacterial transformation of the plasmid. Shake overnight at 300 rpm at 37 °C. While best if timed to use fresh, the bacteria can be flash frozen and used upon thawing if needed.

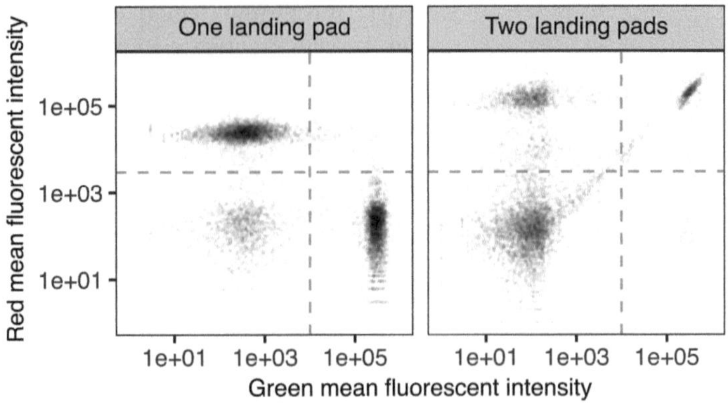

Fig. 3 Example results of the two-color recombination test. When there is only one landing pad in the cell, transfection with a mixture of attB-EGFP and attB-mCherry plasmids should yield green or red cells, but no clear population of green and red cells (left). In contrast, cells with two landing pads will readily yield a discernable population of green and red cells (right). Dotted lines denote green and red fluorescence threshold values separating fluorescence positive from negative cells

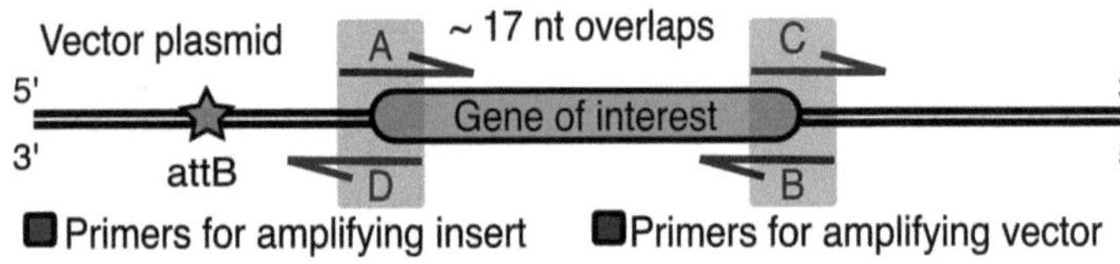

Fig. 4 Schematic showing design of molecular cloning primer pairs with Gibson-compatible overhangs: Gibson Assembly requires primers flanking the backbone plasmid and insert at the 5′ and 3′ end with ~17 nucleotide overhangs. This allows for the insertion of the transgene into the backbone plasmid in one isothermal reaction. Primer pairs A and B contain the hybridizing region of the insert and overhang sequence of the vector plasmid. C and D contain the hybridizing region of the vector plasmid and overhang sequence of the insert

3. Dilute the primer pairs into a single mixture containing 10 μM of each with either nuclease-free water or Tris buffer. Do not use TE buffer to prevent the transfer of EDTA to the PCR product.
4. Dilute the template plasmid (AttB_ACE2_IRES-mCherry-H2A-P2A-PuroR) to 10 ng/μL. In a separate tube, dilute the DNA encoding the transgene of interest to 10 ng/μL.
5. Set up the PCR reactions. We will set up two reactions, one with the template plasmid and one with the transgene of interest, each with appropriate primer pairs. In a total volume of 30 μL, add 10 μL of nuclease-free water, 4 μL of diluted plasmid, 1 μL of diluted primer mix, and 15 μL of 2× KAPA HiFi mastermix.

6. Perform PCR with the following conditions: 95 °C for 5 min, 98 °C for 20 s, 65 °C for 15 s, 72 °C for 8 min, repeat the first four steps six more times (*see* **Note 7**), 72 °C for 5 min, 4 °C hold.
7. Add 1 μL DpnI enzyme to each reaction tube and incubate at 37 °C for 2 h.
8. Mix the two reactions and clean the PCR product using a kit such as the Zymo Research DNA Clean & Concentrator-5 kit.
9. In a new PCR tube, add 1 μL of the cleaned PCR product to 1 μL of Gibson Assembly 2× Enzyme. Incubate for 1 h at 50 °C.
10. Record the OD600 of the overnight culture. Add the starter culture to 200 mL LB media without antibiotics at a dilution of 1:1000 and place into the bacterial incubator at 300 rpm at 37 °C. The culture should be stopped when the OD600 reaches 0.2–0.4, so check the OD600 periodically during incubation. This usually takes about 3.5–4 h.
11. Pre-chill centrifuge and all tubes that will be used for holding the bacterial culture in wet ice. It is important to keep the bacterial culture cold at all times prior to transformation.
12. Once the culture reaches the desired OD600, spin the culture at 4200 rpm at 4 °C for 20 min. Remove supernatant and resuspend bacteria pellet in 5 mL of cold 100 mM calcium chloride. Centrifuge at 4200 rpm at 4 °C for 5 min to wash the pellet.
13. Remove supernatant and resuspend pellet in 1 mL 100 mM calcium chloride.
14. Add 100 μL of competent bacteria to each reaction tube. Make sure to prepare a positive control of intact circular plasmid DNA to determine transformation efficiency. Include a negative control without DNA to make sure it does not produce colonies. Let sit on ice for 30 min.
15. Heat shock at 42 °C for 30–45 s and let sit on ice for at least 2 min before plating.
16. Plate the bacteria on an LB agar plate infused with the appropriate selection antibiotic. The template plasmid we used in this protocol contains an ampicillin resistance gene, so we will plate on LB agar plates containing 150 μg/mL ampicillin.
17. The next day, pick at least three colonies of candidate plasmids and clonally grow each in a 2 mL culture of LB mixed with the appropriate antibiotic (150 μg/mL ampicillin in this case).
18. The next day, perform a miniprep DNA extraction to harvest the candidate plasmids from each clonal bacterial culture.

19. Perform Sanger sequencing to determine which candidate plasmid clones have incorporated the transgene of interest. This is best done by choosing a primer like KAM2199 (5′ TT ACGCCAGCTGGCGAAAGG 3′), which binds to a sequence slightly 5′ of the attB site and reads into the sequence encoding the transgene of interest.
20. For plasmids that seem to have incorporated the transgene of interest, sequence the plasmid in full to ensure it does not have unintended mutations or, more rarely, have formed a concatemer of plasmid sequences. We recommend a combination of full-plasmid nanopore sequencing (such as through Plasmidsaurus) followed by targeted Sanger sequencing to resolve any sites exhibiting a discrepancy between the returned nanopore consensus sequence and the originally expected sequence.
21. Once the original WT construct is generated, create a small set of control constructs that can be used to pilot the phenotypic assay that will be used in Subheading 3.5 (*see* **Note 8**).
22. Generate the plasmid library of interest using site-saturation mutagenesis (*see* **Notes 9** and **10**).

3.5 Plasmid Recombination into Landing Pad Cells

1. Follow the plasmid recombination transfection protocol as described in Subheading 3.3 but replacing the attB-EGFP and attB-mCherry plasmids with the control plasmids (encoding the WT protein or key variants) or plasmid library encoding the transgene of interest.
2. Set aside unselected cells for endpoint analysis. To calculate recombination rate and number of recombinants, set some cells aside before selection steps and run on flow cytometry to determine percentage of mCherry expression. Multiply this percentage with the number of cells transfected (e.g., 120,000 cells) to estimate the number of independently recombined cells in the culture.
3. Perform negative selection with iCasp9. Add 10 nM final concentration of AP1903 about 72 h after recombination. Apoptosis of unrecombined, iCasp9-expressing cells will occur within a few hours. Change media with fresh DMEM + dox to remove the dead cells.
4. Perform positive selection using puromycin. Depending on the ratio of recombined cells to silenced landing pad cells present in the culture well, a positive selection step may be necessary to achieve a pure population of recombined cells. To do this, add puromycin to the well of AP1903-treated cells once they have reached greater than 10% confluence.

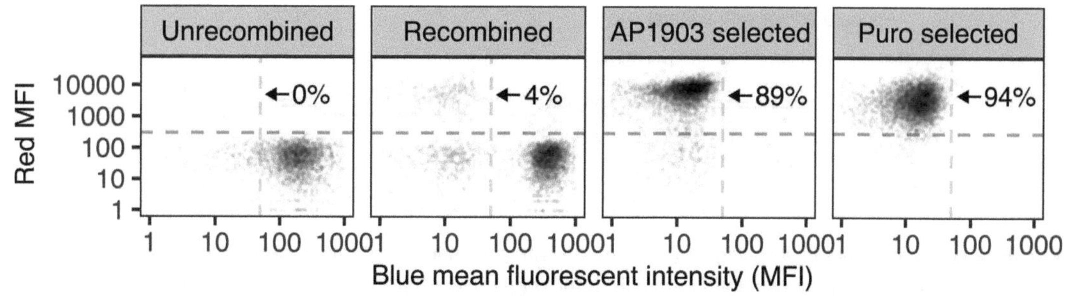

Fig. 5 Example flow cytometry profiles of recombined cells before and after selection. Before transfection with an attB plasmid encoding an IRES-mCherry-2A-PuroR cassette, there are no red cells present, and almost all of the cells are blue fluorescent (**far left**). Upon transfection of the attB plasmid, a small fraction of the blue cells will lose their blue fluorescence while gaining red fluorescence (**middle left**). Addition of AP1903 kills the blue un-recombined cells, drastically increasing the fraction of recombined red cells in the population (**middle right**). The generally small fraction of silenced, nonfluorescent cells that were present following negative selection can be removed by using positive selection for an antibiotic resistance marker expressed from the attB plasmid, like puromycin, yielding the purest population of recombined cells possible through selection (**far right**). Dotted lines denote blue and red fluorescence threshold values separating fluorescence positive from negative cells

5. Perform flow cytometry to quantitate the recombination rate and to access the purity of recombined cells within the culture following the selection steps (Fig. 5). While the recombination rate is not as important for clonal control samples, it is critical for library recombinations to be able to estimate how many independently recombined cells there are per variant and determine if any variants from the library may have been lost due to bottlenecking at the recombination step. We recommend this flow cytometry step to be performed at or later than 5 days following transfection to avoid confounding fluorescent signals from the "promoterless" plasmids themselves (*see* **Note 11**).
6. Perform a phenotype assay specific to the function of the protein of interest to score protein variants for their activity of interest (*see* **Note 12**).

3.6 Illumina Sequencing of Genomically Integrated DNA

1. Create primers to amplify the region of interest. The Amplicon-EZ service through Azenta performs sequencing within 500 bp and requires the use of partial adapter sequences at the end of the primer hybridizing region (Fig. 6) (*see* **Note 13**). Primer pairs should be designed so that the forward primer binds a sequence present only within the landing pad, while the reverse primer binds a sequence only present within the recombination plasmid. Thus, only the singular successfully integrated attB plasmid molecule can be exponentially amplified through

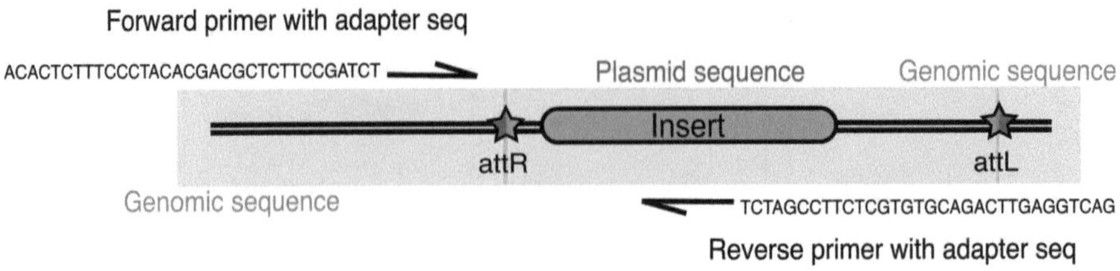

Fig. 6 Schematic showing Amplicon-EZ primer design and binding: Illumina sequencing through Azenta's Amplicon-EZ requires the presence of partial adapter sequences after the hybridizing region (~20 nucleotides) of the primers. The adapter sequence for the forward primer is ACACTCTTTCCCTACACGACGCTCTTCCGATCT and the adapter sequence for the reverse primer is GACTGGAGTTCAGACGTGTGCTCTTCCGATCT. We recommend sequencing across the attR site to ensure coverage of only recombined variants

polymerase chain reaction (*see* **Note 14**). In most cases, we recommend the user utilizes the KAM499 primer binding site (5′ GAGAACGTATGTCGAGGTAGGC 3′) present toward the 3′ end of the Tet-inducible promoter.

2. Extract genomic DNA from the landing pad cell pellet collected at each time point using a genomic DNA extraction kit such as the Qiagen DNeasy Blood and Tissue Kit.
3. Prepare three reactions of 40 μL of each time point sample to amplify the genomic DNA by PCR using Phusion plus high-fidelity PCR mix and appropriate primers. For each reaction, add 1–2 μg of genomic DNA, 1.25 μL each of forward and reverse primers (from 10 uM stock), 20 μL of Phusion plus high-fidelity PCR master mix, and add sufficient volume of nuclease-free water to bring the total volume to 40 μL.
4. Perform PCR with the following conditions: 95 °C for 3 min, 95 °C for 15 s, 60 °C for 15 s, 72 °C for 30 s, repeat first four steps for 23 additional times, 72 °C for 1 min, 4 °C hold.
5. Load PCR product with appropriate amount of SYBR Safe dye on a 1% agarose gel along with a suitable DNA molecular weight ladder, such as a 100 bp DNA Ladder.
6. Extract bands of interest at appropriate fragment size using Freeze 'N Squeeze DNA Gel Extraction Spin Columns.
7. Purify the extracted DNA using the Zymo DNA clean and concentrator kit. Elute in 8 uL of elution buffer.
8. Quantify DNA using Qubit. Normalize concentration of DNA to 20 ng/μL in 25 μL of either elution buffer or Tris buffer (500 ng total).
9. Complete the order form with the required information for Azenta's Amplicon-EZ sequencing service.
10. Analyze the sequencing data to determine calculate phenotype scores (*see* **Note 15**).

4 Notes

1. Transfection of 293T cells for lentiviral production can be performed with various transfection methods, including electroporation, calcium phosphate, and PEI, or with cationic lipids. We describe the method with PEI, as we find this is a highly consistent and cost-effective transfection method for the production of lentiviral particles, when performing this process regularly.
2. We suggest use of complete media without doxycycline during lentivector production, as subgenomic transcription from the internal Tet-inducible promoter may interfere with generation of genome-length lentivector transcripts necessary to produce landing pad lentiviral particles.
3. The titer of lentiviral particles produced with the pLenti-Tet-coBxb1-2A-BFP_IRES-iCasp9-2A-Blast_rtTA3 transfer vector is much lower than for other transfer vectors, likely due to the large lentivector length and the inclusion of a RBglobin transcriptional terminator sequence within the lentivector sequence. Thus, it is advised that the user uses much more lentivector-containing supernatant than normally used.
4. While we originally used homology-directed repair to genomically integrate the landing pad into a precise genomic location [1], we found this method to be prohibitively inefficient, with lengthy validation steps needed to confirm the introduction of a single copy of the landing pad sequence. We have since adopted low multiplicity of infection (MOI) lentivector transduction to introduce a single copy of the landing pad cassette into the cell genome [2, 7].
5. Not every sorted cell will establish a clonal outgrowth; with 293T cells, we have historically observed roughly 1/3 of all sorted cells growing out. Thus, to ensure the outgrowth of sufficient clonal lines to validate for downstream use, we suggest sorting one or two full 96-well plates for each cell line being created. The rate of clonal outgrowth will likely differ depending on the parental cell line. Furthermore, some cell line are highly refractory to outgrowth from single cells (e.g., U2OS cells). In these cases, we suggest adding conditioned media into the 96-well plates prior to sorting, or taking an alternative approach such as the picking of individual colonies from sorted cells sparsely plated in 100 mm plates.
6. The precise characteristics of each landing pad clone may slightly differ and are worth comparing before choosing a clone to move forward with. Furthermore, while low MOI lentiviral transduction should make it highly unlikely that any given clone harbors two landing pad copies, this should be

experimentally validated with the chosen clone before moving forward with subsequent experiments. We have found the "two-color recombination test" to be the most effective method to determine this. This is done by transfecting a mixture of EGFP and mCherry—encoding attB recombination plasmids. If there is only a single landing pad site, only green or red singly fluorescent cells should appear roughly a week following the transfection. If there is more than one landing pad site, a distinct population of dually green and red cells will appear (for an example, *see* Fig. 3).

7. Seven total amplification cycles are typically enough to yield sufficient DNA amplification for success in subsequent molecular cloning steps while keeping the amount of unintended polymerase-introduced sequence errors to a minimum. If struggling to get sufficient transformants with the intended construct (as compared to template plasmids), the number of amplification cycles can be increased, although extra care will be necessary to ensure that resulting plasmid constructs do not harbor problematic mutations.

8. Before performing any multiplexed, library-based experiments, it is worth creating a small panel of constructs encoding phenotypically distinct protein variants to ensure that the phenotypic assay that will be harnessed exhibits sufficient signal. In the case of assays that assess protein steady-state abundance using green fluorescence such as VAMP-seq, this includes creating constructs encoding known loss-of-abundance variants to determine whether the amount of green fluorescence is sufficiently distinct from that elicited by the WT protein [10–12]. In the case of assays that measure the relative depletion or enrichment of various protein-coding sequences over multiple time points, this could involve inclusion of known cell cycle arresting genes like CDKN1A, or gene products known to have little effect on cell division, such as EGFP [2].

9. Potential site-saturation library generation methods that can be performed in house include arrayed inverse PCR [13], such as used in studies of PTEN [10, 14], TPMT [10], and VKORC1 [11], one-pot saturation nicking mutagenesis [15], Precision Oligo-Pool based Code Alteration (POPCode) [16], or purchase through a vendor specializing in generating side-saturation mutagenesis libraries such as Twist Bioscience [12]. Regardless of the library generation method, three key characteristics worth assessing for every library is (1) library coverage/presence of all intended variants, (2) the amount of template contamination of the library, and (3) the frequency distribution of the library members, with the more uniform the frequencies the better.

10. Once fully sequence confirmed, the recombination plasmid that was created in the preceding step can be used to generate the plasmid library. The nature of the plasmid library is project- and user-specific, so we will not get into the details of this step in this protocol, although we list some possible library generation methods above. In this protocol, we harness Amplicon-EZ, a service that allows for 500 nucleotides of returned Illumina sequencing (250 nt per forward and reverse reads). Thus, regions of variable sequence shorter than ~150 amino acids can be analyzed directly from the sequencing reads, while larger regions would require a barcode subassembly method [17, 18] to assess the variant identity of each sequencing read.
11. Despite the attB recombination vectors lacking an annotated promoter, the collective accumulation of cryptic transcripts generated from promoter-like sequences within the large number of transfected plasmids can result in transient expression of mCherry at early time points following transfection. The magnitude of this effect can change based on the transfection method, wherein Lipofectamine 3000 typically elicits a more pronounced effect as compared to Fugene6. We recommend potentially creating an attB-mutated version of your recombination vector to help quantify this effect within your specific experimental conditions.
12. After recombination of the plasmid library into cells, a phenotype assay needs to be developed to demultiplex the variants for further analysis. This step is user-specific, as the assay will need to reflect the function of the protein of interest. For example, if one is interested in determining the change in frequency of a protein variant over time, two time points of the sample could be taken. The first time point can be taken right after selection with AP1903 and puromycin, and then a second time point could be taken a few days after. The following steps in this protocol will describe the procedure after separating the variants and will take into account an example with two time points. Potential library-compatible phenotypic assays performed using this system include assays for protein steady-state abundance [1, 10–12, 14, 19] using Sort-seq [20], functional assays for sensitivity to cytotoxic agents [12, 21, 22], and immunostaining for assessing cell surface localization [23–27].
13. The Azenta Amplicon-EZ service performs 2 × 250 bp Illumina sequencing and provides 50,000 or more reads. This is helpful for pilot experiments, as the company multiplexes samples from multiple users on the back end, providing sufficient reads to assess a handful of conditions for a simple library-based experiment without requiring an excessively large up-front cost. Once the pilot experiments are completed and one now

desires to have a large panel of conditions or replicate samples, it becomes much more cost-effective to forgo Amplicon-EZ and multiplex your own samples on a dedicated Illumina sequencing kit for those series of experiments. This can be done by replacing the Amplicon-EZ forward and reverse adaptor sequences with the Illumina p7 and p5 sequences, taking care to design an indexing scheme to allow for demultiplexing of different samples within the same sequencing run.

14. When designing primers for sequencing, it is important to choose a primer pair that selectively amplifies the recombined DNA. This is because there is only a single recombined plasmid per cell, while there are likely hundreds to dozens of unrecombined plasmid lingering in the cell following transfection. Thus, amplification with primers directly flanking the transgenic sequence of interest encoded in the attB recombination plasmid will preferentially amplify irrelevant DNA, confounding downstream results. We suggest using a pair of primers that spans a recombination junction. In most cases, this will be an amplicon spanning the attR recombination junction (attP left side with attB right side), spanning the Tet-inducible promoter through to the 5′ end of the transgene of interest, although amplicons spanning the attL junction are possible if a nucleotide barcode is engineered to the 5′ of the attB site [2] (Fig. 1b).

15. The turn-around time for Amplicon-EZ is currently between 2 and 3 weeks. They will provide raw FASTQ files of greater than 50,000 reads for each sample. If reads are paired-end, pair the reads using software such as PEAR [28] or fastq-join [29]. The reads can then be trimmed to the relevant sequence and translated from DNA nucleotides to amino acid residues. We will take the example of observing the change in frequency of protein variants over the two time points by calculating phenotype scores between the two time points. The raw count of each protein variant can be done, and frequency of each variant at each time point can be determined by dividing the count of the protein variant over the total number of counts. The phenotype score can be calculated by taking the ratio of the frequency of the variant at the final time point over frequency at the initial time point. We suggest performing these calculations with your own analysis scripts to ensure you understand the analysis process, although convenient analysis software exists, and likely improve throughput and reproducibility for more routine analysis [30].

References

1. Matreyek KA, Stephany JJ, Fowler DM (2017) A platform for functional assessment of large variant libraries in mammalian cells. Nucleic Acids Res 45:e102
2. Matreyek KA, Stephany JJ, Chiasson MA, Hasle N, Fowler DM (2019) An improved platform for functional assessment of large protein libraries in mammalian cells. Nucleic Acids Res 48:1
3. Xu Z, Thomas L, Davies B, Chalmers R, Smith M, Brown W (2013) Accuracy and efficiency define Bxb1 integrase as the best of fifteen candidate serine recombinases for the integration of DNA into the human genome. BMC Biotechnol 13:87
4. Russell JP, Chang DW, Tretiakova A, Padidam M (2006) Phage Bxb1 integrase mediates highly efficient site-specific recombination in mammalian cells. Biotechniques 40:460, 462, 464
5. Ghosh P, Pannunzio NR, Hatfull GF (2005) Synapsis in phage Bxb1 integration: selection mechanism for the correct pair of recombination sites. J Mol Biol 349:331–348
6. Duportet X, Wroblewska L, Guye P, Li Y, Eyquem J, Rieders J et al (2014) A platform for rapid prototyping of synthetic gene networks in mammalian cells. Nucleic Acids Res 42:13440–13451
7. Shukla N, Roelle SM, Suzart VG, Bruchez AM, Matreyek KA (2021) Mutants of human ACE2 differentially promote SARS-CoV and SARS-CoV-2 spike mediated infection. PLoS Pathog 17:e1009715
8. Wang X, Le N, Denoth-Lippuner A, Barral Y, Kroschewski R (2016) Asymmetric partitioning of transfected DNA during mammalian cell division. Proc Natl Acad Sci U S A 113:7177–7182
9. Gibson DG, Young L, Chuang R-Y, Venter JC, Hutchison CA 3rd, Smith HO (2009) Enzymatic assembly of DNA molecules up to several hundred kilobases. Nat Methods 6:343–345
10. Matreyek KA, Starita LM, Stephany JJ, Martin B, Chiasson MA, Gray VE et al (2018) Multiplex assessment of protein variant abundance by massively parallel sequencing. Nat Genet 50:874–882
11. Chiasson MA, Rollins NJ, Stephany JJ, Sitko KA, Matreyek KA, Verby M et al (2020) Multiplexed measurement of variant abundance and activity reveals VKOR topology, active site and human variant impact. elife 9:1–25
12. Suiter CC, Moriyama T, Matreyek KA, Yang W, Scaletti ER, Nishii R et al (2020) Massively parallel variant characterization identifies NUDT15 alleles associated with thiopurine toxicity. Proc Natl Acad Sci U S A 117:5394–5401
13. Jain PC, Varadarajan R (2014) A rapid, efficient, and economical inverse polymerase chain reaction-based method for generating a site saturation mutant library. Anal Biochem 449:90–98
14. Matreyek KA, Stephany JJ, Ahler E, Fowler DM (2021) Integrating thousands of PTEN variant activity and abundance measurements reveals variant subgroups and new dominant negatives in cancers. Genome Med 13:165
15. Wrenbeck EE, Klesmith JR, Stapleton JA, Adeniran A, Tyo KEJ, Whitehead TA (2016) Plasmid-based one-pot saturation mutagenesis. Nat Methods 13:928–930
16. Weile J, Sun S, Cote AG, Knapp J, Verby M, Mellor JC et al (2017) A framework for exhaustively mapping functional missense variants. Mol Syst Biol 13:957
17. Hiatt JB, Patwardhan RP, Turner EH, Lee C, Shendure J (2010) Parallel, tag-directed assembly of locally derived short sequence reads. Nat Methods 7:119–122
18. Yeh C-LC, Amorosi CJ, Showman S, Dunham MJ (2022) PacRAT: a program to improve barcode-variant mapping from PacBio long reads using multiple sequence alignment. Bioinformatics 38:2927–2929
19. Amorosi CJ, Chiasson MA, McDonald MG, Wong LH, Sitko KA, Boyle G et al (2021) Massively parallel characterization of CYP2C9 variant enzyme activity and abundance. Am J Hum Genet 108:1735–1751
20. Peterman N, Levine E (2016) Sort-seq under the hood: implications of design choices on large-scale characterization of sequence-function relations. BMC Genomics 17:206
21. Reisman BJ, Guo H, Ramsey HE, Wright MT, Reinfeld BI, Ferrell PB et al (2022) Apoptolidin family glycomacrolides target leukemia through inhibition of ATP synthase. Nat Chem Biol 18:360–367
22. Glazer AM, Kroncke BM, Matreyek KA, Yang T, Wada Y, Shields T et al (2020) Deep mutational scan of an SCN5A voltage sensor. Circ Genom Precis Med 13:e002786
23. McKee AG, Kuntz CP, Ortega JT, Woods H, Most V, Roushar FJ et al (2021) Systematic profiling of temperature- and retinal-sensitive rhodopsin variants by deep mutational scanning. J Biol Chem 297:101359

24. Roushar FJ, McKee AG, Kuntz CP, Ortega JT, Penn WD, Woods H et al (2022) Molecular basis for variations in the sensitivity of pathogenic rhodopsin variants to 9-cis-retinal. J Biol Chem 298:102266
25. Coyote-Maestas W, Nedrud D, Suma A, He Y, Matreyek KA, Fowler DM et al (2021) Probing ion channel functional architecture and domain recombination compatibility by massively parallel domain insertion profiling. Nat Commun 12:7114
26. Coyote-Maestas W, Nedrud D, He Y, Schmidt D (2022) Determinants of trafficking, conduction, and disease within a K+ channel revealed through multiparametric deep mutational scanning. elife 11. https://doi.org/10.7554/eLife.76903
27. Ouyang WO, Tan TJC, Lei R, Song G, Kieffer C, Andrabi R et al (2022) Probing the biophysical constraints of SARS-CoV-2 spike N-terminal domain using deep mutational scanning. Sci Adv 8:eadd7221
28. Zhang J, Kobert K, Flouri T, Stamatakis A (2014) PEAR: a fast and accurate illumina paired-end reAd mergeR. Bioinformatics 30: 614–620
29. Aronesty E (2013) Comparison of sequencing utility programs. Open Bioinf J. Available: https://benthamopen.com/ABSTRACT/TOBIOIJ-7-1
30. Rubin AF, Gelman H, Lucas N, Bajjalieh SM, Papenfuss AT, Speed TP et al (2017) A statistical framework for analyzing deep mutational scanning data. Genome Biol 18:150

Chapter 11

Flow Cytometry Quantification of Transient Transfections in Mammalian Cells

Jacob Beal

Abstract

Flow cytometry is a powerful quantitative assay supporting high-throughput collection of single-cell data with a high dynamic range. For flow cytometry to yield reproducible data with a quantitative relationship to the underlying biology, however, requires that (1) appropriate process controls are collected along with experimental samples, (2) these process controls are used for unit calibration and quality control, and (3) data are analyzed using appropriate statistics. To this end, this chapter describes methods for quantitative flow cytometry through the addition of process controls and analyses, thereby enabling better development, modeling, and debugging of engineered biological organisms. The methods described here have specifically been developed in the context of transient transfections in mammalian cells but may in many cases be adaptable to other categories of transfection and other types of cells.

Key words Synthetic biology, Metrology, Flow cytometry, Fluorescence measurements, Single-cell biology

1 Introduction

Flow cytometry is a long-established method for assaying the optical properties of large numbers of single cells, in use for more than half a century [1, 2]. Much of the development and use of flow cytometry has been in clinical applications focused on the categorization of cells into classes, such as immunoassays and diagnosis of blood diseases. In recent years, however, flow cytometers have also been used to great effect in synthetic biology for measuring the properties of individual cells, e.g., measuring the expression level of a gene by estimating the number of molecules of a proxy fluorescent reporter.

In a typical flow cytometer, a stream of cells are passed in front of one or more lasers, with an arrangement of optical filters and detectors collecting information on the light scattering and fluorescence properties of each cell-like particle as it passes. Flow

Francesca Ceroni and Karen Polizzi (eds.), *Mammalian Synthetic Systems*, Methods in Molecular Biology, vol. 2774,
https://doi.org/10.1007/978-1-0716-3718-0_11,

	Fluorimeter	Plate Reader	Flow Cytometer	Fluorescence Microscope
Throughput	Low	High	High	Low - High
Resolution	Population	Population	Single Cell	Subcellular
Time Series	Yes	Yes	No	Yes
Dynamic Range	2-3 logs	2-3 logs	3-6 logs	1-2 logs
# cells	n/a	n/a	High	Low

Fig. 1 Flow cytometry provides high dynamic range data from large numbers of single cells at a single time point, complementary to the properties of other common fluorescence-based assays

cytometer optical detectors often have high sensitivity and range, with signals distinguishable on the order of a 10^3 to 10^6 ratio between the highest and lowest distinguishable values. Data are typically collected from on the order of 10^3 to 10^5 particles per sample over the time span from seconds to minutes, and automated samplers allow high-throughput processing of dozens to hundreds of samples from microplates.

Flow cytometers can thus be prodigious sources of single-cell data with a high dynamic range and form an important part of the synthetic biology engineering toolkit. Compared to other broadly accessible fluorescence assays (Fig. 1), flow cytometry shines in the number of cells assayed and sensitivity (i.e., dynamic range). Because cells must be suspended in sheath fluid and passed through the instrument, however, the assay is typically disruptive to cell state and thus generally not suitable for collecting time series data. Plate readers provide a useful complementary capability for flow cytometry, since they are effective at collecting time series data and can be calibrated to equivalent units [3, 4]. Likewise, microscopy provides the complementary capabilities of subcellular resolution and time series data. Indeed, imaging flow cytometry can fuse microscopy and flow cytometry by adding a microscopy stage to a flow cytometer [5], though it is still disruptive to cell states.

The relationship between the properties of a cell and the optical signal measured in a channel of a flow cytometer, however, is highly sensitive to specifics of the instrument's configuration, current settings, and the level of optical wear, as well as to the choice of fluorescent reporter, cell state, and interference from other measurement channels [6–8]. As a consequence, interpreting flow cytometry data to produce precise, reproducible, and biologically relevant estimates of molecule counts requires careful use of process controls linked to appropriate quality control and analytical methods [9]. When appropriate process control, calibration, and analysis protocols are applied, however, flow cytometry can reliably produce reproducible measurements with a quantitative relationship to the underlying biology.

This chapter aims to increase the accessibility of quantitative flow cytometry by summarizing methods for:

1. Enhancing a flow cytometry study by adding process controls to support calibration and quantification
2. Using data from these process controls for unit calibration and quality control
3. Analysis of the resulting calibrated data using appropriate statistics

More specifically, this chapter focuses on methods that were developed in the context of transient transfections in mammalian cells over the course of a number of studies that used these methods to develop, model, refine, and predict the behavior of a variety of genetic devices [10–18]. Note that these methods are modular with respect to most aspects of cell type, transfection, and culturing, so the actual preparation of the experimental samples will be largely abstracted, while the methods focus on the process controls necessary for effective calibration and the analysis used to make use of those process controls. Likewise, many aspects of the methods should be adaptable to non-transient transfections and to non-mammalian cells, and to this end the methods will explicitly identify the points where enabling assumptions are made based on mammalian transient transfection in order to better understand limits and enable adaptation to other contexts.

2 Materials

The description of these materials and their potential substitutions are provided in Subheading 4.1.

1. SpheroTech Ultra Rainbow Quantitative Particle Kit (URQP-38-6K)
2. SpheroTech Polystyrene Particle Size Standard Kit (PPS-6K)
3. Wild-type negative control
4. Single-color controls (one per fluorescence channel)
5. Multi-color control
6. Experimental samples per experimenter's study design.

3 Methods

The analytical methods below are described with sufficient detail to allow them to be understood and implemented by the reader. Note, however, that application of these analysis methods can be simplified by using the TASBE Flow Analytics Matlab software

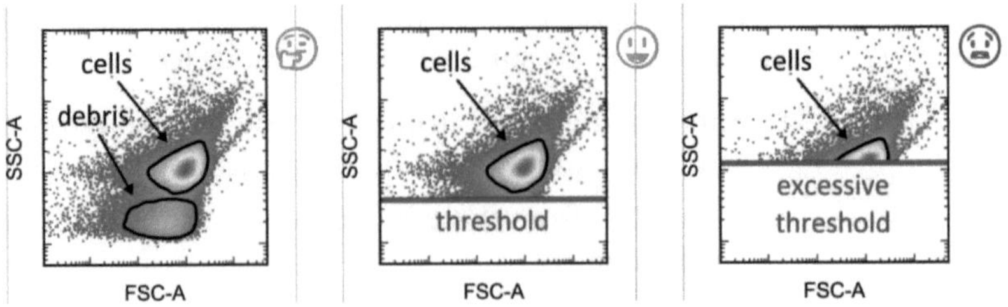

Fig. 2 Examples of data collection gating with size thresholds that are too permissive (left), appropriate (center), and too restrictive (right), adapted from protocol in [22]

package [19], an existing implementation of methods and workflows to automate all of the steps described in this section, as well as to support the debugging steps listed below in Subheading 4. Many of these steps are also automatable using existing Python implementations in FlowCal [20] or CytoFlow [21].

3.1 Data Collection

1. Set a conservative forward scatter (particle size) trigger, such that the number of small non-cell particle events is minimized while not removing smaller cells (Fig. 2). Other instrument gating should not be used, as it is generally better to use post-facto filtering to ignore unneeded events, rather than to discard cell data that may later prove valuable.
2. Record all forward scatter (FSC) and side scatter (SSC) channels and the area channel for each fluorescent reporter.
3. Ensure that the instrument is putting laser and filter information in FCS files and that the values for these are correct. This is valuable for ensuring that instrument settings are kept consistent between samples.
4. Gather between 50,000 and 100,000 events per sample.

Explanatory notes and information on alternatives may be found in Subheading 4.3.

3.2 Log-Scale Statistical Analyses

All of the subsequent stages of this protocol are analytical, and most depend on log-scale statistical analyses of data, so here we will discuss the reasons for this analysis and provide their formulations for those who may be unfamiliar.

Log-scale analysis is motivated by the fact that strong gene expression typically has a log-normal distribution [23]. Transient transfections in mammalian cells typically deliver many copies of a genetic construct to each cell, and fluorescent reporter signals are typically engineered to be as strong as an experimenter can obtain, both of which mean that strong overall expression is the typical case. For this reason, cell behavior often shows distributions that are better suited to geometric statistics on the log scale, rather than arithmetic statistics on the linear scale, as illustrated in the example in Fig. 3 (other statistical alternatives are discussed in Subheading 4.5).

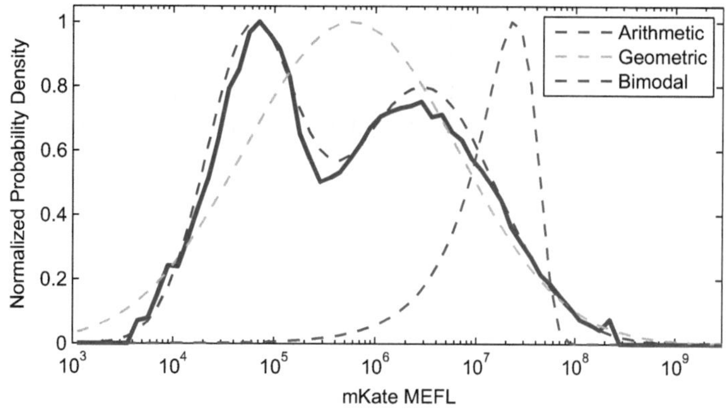

Fig. 3 Cell behavior often shows distributions that are better suited to geometric statistics on the log scale, rather than arithmetic statistics on the linear scale. In many cases, fits with bimodal or multi-modal log-normal distributions are appropriate. Figure adapted from [23]

Geometric mean and geometric standard deviation may be computed by computing the arithmetic mean and standard deviation of a variable that has been transformed onto the log scale:

- Geometric mean of variable X:

$$\mu_g(X) = 10^{\mu(log_{10}(X))}$$

- Geometric standard deviation of variable X:

$$\sigma_g(X) = 10^{\sigma(log_{10}(X))}$$

The comparison of these statistics should also be discussed with respect to geometric (fold) differences. For example, a decrease from 4000 MEFL to 1000 MEFL or 250 MEFL should be interpreted as a 4-fold or 16-fold decrease, respectively, and not 75% or 94%.

Histograms should likewise generally be computed on the log scale. For purposes of the analyses discussed here, histograms should be computed using a granularity of 10 bins per decade: for example, in the range between 100 and 1000, there is a bin for values from 100 to 125.9, from 125.9 to 158.5, etc., until the last bin holds values from 794.3 to 1000. This may be computed by taking a histogram of data transformed onto the log scale. This granularity matches well with typical flow cytometer precision and the number of data points recommended for capture in Subheading 3.1. If the bin-to-bin variability is too high, however, the granularity may be reduced, e.g., from 10 bins per decade to 5 bins per decade.

The parametric analysis combines histograms on the log scale with geometric statistics. This analysis is computed by grouping flow cytometry events into log-scale bins with respect to a selected fluorescent channel, at a granularity of 10 bins per decade. Then, for each bin, the geometric mean and the standard deviation of other fluorescent channels are calculated over all of the events in the bin. Bins with less than a minimum count of 10 events are excluded from the parametric results, due to the uncertainty of values computed from small numbers of data points. The result is a function capturing the relationship between the selected channel and the other fluorescence channels. Examples of its use are measuring the spillover fluorescence in a spectrally overlapping channel as a function of fluorescence in the intended channel (Subheading 3.6) or the value of an experimental variable as a function of transfection marker expression (Subheading 3.8).

3.3 Bead-Based Fluorescence Channel Calibration

For each channel for a fluorescent reporter, compute the unit scaling between arbitrary units on the fluorescence channel and units of Equivalent Reference Fluorophore (ERF) for that channel:

1. Compute a log-scale histogram of the channel values for the rainbow bead sample.
2. Identify the peaks in the channel and compute the geometric mean for each peak region, as shown in the example in Fig. 4a.
3. Exclude low-valued peaks and doublets.
4. Perform a constrained linear fit (slope=1) on the log scale between peak values and manufacturer supplied peak ERF values, as shown in the example in Fig. 4b.

Explanatory notes and information on alternatives may be found in Subheading 4.6.

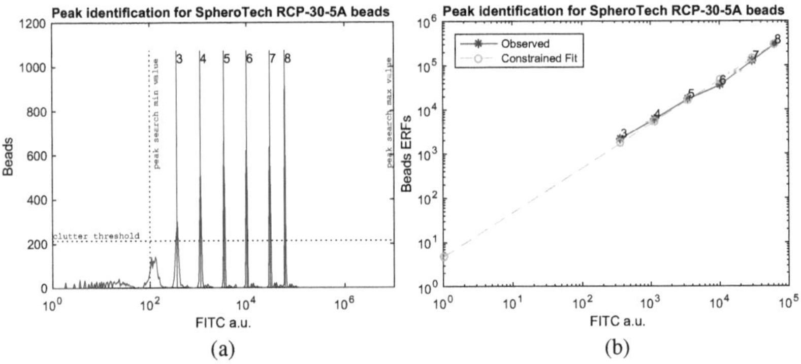

Fig. 4 Example of color bead analysis, showing (**a**) peak identification for the fluorescein channel against an alternate calibration bead, SpheroTech RCP-30-5A. This is an 8-peak calibrant, and in this case only the top six peaks are being used. Note the small doublet peaks slightly above each identified peak. (**b**) Scaling fit for a unit conversion function mapping FITC area channel arbitrary units to Molecules of Equivalent FLuorescein (MEFL). Figure adapted from [24]. (**a**) MEFL bead peaks. (**b**) MEFL bead fit

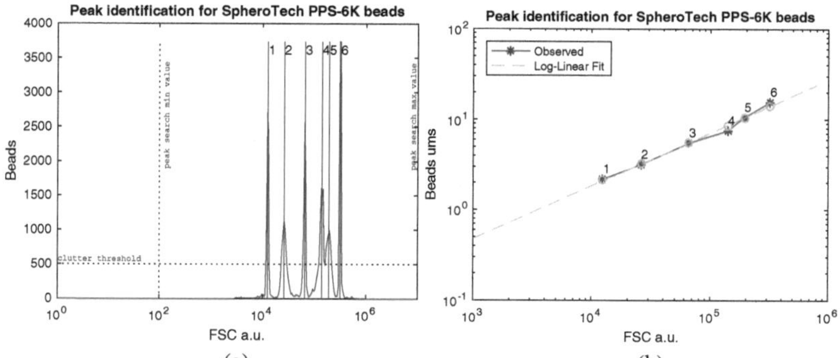

(a) (b)

Fig. 5 Example of size bead analysis, showing (**a**) peak identification and (**b**) log–log fit for unit conversion function mapping forward scatter area (FSC-A) channel arbitrary units to equivalent micro-meter (Eμm) diameter. Note that the fit function does not have a 1:1 slope in the log scale, meaning that it is not a linear scaling relationship. Figure adapted from [24]. (**a**) Size bead peaks. (**b**) Size bead fit

When channel conversion is used (Subheading 3.7), only the Molecules of Equivalent FLuorescein (MEFL) scaling factor will actually be used. Nevertheless, unit scaling should be computed for all fluorescence channels, as the ability to compute a reliable unit scaling all channels is a process control indicating correct instrument operation.

3.4 Bead-Based Particle Size Calibration

1. Compute a log-scale histogram of the forward scatter area (FSC-A) channel values for the size bead sample.
2. Identify the peaks in the channel and compute the geometric mean for each peak region, as shown in the example in Fig. 5a.
3. Exclude low-valued peaks.
4. Compute the unit scaling factor from FSC-A arbitrary units to equivalent micro-meter (Eμm) diameter by performing a linear fit on the log scale between peak values and manufacturer supplied peak diameter values, as shown in the example in Fig. 5b.

Explanatory notes and information on alternatives may be found in Subheading 4.7.

3.5 Cell Gating

1. Load events from the negative control (or null transfection if one was used per Subheading 4.2).
2. Compute a two-component n-dimensional Gaussian mixture model (GMM) fit jointly on all the forward and side scatter channels.
3. To gate for cells, select all events that are within two standard deviations of the mean of the GMM component with the smallest standard deviation and also not closer to the mean of any other component. These are candidate cell events.

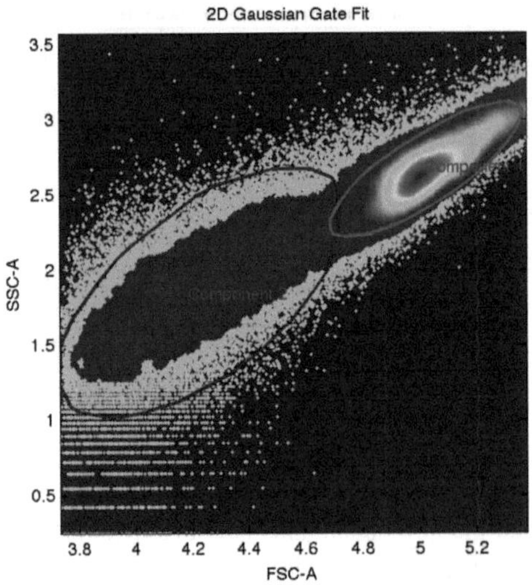

Fig. 6 Example of gate to select cell events via a Gaussian mixture model (GMM), in this case a two-component GMM built against forward and side scatter areas. The smaller, bright red, component selects events likely to be from cells. Figure adapted from [24]

4. Plot a two-dimensional density map of all events, with a convex hull around the selected events. Optionally, also plot the convex hulls of the events that would be selected for each other component. Figure 6 shows an example of selecting the cell component for a two-component GMM.
5. Compare selected component with size beads to determine is estimated diameter of cell-like events is consistent with experimental expectations.
6. If the distribution of events is such that cell-like events are not well identified by two standard deviations from the smaller standard deviation component of a two-component GMM, then select a different component, adjust the number of components, and/or adjust the number of standard deviations until a reasonable selection is achieved. Adjust components before adjusting the number of standard deviations, increasing the number incrementally until the best fit is identified.

Explanatory notes and information on alternatives may be found in Subheading 4.9.

3.6 Fluorescence Compensation

In order to quantify the fluorescence due to the presence of a fluorescent reporter, two other sources of fluorescence must be removed from its channel: cell autofluorescence (estimated from the negative control or null control) and spillover fluorescence from other fluorescent reporters due to spectral overlap (estimated from single-color controls). Together, this removal process is known as compensation and is handled via linear transformations. Here we

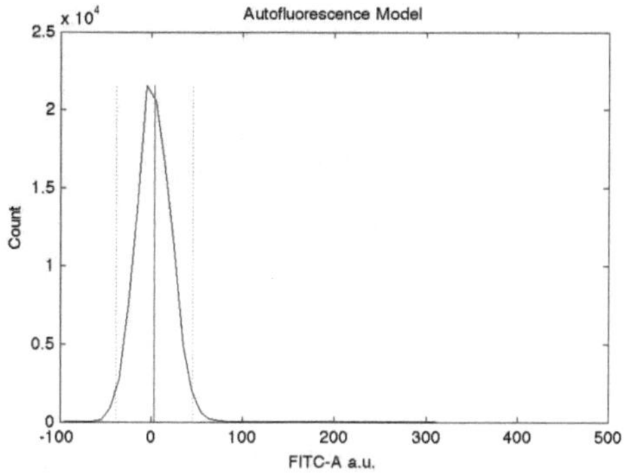

Fig. 7 Example of an autofluorescence model computed from a negative control (the solid red line shows mean, and dotted red lines show ± 2 std. dev.), plotted against a linear scale histogram of events from the sample. Note the use of a linear scale and arithmetic statistics rather than logarithmic in this case. Figure adapted from [24]

present a minimal protocol for compensation for this context; a more general presentation of compensation and the principles behind it may be found in [6, 7] (Fig. 7).

For each channel for a fluorescent reporter, compute an autofluorescence model for that channel:

1. Load events from the negative control (or null transfection if one was used per Subheading 4.2)
2. Gate events to retain only cell events, using the gate computed in Subheading 3.5.
3. Compute the arithmetic mean and standard deviation for all cell events in the sample.

For each channel for a fluorescent reporter, compute the spectral overlap model from that channel into other channels:

1. Load events from the single-color control for the channel.
2. Gate events to retain only cell events, using the gate computed in Subheading 3.5, and subtract autofluorescence means to convert to net fluorescence.
3. Compute a parametric analysis of all other fluorescence channels (passive channels) with respect to the expressed fluorescent reporter (driven channel).
4. For each passive channel, select the significant bins by selecting only those where the passive channel mean net value is above 2 std. dev. of channel autofluorescence.
5. If there is at least one significant bin, the spectral overlap is computed as a constrained linear fit (slope=1) for passive versus driven values for the significant bins. Otherwise, spectral overlap is zero. An example of computing spectral overlap is shown in Fig. 8a.

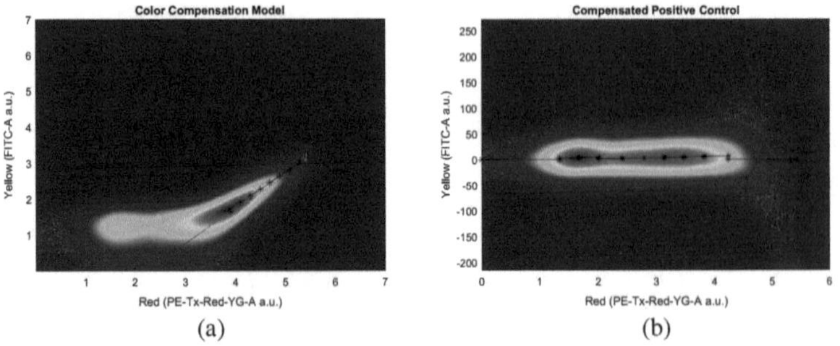

Fig. 8 (**a**) Example of calculating spectral overlap, in this case from the PE-Tx-Red-YG channel (driven) into the FITC-A channel (passive), finding a spillover rate of a little under 1%. Black dots are the parametric means of significant bins, and the red line is the constrained linear fit for spectral overlap. (**b**) Example of compensation verification plot using the compensated values for the same sample and channels. Black dots are means of selected bins, and black lines are visualizing bin levels. Figure adapted from [24]. (**a**) Single-Color Control analysis. (**b**) Compensated Single-Color Control

6. Compensation for spectral overlap is computed by dividing a vector of net fluorescence values by the spectral overlap matrix.
7. Verify correct compensation with a log-linear plot of compensated event values for each single-color control with respect to each driven channel, which should produce a balanced distribution centered on zero. An example is shown in Fig. 8b.

Given the autofluorescence and spectral overlap model, compensated fluorescence values are produced by first subtracting autofluorescence mean from each channel and then dividing by the spectral overlap matrix.

Explanatory notes and information on alternatives may be found in Subheading 4.10.

3.7 Fluorescent Channel Conversion

1. Load events from the multi-color control.
2. Gate events to retain only cell events, using the gate computed in Subheading 3.5, and compensate to net fluorescence using the compensation model computed in Subheading 3.6.
3. For each fluorescent channel C other than the Molecules of Equivalent FLuorescein (MEFL) channel, compute a parametric analysis of channel C and the MEFL channel with respect to a third channel.
4. Select the significant bins by selecting only those where the channel mean net value for both the C and MEFL channels is above 2 std. dev. of channel autofluorescence.
5. The unit conversion factor from C to MEFL is computed as a constrained linear fit (slope=1) for C mean net values versus MEFL mean net values for the significant bins. Figure 9 shows

Fig. 9 Example of multi-color control being used to build a channel unit conversion model. Colors show the density of data; black marks show parametric geometric mean of blue (Pacific Blue-A) and yellow (FITC-A) fluorescence with respect to red fluorescence (parametric axis not shown) for significant bins at 10 bins per decade. The red line shows the mean scaling factor for blue arbitrary units to yellow arbitrary units. Figure adapted from [24]

an example of computing a color translation model from the Pacific Blue channel to the FITC channel, parameterized by values on a Texas Red channel.

Explanatory notes and information on alternatives may be found in Subheading 4.13.

3.8 Experimental Sample Analysis

1. Load events for each experimental sample.
2. Gate events to retain only cell events, using the gate computed in Subheading 3.5.
3. Compensate to net fluorescence using the compensation model computed in Subheading 3.6.
4. Translate all fluorescence channels to the MEFL channel using the conversion factor computed in Subheading 3.7.
5. Convert to units of MEFL (fluorescence channels) and Eμm (FSC-A) using the conversion factors calculated in Subheadings 3.3 and 3.4, respectively.

Figure 10 shows an example of a sample's raw event distribution converted to a calibrated cell event distribution.

Once data have been gated and converted to MEFL and Eμm, the further analyses that will be of value will depend on the specifics of the experiment. The same histogram and parametric analyses as have been used above, however, are also typically useful for analyzing some common experimental properties of interest.

1. Distribution of transfection levels can be analyzed by fitting a Gaussian mixture model (GMM) to a log-scale histogram analysis of a constitutive fluorescent reporter. Figure 11a shows a typical transient transfection distribution from lipofection.

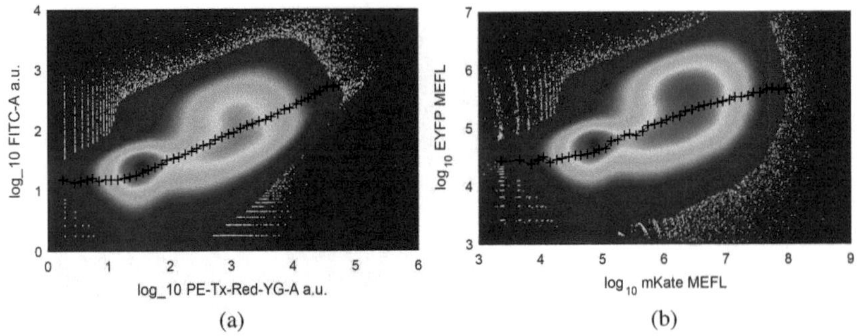

Fig. 10 Example of data from a flow cytometry file transformed by the full gating and calibration model (gating, compensation color translation, and ERF units). Colors show the density of data; black marks show parametric geometric mean of yellow fluorescence with respect to red fluorescence at 10 bins per decade. Note that the transformed data have an enriched relative density of strongly transfected cells, that these cells have a more balanced distribution, and that the two axes have the same MEFL units. Figure adapted from [24]. (**a**) Raw FCS data distribution. (**b**) Calibrated cell data distribution

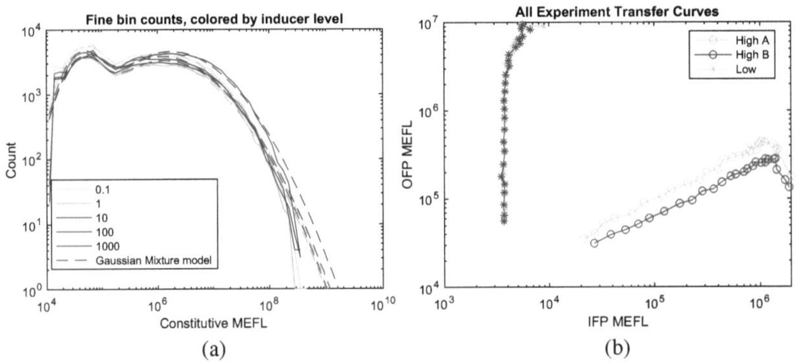

Fig. 11 Examples of log-scale statistical analysis, as applied to transient transfection samples: (**a**) log-scale histogram analysis of transfection marker expression, showing a bimodal distribution of transfection levels. (**b**) Parametric analysis of two test conditions of a repressor under high or low induction, comparing input fluorescence protein (IFP) expression levels versus output fluorescence protein (OFP) expression levels, binned with respect to the strongly transfected range of a constitutive fluorescent protein. Figure adapted from [24]. (**a**) Histogram Analysis. (**b**) Parametric Analysis

2. The behavior of transfected constructs can be analyzed by a parametric analysis of the experimental fluorescent reporters with respect to a constitutive fluorescent reporter. The parametric range should be limited to only strongly transfected cells. Figure 11b shows an example of comparing test conditions using parametric analysis with respect to a constitutive fluorescent reporter.

Explanatory notes and information on alternatives may be found in Subheading 4.14.

4 Notes

This section presents common issues that can occur in quantifying transient transfections in mammalian cells with flow cytometry and adjustments to the method that can be used to resolve these issues.

4.1 Notes on Reagents

For calibration and quality control on a flow cytometry study, two classes of calibrants are needed: calibrant reagents and cellular calibrants.

Calibrant reagents are materials with known optical properties that are used to compute a unit conversion factor between the arbitrary unit values output by a flow cytometer and the physical properties of size and molecule count. Failures in calibrant reagents are typically uncorrelated with failures in the preparation of the experimental samples to be assayed, so these calibrants serve as a process control testing for correct operation of the flow cytometer.

Cellular calibrants, on the other hand, are positive and negative controls that establish the expected dynamic range of cellular behaviors in each fluorescent channel and the relationships between measurements in different channels. Failures in cellular calibrants are typically correlated with failures in the preparation of experimental samples, so these calibrants serve as a process control testing for correct execution of the preparation of experimental samples.

4.1.1 Calibrant Reagents

- **SpheroTech Ultra Rainbow Quantitative Particle Kit (URQP-38-6K)**
  - This kit provides beads with six different intensities of fluorescence and NIST-certified Equivalent Reference Fluorophore (ERF) value that map each intensity to an equivalent number of fluorescent molecules on four channels: fluorescein, Nile Red, APC, and Coumarin 30. Calibrant should be stored, handled, and dispensed according to the manufacturer directions. One kit provides sufficient materials for many assays.
  - These beads are used to compute a linear unit conversion factor that maps fluorescence channel values to molecules of equivalent fluorophore. The specific units depend on the channel, e.g., Molecules of Equivalent FLuorescein (MEFL).
  - Alternate beads may be substituted but must have multiple separable intensities and certified ERF values, one set of which must be for fluorescein.
- **SpheroTech Polystyrene Particle Size Standard Kit (PPS-6K)**
  - This kit provides beads with six different diameters. Calibrant should be stored, handled, and dispensed according to the manufacturer directions. One kit provides sufficient materials for many assays.

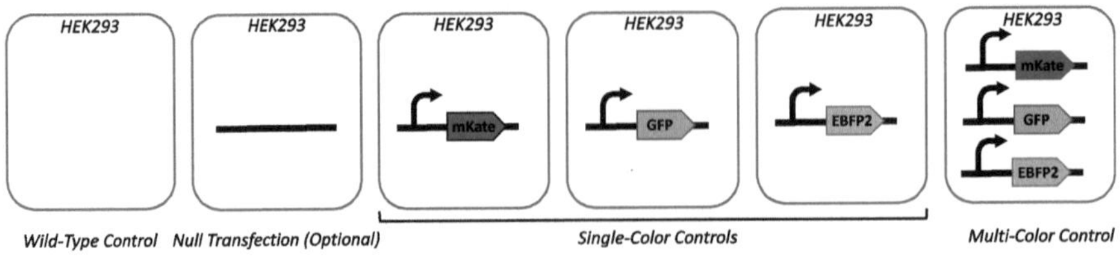

Fig. 12 Example genetic constructs for cellular controls, for an experiment with red and green fluorescence, adding blue fluorescence as a transfection marker, plus the optional null transfection control (*see* Subheading 4.2)

- These beads are used to compute a nonlinear unit conversion function that maps forward scatter area (FSC-A) values to estimated diameter in units of equivalent micron particle diameter (Eμm).
- Alternate beads may be substituted but must be monodisperse and must have multiple separable sizes.
- Certified size beads are not currently available but should be preferred if they become available in the future.

4.1.2 Cellular Calibrants

The set of cellular calibrants to be used is shown in Fig. 12.

- **Wild-Type Negative Control**
  - For each strain of cell used, at least one replicate with a sample of wild-type cells, i.e., not transformed with any construct. For example, if CHO and HEK293 cells are included in the experiment, there should be at least one sample of wild-type CHO and at least one sample of wild-type HEK293.
  - The wild-type control is used to estimate autofluorescence and to assess strain response to culturing and/or other sample preparation activities.
  - The wild-type control should be cultured and/or otherwise prepared alongside the experimental samples, but not given any other experimental treatment.
- **Single-Color Controls**
  - For each fluorescent reporter used, at least one replicate transiently transfected with a construct for strong constitutive expression of the fluorescent reporter (and no other genetic function in its design).
  - Transient transfection is preferable to integration, as it is both faster and produces a wider range of expression levels, including some that are much stronger.
  - The single-color controls are used to estimate spectral overlap between channels and to assess strain response to culturing.

- Fluorescent reporters should be selected such that no pair of channels has more than a 10% spectral overlap. This typically allows three fluorescent reporters (red/green/blue) or four if deep red/infrared fluorescence is used.
- The single-color controls should be cultured and/or otherwise prepared alongside the experimental samples, but not given any other experimental treatment.

- **Multi-color Control**
  - At least one replicate transiently co-transfected with equal doses of single-color control constructs for all of the fluorescent reporters used, where the constructs differ only in their fluorescent reporter (i.e., the vector and regulatory regions are identical) and have identical dosages in the co-transfection.
  - The multi-color control is used to estimate unit conversions from fluorescent channels, such that the expression levels of different fluorescent reporters can be directly compared in the same molecular units. If only reference fluorophores are used, then ERF values can be compared directly, but reference fluorophores are typically small-molecule dyes rather than biologically expressed reporter constructs. With biologically expressed reporters, ERF values for different channels are generally incommensurate due to the differences in relationship between fluorescent reporter and reference fluorophore. For example, the relationship between fluorescein and GFPmut3 in the FITC-A channel is expected to be different from the relationship between Coumarin 30 and EBFP2 in the Pacific Blue-A channel.
  - The recommended unit to convert to is Molecules of Equivalent FLuorescein (MEFL), as fluorescein channels are common and often have similar spectral properties and as fluorescein has similar spectral properties to those of many green fluorescent reporters.
  - It is preferable that the multi-color control has at least three colors, as this will allow better segmentation into subpopulations for analysis.
  - The multi-color control should be cultured and/or otherwise prepared alongside the experimental samples, but not given any other experimental treatment.
  - This control may be omitted if the only experimental fluorescent reporter is measured in the fluorescein channel (e.g., GFP and YFP). If there is no experimental reporter measured in the fluorescein channel, then a strong constitutive expression construct for such a reporter should be added to the multi-color control and also as an additional single-color control.

- This control depends on the relatively low sensitivity of mammalian promoter transcription rates to the contents of the coding sequences being transcribed, such that the transcriptional activity of each co-transfected construct can be expected to be equivalent. This control also depends on co-transfection introducing a large number of plasmids to each cell (typical of many protocols for mammalian transient transfection, e.g., lipofection), which means that strongly transfected cells should have a ratio between plasmids tightly distributed around 1:1. These assumptions often do not hold for other types of cells or transformations.

4.1.3 Experimental Samples

- Experimental samples should be prepared according to the experimenter's study design.
- The single-color control for a non-conflicting fluorescent reporter should be co-transfected along with the experimental construct or constructs. During analysis, this fluorescence channel will be used as a transfection marker, enabling parametric analysis of behavior with respect to transfection copy count.
- Experimental samples should be cultured along with the cellular calibrants and should be measured at the same time as both the calibrant reagents and cellular calibrants, in order to maximize the likelihood that issues with the protocol or instrument can be detected. Note that single-color and multi-color controls may simultaneously server as experimental samples, if appropriate for the experiment design.

4.2 Null Transfection Control

Some strains of cell respond to some transfection protocols or some vector backbones with large increases in autofluorescence. This is most often in red channels and often indicates that cell viability is degraded. If this is suspected to be the case, then a null transfection control should be added, in which cells are transfected with a vector that contains none of the constructs under study. This construct will then be used for estimating autofluorescence instead of the wild-type, but wild-type will still be used for assessing strain response to culturing.

4.3 Data Collection

- The minimum size threshold trigger is needed because flow cytometers typically have limits in the rate at which events can be recorded, and if the instrument's recording capability is saturated, then many cell measurements will be lost.
- Forward scatter and size scatter will be used to gate out non-cell particles and multi-cell clumps, and thus the height and width channels, if available, are useful in addition to the area channels.
- In fluorescent channels only area is needed, as that is the channel that will be used to estimate equivalent molecule counts.

- In a transient transfection, cells generally receive a widely varied number of copies of the plasmid or a co-transfected set of plasmids. Collecting a larger number of events allows a parametric analysis in which events are divided into subpopulations by transfection marker (*see* Subheading 3.2).
- If the subpopulation analysis is not needed (e.g., for integrated constructs), the number of events can be reduced to 10,000 to 30,000.

4.4 Small Fraction of Cell-Like Particles

If the number of cell-like particles making it through gating is small (under 50%), this may indicate problems with data collection. This may be diagnosed with a plot of forward versus side scatter for the wild-type negative control.

- If the distribution of cell-like particles is truncated against a low threshold, the instrument's forward scatter trigger for event capture likely needs to be lowered.
- If there are smaller particles than cell-like particles, the instrument's forward scatter trigger for event capture likely needs to be raised.
- If shifting triggers does not correct the problem, then the flow rate may need to be lowered in order to decrease the rate at which events occur.

4.5 Alternative Statistics Analyses

Many flow cytometry analyses recommend the use of median or mode values instead of geometric means. The median often gives values that are very similar to the geometric mean, but it does not have a biological distribution theory supporting its use. As a result, although other percentiles can provide some notion of deviation, there is no principled equivalent of geometric standard deviation of multi-modal log-normal fits to guide the interpretation of median-based analyses.

Mode-based analysis is typically presented as finding a peak or maximum in a histogram. While this works well in some cases, biological variability and the inherent quantization associated with binning to produce a histogram mean that there is often significant uncertainty in the location of the mode, and the value can be strongly affected by the details of how a histogram is computed. The same qualitative goals can thus often be better satisfied by use of the multi-modal geometric statistics.

Finally, it is useful to note that log-scale statistics are ill-defined for non-positive numbers and thus do not provide correct results for small values, where error tends to be dominated by arithmetic contributions from instrument noise or fluorescence compensation. This is often addressed in visualization by using a bi-exponential or "logicle" remapping [25], which provides a clean visual transition between regions of geometric and arithmetic error dominance, but which involves an arbitrary cutoff and does not shed light on the underlying question of biological activity. Better methods for incorporation of low and negative flow cytometry values into analysis remains an open question.

4.6 Bead-Based Fluorescence Channel Calibration

- Channel range (e.g., voltage) should be set based on the cells to be measured, not the beads. It is acceptable for some beads to be saturated high, lost to low noise, or otherwise not have usable peaks. For purposes of fit quality, however, it is preferable to have at least three usable peaks.
- Low-valued peaks, typically below around 10^2 arbitrary units, will often be blurred due to instrument noise or other issues. It is better to exclude these peaks and use only the better-defined peaks at higher values.
- Bead doublets will often form a small peak with a value twice that of the primary peak for each expected bead intensity. Sometimes, a third triplet peak with a three-times value will also be distinguishable. Data from these doublet and triplet peaks should be excluded from peak calculations.
- Mean fit error should be under 10%. Higher error typically indicates that peaks have not been correctly identified: *see* Subheading 4.8 for likely issues.
- Remember that the ERF values are not actually the number of molecules of the fluorescent reporter, but the amount of fluorescence that would be produced by that number of molecules of the reference fluorophore. With good spectral matching between the fluorescent reporter, reference fluorophore, and measurement channel, however, the impact of this difference is likely to be less than the magnitude of error from other sources [9]. Fluorescent reporter molecules also may not properly mature to give full fluorescence due to biological issues such as incomplete folding, lack of oxygen, pH range, etc.

4.7 Bead-Based Particle Size Calibration

- Channel range (e.g., voltage) should be set based on the cells to be measured, not the beads. Due to the smaller range of typical cell sizes, however, it is typically possible to set FSC-A range to be appropriate for cell measurement and also to keep all beads in a range where peaks can be clearly distinguished.
- If the channel cannot be adjusted to make all bead peaks distinguishable, it is acceptable for some beads to be saturated high, lost to low noise, or otherwise not have usable peaks. For purposes of fit quality, however, it is preferable to have at least three usable peaks.
- Likewise, if there are low-valued peaks, typically below around 10^2 arbitrary units, that are blurred due to instrument noise or other issues. It is better to exclude these peaks and use only the better-defined peaks at higher values.

- At present there is generally a higher degree of variability in size beads than rainbow beads due to differences in the available calibrant materials. Doublet and triplet peaks are typically less of an issue with size beads, however, due to the lower amount of separation between peaks.
- Unlike with color beads, this conversion should not have a unit slope, as there is a nonlinear relationship between diameter and forward scatter [26, 27]. Unit transformation will thus be a nonlinear function.
- Mean fit error should be under 10%. Higher error typically indicates that peaks have not been correctly identified: *see* Subheading 4.8 for likely issues.

4.8 Bead Peak Problems

In addition to the notes already presented in Subheadings 3.3 and 3.4, other common issues in identifying bead peaks are small counts, high background, blurred peaks, and offset peaks.

- If the number of events per peak is small, on the order of 200 or less per peak, then the precision of unit conversion is likely to be poor. This issue also typically indicates problems with the FSC threshold for event capture.
  - If the total number of events in the bead file is low, this typically indicates that the event capture threshold is too high.
  - If the total number of events is as expected, but most are low values not forming clean peaks, this typically indicates that the event capture threshold is too low.
- High levels of background events between peaks can make peaks difficult to identify or separate. If this occurs for values higher than around the 10^2 arbitrary units, this typically indicates an instrument that is dirty or has other service issues that are creating the high rates of background events.
- If there are peaks, but they do not generally have a single highly distinct maximum (e.g., wide peaks or doubled peaks), then this generally indicates an instrument hardware issue that requires service, such as lasers having come out of alignment.
- If there are clear peaks, but they do not produce a good linear fit, this typically indicates that the peak identities have been offset from their true identities, e.g., peaks 4–6 have been mis-labeled as peaks 2–4.

4.9 Cell Gating

- Forward scatter area (FSC-A) and side scatter area (SSC-A) channels are often sufficient for gating.
- Including height and width channels can help further refine cell selection, particularly with respect to excluding doublet events.

4.10 Fluorescence Compensation

- The autofluorescence mean will be used for compensation, subtracted from each event to convert total fluorescence to net fluorescence.
- The autofluorescence standard deviation will be used to determine where values cannot be distinguished from autofluorescence. Any value less than 2 std. dev. net fluorescence should be considered equivalent to autofluorescence. Truncating such values (e.g., to a minimum net fluorescence of 1) will also remove non-positive net fluorescence values, which are problematic for log-scale statistics.
- Linear statistics are used with autofluorescence because the instrument measurement noise is typically expected to dominate the variability of autofluorescence measurements. If this is not the case, it may be necessary to use null transfections and/or to switch to geometric statistics.
- For single-color controls, the parametric function for each channel should either be consistently near zero (no significant overlap) or a "dog-leg" that is near zero for low values but then switches to a tight linear relation with unit slope at the point where spectral overlap begins to exceed autofluorescence.
- For good resolution on low values, spectral overlap should be less than 10%.

4.11 Compensation Problems

- If spectral overlap is more than 10%, the level of effective overlap can be reduced by adjusting the choice of fluorescent reporter and/or channel voltage.
- Expression of the single-color control is too weak if it could not be used to identify at least a 1% spectral overlap. Typically this means that there should be significant numbers of events through at least the 10^4 to 10^5 arbitrary unit range. Weak expression can be caused by a number of issues:
  - The selected channel may not be a close enough match for the excitation and emission properties of the fluorescent reporter, such that only a small fraction of the potential fluorescent energy is being measured, in which case either the channel or the reporter should be changed.
  - The fluorescent reporter may have poor brightness or there may be stronger than usual autofluorescence in the selected channel, in which case the reporter should be changed.
  - The fluorescent reporter may be a poor choice for the protocol (e.g., mismatched chemical properties), in which case the reporter should be changed.
  - If none of these is the case, then it is an indicator that there are problems with the reporter construct (e.g., poor

promoter strength and interference between promoter and reporter) and/or the experimental protocol.

- If compensation verification shows a failure, i.e., the center of the distribution does not stay roughly centered on zero, this is typically caused by problems with one or more of the single-color controls or their analysis.
  - Weak expression can cause compensation failures and should be addressed as discussed above.
  - If the control has a "messy" distribution instead of the expected pattern of autofluorescence followed by a tight linear unit slope, then it typically indicates problems in either the reporter construct, experimental protocol, or cell gating.
  - Instrument problems can also cause compensation failures, but it is highly unlikely for this to be the case without the problem already being identified during peak identification for the color beads.

4.12 Channel Conversion Problems

- If there are no significant bins or there is a poor fit (more than 10% error), first check the single-color controls to see if their expression is weak or "messy," per Subheading 4.11.
- If all of the single-color controls are strong and have clean distributions, then it is likely that there is a problem with the co-transfection protocol or the specifics of the multi-color control.

4.13 Fluorescent Channel Conversion

- If there are only two channels, compute a parametric analysis of the second channel against the MEFL channel.
- If there is no MEFL channel, another ERF-calibrated channel can be substituted, but this is not desirable because it will make it more difficult to compare data with data produced by experiments using MEFL.
- If there are no significant bins or the mean fit error is more than 10%, this indicates a problem with the multi-color control. *See* Subheading 4.12 for likely issues.

4.14 Experimental Sample Analysis

- Some of the cellular process controls are likely to also be used as experimental controls.
- There are typically two major components in a transient transfection distribution: strongly transfected cells and transfection failures. A strong lipofection should typically have more than 50% of the cells in the strongly transfected component.
- If less than 30% of cell events are in the strongly transfected component, this usually indicates a failure of the transfection protocol.

- Biologically, it is not clear whether transfection failure means no transfection or just few copies. The transfection failure component should typically be excluded from experimental analysis in any case, as the behavior of cells in the transfection failure component is generally qualitatively different from those in the strongly transfected component.
- In some cases, the fit for a transfection level distribution is better when a third component is included in the middle, though it is unclear whether this represents a true third population or if the better fit is due to a distribution asymmetry, e.g., due to cellular resource limitations.
- A significant drop in the strongly transfected component of a transfection distribution from run to run or from sample to sample is often an indication of problems in the protocol execution that will render the data from the less-well-transfected samples unreliable.
- With transient transfection, parametric analysis generally provides more insight than bulk statistical analysis because of the high variability in transfection level.

4.15 Transfection Problems

Figure 13 shows examples of experimental samples failing quality control due to differences between expected and observed transfection distributions. In Fig. 13a, the lowest three samples have significantly lower strongly transfected components than the rest, which are tightly bunched. Those three samples were thus considered to be likely protocol failures and excluded, even though they have more than 50% of their events in the strongly transfected component. In Fig. 13b, the entire collection of samples is showing a strongly transfected component containing less than 30% of the cell events, and thus the entire experimental run was considered likely to have been affected by a protocol failure and was rerun.

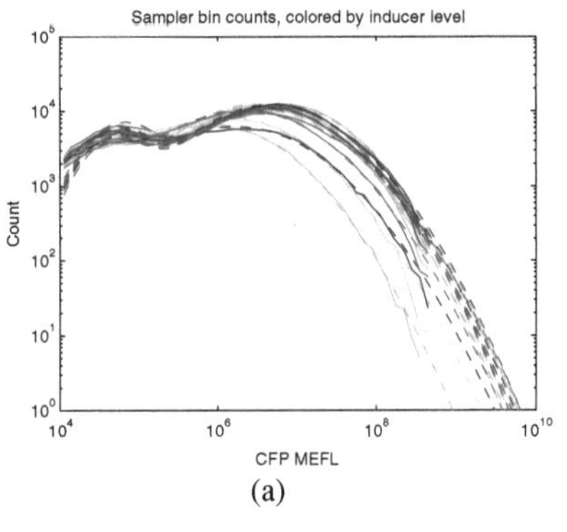

Fig. 13 Example of distribution expectations allowing identification of individual bad samples (**a**) and a full batch of samples gone wrong from a badly executed protocol (**b**). (**a**) Problematic Samples. (**b**) Problematic Experiment

Acknowledgements

This chapter does not contain technology or technical data controlled under either the U.S. International Traffic in Arms Regulations or the U.S. Export Administration Regulations.

References

1. Fulwyler MJ (1965) Electronic separation of biological cells by volume. Science 150(3698):910–911
2. Kamentsky LA (1973) Cytology automation. Adv Biol Med Phys 14:93–161
3. Beal J, Baldwin GS, Farny NG, Gershater M, Haddock-Angelli T, Buckley-Taylor R, Dwijayanti A, Kiga D, Lizarazo M, Marken J, et al (2021) Comparative analysis of three studies measuring fluorescence from engineered bacterial genetic constructs. PloS One 16(6): e0252263
4. Beal J, Farny NG, Haddock-Angelli T, Selvarajah V, Baldwin GS, Buckley-Taylor R, Gershater M, Kiga D, Marken J, Sanchania V, et al (2020) Robust estimation of bacterial cell count from optical density. Commun Biol 3(1): 512
5. Barteneva NS, Fasler-Kan E, Vorobjev IA (2012) Imaging flow cytometry: coping with heterogeneity in biological systems. J Histochem Cytochem 60(10):723–733
6. Roederer M (2001) Spectral compensation for flow cytometry: visualization artifacts, limitations, and caveats. Cytometry 45(3):194–205
7. Roederer M (2002) Compensation in flow cytometry. Curr Protoc Cytom 22(1):1–14
8. Gaigalas AK, Wang L, Schwartz A, Marti GE, Vogt Jr RF (2005) Quantitating fluorescence intensity from fluorophore: assignment of MESF values. J Res Natl Inst Stand Technol 110(2):101
9. Beal J, Teague B, Sexton JT, Castillo-Hair S, DeLateur NA, Samineni M, Tabor JJ, Weiss R, Consortium CFCS (2022) Meeting measurement precision requirements for effective engineering of genetic regulatory networks. ACS Synth Biol 11(3):1196–1207
10. Beal J, Weiss R, Yaman F, Davidsohn N, Adler A (2012) A method for fast, high-precision characterization of synthetic biology devices. Tech. Rep. MIT-CSAIL-TR-2012-008, MIT CSAIL. http://hdl.handle.net/1721.1/69973
11. Davidsohn N, Beal J, Kiani S, Adler A, Yaman F, Li Y, Xie Z, Weiss R (2014) Accurate predictions of genetic circuit behavior from part characterization and modular composition. ACS Synth Biol 4(6):673–681. https://doi.org/10.1021/sb500263b
12. Kiani S, Beal J, Ebrahimkhani MR, Huh J, Hall RN, Xie Z, Li Y, Weiss R (2014) Crispr transcriptional repression devices and layered circuits in mammalian cells. Nat Methods 11(7): 723–726
13. Kiani S, Chavez A, Tuttle M, Hall RN, Chari R, Ter-Ovanesyan D, Qian J, Pruitt BW, Beal J, Vora S, et al (2015) Cas9 gRNA engineering for genome editing, activation and repression. Nat Methods 12(11):1051–1054
14. Beal J, Wagner TE, Kitada T, Azizgolshani O, Parker JM, Densmore D, Weiss R (2015) Model-driven engineering of gene expression from RNA replicons. ACS Synth Biol 4(1): 48–56
15. Wagner TE, Becraft JR, Bodner K, Teague B, Zhang X, Woo A, Porter E, Alburquerque B, Dobosh B, Andries O, et al (2018) Small-molecule-based regulation of RNA-delivered circuits in mammalian cells. Nat Chem Biol 14(11):1043–1050
16. Siciliano V, DiAndreth B, Monel B, Beal J, Huh J, Clayton KL, Wroblewska L, McKeon A, Walker BD, Weiss R (2018) Engineering modular intracellular protein sensor-actuator devices. Nat Commun 9(1):1881
17. Weinberg BH, Cho JH, Agarwal Y, Pham NH, Caraballo LD, Walkosz M, Ortega C, Trexler M, Tague N, Law B, et al (2019) High-performance chemical-and light-inducible recombinases in mammalian cells and mice. Nat Commun 10(1):4845
18. Kiwimagi KA, Letendre JH, Weinberg BH, Wang J, Chen M, Watanabe L, Myers CJ, Beal J, Wong WW, Weiss R (2021) Quantitative characterization of recombinase-based digitizer circuits enables predictable amplification of biological signals. Commun Biol 4(1):875
19. Beal J, Overney C, Adler A, Yaman F, Tiberio L, Samineni M (2019) TASBE flow analytics: a package for calibrated flow cytometry analysis. ACS Synth Biol 8(7):1524–1529
20. Castillo-Hair SM, Sexton JT, Landry BP, Olson EJ, Igoshin OA, Tabor JJ (2016)

FlowCal: a user-friendly, open source software tool for automatically converting flow cytometry data from arbitrary to calibrated units. ACS Synth Biol 5(7):774–780

21. Teague B (2022) Cytoflow: A python toolbox for flow cytometry. bioRxiv pp 2022–07

22. Beal J, Teague B, Sexton JT, Castillo-Hair S, DeLateur NA, Samineni M, Tabor JJ, Weiss R, the Calibrated Flow Cytometry Study Consortium (2021) Meeting measurement precision requirements for effective engineering of genetic regulatory networks. ACS Synth Biol 11(3):1196–1207. bioRxiv https://doi.org/10.1101/2021.10.10.460840. https://www.biorxiv.org/content/early/2021/10/10/2021.10.10.460840. https://www.biorxiv.org/content/early/2021/10/10/2021.10.10.460840.full.pdf

23. Beal J (2017) Biochemical complexity drives log-normal variation in genetic expression. Eng Biol 1(1):55–60

24. Overney C, Beal J (Retrieved September 18, 2020) TASBE flow analytics tutorial. https://github.com/TASBE/TASBEFlowAnalytics-Tutorial

25. Tung JW, Heydari K, Tirouvanziam R, Sahaf B, Parks DR, Herzenberg LA, Herzenberg LA (2007) Modern flow cytometry: a practical approach. Clin Lab Med 27(3):453–468

26. Koch AL, Robertson BR, Button DK (1996) Deduction of the cell volume and mass from forward scatter intensity of bacteria analyzed by flow cytometry. J Microbiol Methods 27:49–61

27. Chandler WL, Yeung W, Tait JF (2011) A new microparticle size calibration standard for use in measuring smaller microparticles using a new flow cytometer. J. Thromb Haemost 9: 1216–24

Chapter 12

RNA Switches Using Cas Proteins

Moe Hirosawa and Hirohide Saito

Abstract

Expanding the number of available RNA-binding proteins (RBPs) is vital to establishing posttranscriptional circuits in mammalian cells. We focused on CRISPR-Cas systems and exploited Cas proteins for their versatility as RBPs. The translation of genes encoded in an mRNA becomes regulatable by a Cas protein by inserting a crRNA/sgRNA sequence recognizable by the specific Cas protein into its 5′UTR. These Cas protein-responsive switches vastly expand the available tools in synthetic biology because of the wide range of Cas protein orthologs that can be used as trigger proteins.

Here, we describe the design principle of Cas protein-responsive switches, both plasmid and RNA versions, using *Streptococcus pyogenes* Cas9 (SpCas9) as an example and show an example of its use in mammalian cells, HEK293FT cells.

Keywords CRISPR–Cas system, RBP, Aptamer, Protein-responsive switch, Translational regulation

Abbreviations

crRNA	CRISPR RNA
mRNA	messenger RNA
RBPs	RNA-binding proteins
sgRNA	single guide RNA
SpCas9	*Streptococcus pyogenes* Cas9
tracrRNA	trans-activating crRNA
UTR	untranslated region

1 Introduction

Protein-responsive mRNA switches, which contain a specific protein-binding sequence (aptamer) in the 5′UTR, regulate the translation of its encoded gene in response to a trigger protein [1, 2]. In the absence of the trigger protein, the typical mRNA switch is in the ON state by default and expresses the encoded gene.

Francesca Ceroni and Karen Polizzi (eds.), *Mammalian Synthetic Systems*, Methods in Molecular Biology, vol. 2774, https://doi.org/10.1007/978-1-0716-3718-0_12,

On the other hand, when the trigger protein is present, it binds to the aptamer in the 5′UTR and inhibits translation, hence switching to the OFF state. Such switches can potentially build complex genetic circuits because RBPs can function as both input and output [3, 4]. However, the limited number of RBPs available to control translation has hindered the implementation of such post-transcriptional circuits in mammalian cells.

We focused on Cas proteins as an ideal candidate for constructing protein-responsive mRNA switches. Cas proteins interact with their corresponding crRNA or sgRNA, a chimera of a crRNA and a tracrRNA [5]. Hence, Cas proteins can be used as translational regulators when their corresponding crRNA/sgRNA sequences are inserted into the 5′UTR of mRNAs (Fig. 1a). Many Cas proteins have been characterized to date [6], and indeed we demonstrated their high orthogonality in regulating target mRNA translation [7]. Thus, by employing Cas proteins, we have expanded the diversity of RBPs capable of functioning as the translational regulator of protein-responsive mRNA switches for multiplex applications. In addition, such switches can be combined with existing CRISPR-Cas technologies from the genome editing field to facilitate the construction of an assortment of complex synthetic biological circuits.

In this chapter, we detail the construction of Cas protein-responsive switches (plasmid and mRNA versions) and illustrate their application in mammalian cells.

2 Materials

2.1 Plasmid Construction

1. pAptamerCassette-EGFP plasmid (available from Addgene, #140288).
2. Oligo DNA (*see* Table 1).
3. T4 polynucleotide kinase (TAKARA).
4. 10× kinase buffer (TAKARA).
5. 75 mM ATP (Ambion, supplied with MEGAscript T7 Transcription Kit) or equivalent products.
6. Annealing buffer (20 mM Tris–HCl, 5 mM NaCl).
7. Ligation High ver.2 (TOYOBO).
8. rAPid alkaline phosphatase (Roche).
9. AgeI and BamHI.
10. KOD -Plus- Neo (TOYOBO).
11. MinElute PCR Purification Kit (QIAGEN) or Monarch PCR & DNA Cleanup Kit (NEB).
12. Midiprep kit (QIAGEN or Promega).

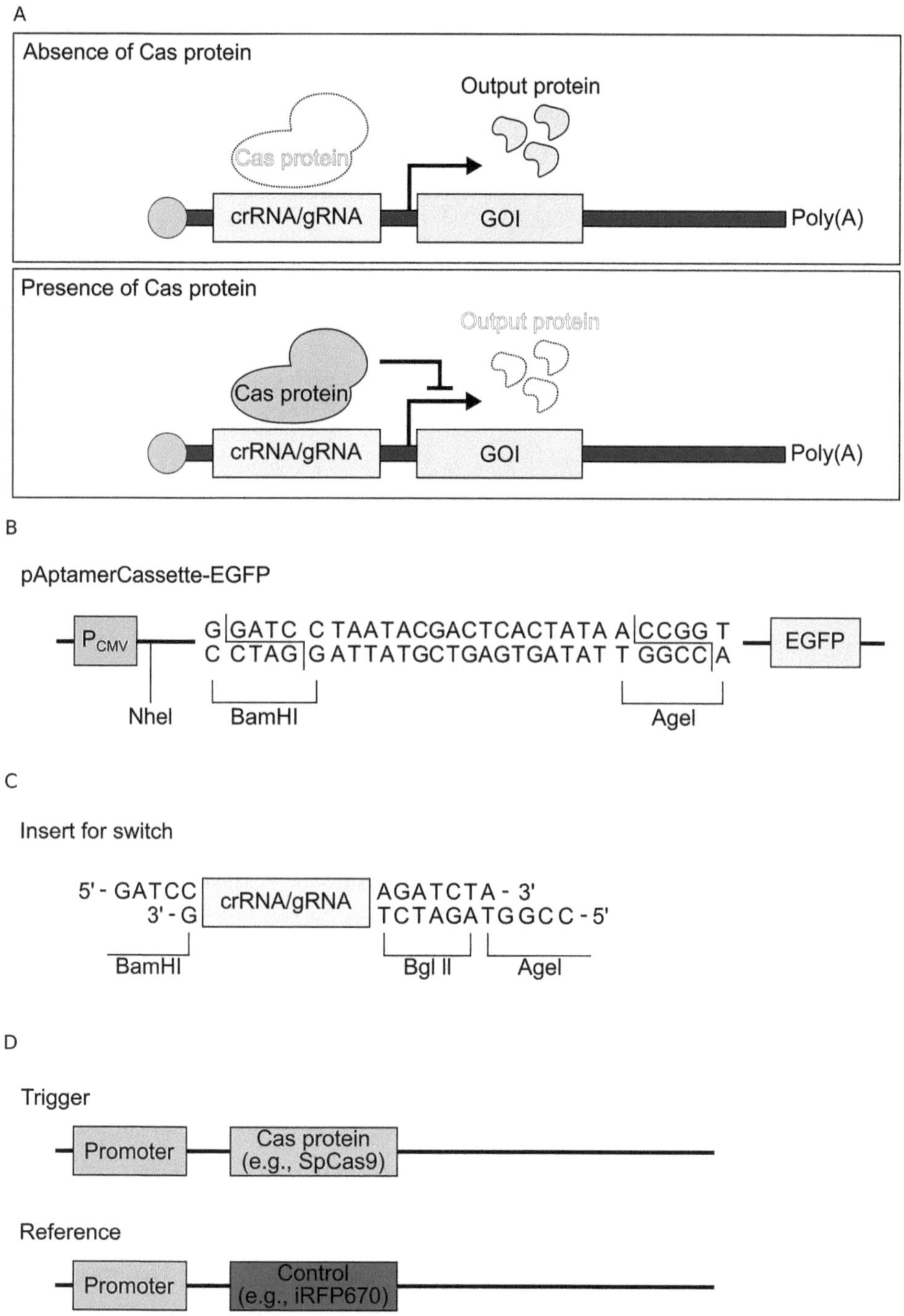

Fig. 1 Overview of the Cas protein-responsive switch. (**A**) Mechanism of the Cas protein-responsive switch. This switch contains a crRNA/gRNA sequence for interaction with the corresponding Cas protein in its 5′UTR. In the absence of Cas protein, output protein is expressed from the switch (top). On the other hand, in the presence of Cas protein, output protein is not expressed from the switch (bottom). (**B**) Configuration around the 5′UTR of pAptamerCassette-EGFP. (**C**) Annealing the product derived from two synthetic oligo DNAs. BamHI and AgeI are used to insert crRNA/gRNA sequences into the pAptamerCassette-EGFP plasmid. (**D**) Configuration of the trigger and reference plasmids

Table 1
Oligo DNAs for switch plasmid

No.	Oligo DNA name	Sequence (5′ -> 3′)
1	BamH1-Gluc-Sp_gRNA-Bgl2-Age1_code	GATCCGAGATCAGGGCAAACAGAACTGTTTTAGAGCTAGAAATAGCAagttaaaataaggctagtccgttatcaacttgaaaaagtggcaccgagtcggtgcAGATCTA
2	BamH1-Gluc-Sp_gRNA-Bgl2-Age1_temp	CCGGTAGATCTgcaccgactcggtgccactttttcaagttgataacggactagccttattttaactTGCTATTTCTAGCTCTAAAACAGTTCTGTTTGCCCTGATCTCG

Underline: Sp_gRNA backbone
GAGATCAGGGCAAACAGAAC: spacer sequence for targeting an exogenous gene, Gaussia luciferase

2.2 RNA Template and RNA Construction

1. SpCas9 coding plasmid (e.g., Addgene, plasmid #60599).
2. Primer and oligo DNA (*see* Table 2).
3. KOD -Plus- Neo (TOYOBO).
4. MinElute PCR Purification Kit (QIAGEN) or Monarch PCR & DNA Cleanup Kit (NEB).
5. Dpn I (TOYOBO).
6. MEGAscript T7 Transcription Kit (Ambion).
7. Pseudouridine-5′-triphosphate (TriLink BioTechnologies).
8. 5-Methylcytidine-5′-triphosphate (TriLink BioTechnologies).
9. ARCA (TriLink BioTechnologies).
10. RNeasy MinElute Cleanup Kit (QIAGEN) or Monarch RNA Cleanup Kit (NEB).
11. Antarctic phosphatase (NEB).
12. MultiNA (SHIMADZU).

2.3 Cell Culture, Plasmid/RNA Transfection, and Assay

1. SpCas9 expression plasmid (e.g., pcDNA3.1-SpCas9 [7] and Addgene, plasmid #60599, etc.).
2. Reference plasmid (e.g., pCMV-tdiRFP670 [8])
3. pcDNA3.1-myc-HisA plasmid (Invitrogen).
4. Cultured cell lines. In this chapter, we use HEK293FT cells (Invitrogen, #R70007).
5. DMEM high glucose supplemented with 10% FBS.
6. L-Glutamine (Invitrogen).
7. Nonessential amino acids (Invitrogen).
8. Sodium pyruvate (Sigma).
9. PBS.
10. 0.25% trypsin–EDTA (Thermo Fisher Scientific).

Table 2
Primer and oligo DNAs for RNAs

No.	Primer/Oligo DNA name	Sequence (5′ -> 3′)	Note
1	TAP_T7_G3C fwd primer	CAGTGAATTGTAATACGACTCACTATAGGGC	5′-UTR PCR and Fusion PCR (control)
2	IVT_5prime_ UTR primer	CAGTGAATTGTAATACGACTCACTATAGGGC GAATTAAGAGAGAAAAGAAGAGTAAGAAG AAATATAAGACACCGGTCGCCACCATG	5′-UTR PCR
3	Rev5UTR primer	CATGGTGGCGACCGGTGTCTTATATTTCTT CTTACTC	
4	Fwd3UTR primer	TCTAGACCTTCTGCGGGGC	3′-UTR PCR
5	IVT_3prime_ UTR primer	TCTAGACCTTCTGCGGGGCTTGCCTTCTG GCCATGCCCTTCTTCTCTCCCTTGCACC TGTACCTCTTGGTCTTTGAATAAAGCCTG AGTAGG	
6	Rev3UTR2T20	TTTTTTTTTTTTTTTTTTTTCCTACTCAGGC TTTATTCAAAGACCAAG	
7	SphcCas9 ORF fwd primer	CACCGGTCGCCACCATGGATAAGAAATAC AGCATTGGAC	Cas9 ORF
8	SphcCas9 ORF rev primer	GCCCCGCAGAAGGTCTAGACTATCACACC TTCCTCTTCTTCTTGG	
9	YF771_T7_ 5UTR_fwd	ATTGTAATACGACTCACTATAGGGCGAAT TAAGAGAGAAAAGAAGAGTAAG	iRFP670 Template and EGFP Template
10	KWC00425_T7-5UTR for cassette1	TAATACGACTCACTATAggTCAGATCCGCT AGCGGATCC	Gluc-Sp_gRNA-EGFP ORF
11	KWC00427_ EGFP ORF_Rv	GCCCCGCAGAAGGTCTAGACTACTTGTA CAGCTCGTCCATGCCGAGAG	
12	3UTR120A	TTTTTTTTTTTTTTTTTTTTTTTTTTTTTT TTTTTTTTTTTTTTTTTTTTTTTTTTTT TTTTTTTTTTTTTTTTTTTTTTTTTTTT TTTTTTTTTTTTTTTTTTTTTTTTTTTTT TTTTTCCTACTCAGGCTTTATTCA	Fusion PCR Rev and iRFP670 Template and EGFP Template

11. Opti-MEM I Reduced-Serum Medium (Thermo Fisher Scientific).
12. Lipofectamine 2000 Transfection Reagent (Thermo Fisher Scientific) for DNA transfection.
13. Lipofectamine MessengerMAX Transfection Reagent (Thermo Fisher Scientific) for RNA transfection.
14. Flow cytometer (e.g., BD Accuri C6).
15. FlowJo (BD).

3 Methods

3.1 Construction of the Cas Protein-Responsive Switch (The Principle of the Switch Is Described in Note 1)

3.1.1 SpCas9-Responsive Switch (Gluc-Sp_gRNA-EGFP) (Fig. 1 and See ***Note 2****)*

1. Design two oligo DNAs (Table 1 and *see* **Note 3**):
   - BamH1-Gluc-Sp_gRNA-Bgl2-Age1_code.
   - BamH1-Gluc-Sp_gRNA-Bgl2-Age1_temp.
2. Phosphorylate the oligo DNAs using the following condition: reaction condition—37 °C for 30 min and heat the sample at 95 °C for 5 min to inactivate the enzyme (Table 3).
3. Anneal the oligo DNAs using the following condition (Table 4), and add 102.8 μL MQ to adjust the concentration of oligos to 70 nM.

Table 3
Phosphorylation

Phosphorylation

component		vol.	final conc
	D2W	6.75 μL	up to 10 μL
10 ×	T4 Polynucleotide Kinase Buffer	1 μL	1 x
75 mM	ATP	0.25 μL	1.875 mM
100 μM	Oligo DNA	1 μL	10 μM
	T4 Polynucleotide Kinese	1 μL	20 U
Total		10 μL	

Reaction

37 °C	30 min
95 °C	5 min

Table 4
Annealing

Annealing

component		vol.
10 μM	Oligo DNA fwd	1 μL
10 μM	Oligo DNA rev	1 μL
	Annealing Buffer	38 μL
Total		40 μL

Reaction

95 °C	4 min
80 °C	1 min
73 °C	5 min
25 °C	10 min

Table 5
Ligation

Ligation

component		vol.
70 nM	Annealed Oligo DNAs	1 μL
10 fmol/μL	Digested plasmid	1 μL
	Ligation High ver.2	2 μL
	Total	4 μL

Reaction

16 °C	30 min

4. Digest the pAptamerCassette-EGFP plasmid with AgeI and BamHI and purify the digested plasmid with MinElute PCR Purification Kit or Monarch PCR & DNA Cleanup Kit.
5. Ligate the annealed oligo DNAs and the digested plasmid using the following condition (Table 5).
6. Transform *E. coli* with the product.
7. Check the sequence and purify the product with a Midiprep kit.

3.2 Construction of the Cas Protein-Responsive Switch and Its Trigger mRNA

3.2.1 Construction of the SpCas9 mRNA (Trigger), SpCas9-Responsive mRNA (Gluc-Sp_gRNA-EGFP mRNA, Switch), pAptamerCassette-EGFP mRNA (Control Switch), and iRFP670 mRNA (Reference) Templates

1. Perform PCR for the control 5′-UTR fragment, 3′-UTR fragment, SpCas9 ORF (for trigger), Gluc-Sp_gRNA-EGFP ORF (for switch), EGFP mRNA template (for control switch), and iRFP670 mRNA template (for reference) using the following condition (Table 6).
2. Confirm the PCR products by agarose gel electrophoresis.
3. After the PCR reaction, add 1 μL DpnI at 37 °C for 30 min to remove methylated plasmid DNA. This reaction is performed for the ORF PCR products, EGFP mRNA template, and iRFP670 mRNA template.
4. Purify the PCR product.
5. To construct the full-length mRNA template, perform fusion PCR using the following condition (Table 7).
6. Confirm the PCR products by agarose gel electrophoresis.
7. Purify the PCR product.

3.2.2 Construction of RNA by In Vitro Transcription (IVT) (See Note 4)

1. Perform the IVT reaction (37 °C for 6 h) using the following condition (Table 8).
2. To remove the template DNA, add 1 μL TURBO DNase and incubate at 37 °C for 30 min.
3. Purify the mRNA. (We elute 40 μL.)
4. Add 4 μL 10× Antarctic phosphatase buffer and 1 μL Antarctic phosphatase and incubate at 37 °C for 30 min.

Table 6
First PCR

5'-UTR

component		vol.	final conc
	D2W	28 μL	up to 50 μL
10 ×	Kod-plus-Neo Buffer	5 μL	1 x
2 mM	dNTPs	5 μL	200 mM
25 mM	MgSO4 aq	3 μL	1.5 mM
10 μM	Fwd primer (No.1)	1.5 μL	0.3 μM
10 μM	Rev primer (No.3)	1.5 μL	0.3 μM
100 nM	UTR oligo (No.2)	5 μL	10 nM
1 U/μL	Kod+ polymerase	1 μL	0.02 U/μL
	Total	50 μL	

5'-UTR

94 °C	2 min		
98 °C	10 sec	13 Cycle	
68 °C	10 sec		30 s/Kb
15 °C	∞		

3'-UTR

component		vol.	final conc
	D2W	28 μL	up to 50 μL
10 ×	Kod-plus-Neo Buffer	5 μL	1 x
2 mM	dNTPs	5 μL	200 mM
25 mM	MgSO4 aq	3 μL	1.5 mM
10 μM	Fwd primer (No.4)	1.5 μL	0.3 μM
10 μM	Rev primer (No.6)	1.5 μL	0.3 μM
100 nM	UTR oligo (No.5)	5 μL	10 nM
1 U/μL	Kod+ polymerase	1 μL	0.02 U/μL
	Total	50 μL	

3'-UTR

94 °C	2 min		
98 °C	10 sec	13 Cycle	
68 °C	10 sec		30 s/Kb
15 °C	∞		

(continued)

Table 6
(continued)

SpCas9 ORF, iRFP670 mRNA Template or EGFP mRNA Template

component		vol.	final conc
	D2W	28 μL	up to 50 μL
10 ×	Kod-plus-Neo Buffer	5 μL	1 x
2 mM	dNTPs	5 μL	200 mM
25 mM	MgSO4 aq	3 μL	1.5 mM
10 μM	Fwd primer (No.7 or 9)	1.5 μL	0.3 μM
10 μM	Rev primer (No.8 or 12)	1.5 μL	0.3 μM
10 ng/μL	Plasmid	5 μL	1 ng/μL
1 U/μL	Kod+ polymerase	1 μL	0.02 U/μL
	Total	50 μL	

Plasmids used

- For SpCas9 ORF: pHL-EF1α-SphcCas9-iP-A (Addgene, plasmid #60559)
- For iRFP670 mRNA Template: pUC19-iRFP670 wo T7fwd *(7)*
- For EGFP mRNA Template: pUC19-EGFPv2woT7r *(7)*

Primers used

- For SpCas9 ORF: primers 7 and 8
- For iRFP670 and EGFP mRNA Templates: primer 9 and 12

** iRFP670 and EGFP mRNA Templates are completed in this PCR reaction.

SpCas9 ORF

94 ℃	2 min		
98 ℃	10 sec	20 Cycle	
68 ℃	130 sec		30 s/Kb
15 ℃	∞		

iRFP670 mRNA Template

94 ℃	2 min		
98 ℃	10 sec	25 Cycle	
60 ℃	30 sec		
68 ℃	52 sec		30 s/Kb
15 ℃	∞		

EGFP mRNA Template

94 ℃	2 min		
98 ℃	10 sec	25 Cycle	
50 - 65 ℃	30 sec		
68 ℃	30 sec		30 s/Kb
15 ℃	∞		

(continued)

Table 6 (continued)

Gluc-Sp_gRNA-EGFP ORF (switch)

component		vol.	final conc
	D2W	28 μL	up to 50 μL
10 ×	Kod-plus-Neo Buffer	5 μL	1 x
2 mM	dNTPs	5 μL	200 mM
25 mM	MgSO4 aq	3 μL	1.5 mM
10 μM	Fwd primer (No.10)	1.5 μL	0.3 μM
10 μM	Rev primer (No.11)	1.5 μL	0.3 μM
10 ng/μL	pGluc-Sp_gRNA-EGFP	5 μL	1 ng/μL
1 U/μL	Kod+ polymerase	1 μL	0.02 U/μL
	Total	50 μL	

Gluc-Sp_gRNA-EGFP ORF

94 °C	2 min		
98 °C	10 sec	25 Cycle	
50 - 65 °C	30 sec		
68 °C	30 sec		30 s/Kb
15 °C	∞		

5. Purify the mRNA again.
6. Check the mRNA quality. We usually use a microchip electrophoresis system, MultiNA, but denaturing polyacrylamide gel electrophoresis can be alternatively used to check the size of mRNAs.

3.3 Validation

3.3.1 Preparation of Samples

1. Seed HEK293FT cells in a 24-well plate 24 h before the transfection as follows:
   (a) Aspirate the medium from a dish containing HEK293FT cells.
   (b) Wash the cells with 5 mL PBS and aspirate the PBS.
   (c) Add 1 mL 0.25% trypsin–EDTA and incubate at 37 °C for 5 min.
   (d) Add 9 mL DMEM medium and detach the cells by pipetting.
   (e) Harvest the cells in a 15 mL conical tube.
   (f) Count the cell number using a cell counter.
   (g) Dilute the cells to a concentration of 2×10^5 cells/mL, and seed 500 μL of the cell suspension per well of a 24-well plate.

Table 7
Second PCR

Fusion PCR: SpCas9 mRNA Template

component		vol.	final conc
	D2W	22 μL	up to 50 μL
10 ×	Kod-plus-Neo Buffer	5 μL	1 x
2 mM	dNTPs	5 μL	200 mM
25 mM	MgSO4 aq	3 μL	1.5 mM
10 μM	Fwd primer (No.1)	1.5 μL	0.3 μM
10 μM	Rev primer (No.12)	1.5 μL	0.3 μM
10 ng/μL	Cas9 mRNA ORF	1 μL	0.2 ng/μL
100 nM	5'-UTR	5 μL	10 nM
100 nM	3'-UTR	5 μL	10 nM
1 U/μL	Kod+ polymerase	1 μL	0.02 U/μL
Total		50 μL	

SpCas9 mRNA Template

94 °C	2 min		
98 °C	10 sec	20 Cycle	
68 °C	140 sec		30 s/Kb
15 °C	∞		

Fusion PCR: Gluc-Sp_gRNA-EGFP mRNA Template

component		vol.	final conc
	D2W	27 μL	up to 50 μL
10 ×	Kod-plus-Neo Buffer	5 μL	1 x
2 mM	dNTPs	5 μL	200 mM
25 mM	MgSO4 aq	3 μL	1.5 mM
10 μM	Fwd primer (No.10)	1.5 μL	0.3 μM
10 μM	Rev primer (No.12)	1.5 μL	0.3 μM
10 ng/μL	Gluc-Sp_gRNA-EGFP ORF	1 μL	0.2 ng/μL
100 nM	3'-UTR	5 μL	10 nM
1 U/μL	Kod+ polymerase	1 μL	0.02 U/μL
Total		50 μL	

Gluc-Sp_gRNA-EGFP mRNA Template

94 °C	2 min		
98 °C	10 sec		
50 - 65 °C	30 sec	25 Cycle	
68 °C	30 sec		30 s/Kb
15 °C	∞		

Table 8
IVT reaction for mRNAs

SpCas9 and iRFP670 mRNAs

component		vol.	final conc
D2W		μL	up to 10 μL
100-400 ng/μL	Template DNA	μL	10-40 ng/μL
10 ×	Buffer	1 μL	1 x
75 mM	ARCA : GTP = 4 : 1	1 μL	6 mM 1.5 mM
75 mM	ATP	1 μL	7.5 mM
75 mM	5mC	1 μL	7.5 mM
75 mM	Ψ	1 μL	7.5 mM
	T7 enzyme	1 μL	1 x
	Total	10 μL	

Template DNAs are full-length mRNA templates made in section 3.2.1.
All templates contain T7 promoter at 5' region.

Gluc-Sp_gRNA-EGFP (switch) and EGFP (control switch) mRNAs

component		vol.	final conc
D2W		μL	up to 10 μL
100-400 ng/μL	Template DNA	μL	10-40 ng/μL
10 ×	Buffer	1 μL	1 x
75 mM	ARCA : GTP = 4 : 1	1 μL	6 mM 1.5 mM
75 mM	ATP	1 μL	7.5 mM
75 mM	CTP	1 μL	7.5 mM
75 mM	UTP	1 μL	7.5 mM
	T7 enzyme	1 μL	1 x
	Total	10 μL	

Template DNAs are full-length mRNA templates made in section 3.2.1.
All templates contain T7 promoter at 5' region.

3.3.2 Transfection of Plasmid (See ***Note 5****)*

1. Mix 100 ng Cas protein-responsive mRNA switch plasmid, 400 ng trigger plasmid, and 100 ng reference plasmid in Opti-MEM. We use pCMV-tdiRFP670 [8] as a reference plasmid.
2. Add 2 μL Lipofectamine 2000 reagent to 48 μL Opti-MEM.
3. Mix the two solutions above and incubate at room temperature for 15–20 min.
4. Add 100 μL mixture per well of a 24-well plate and gently shake the plate to mix.

5. Incubate the cells until the analysis. (We incubate the cells for 24 h.)
6. If needed (in case of cell toxicity is observed), change the medium 4 h after transfection.

3.3.3 Transfection of mRNA (See ***Note 6****)*

1. Mix 100 ng spCas9 mRNA, 100 ng switch mRNA, and 100 ng reference mRNA in Opti-MEM.
2. Add 1 μL Lipofectamine MessengerMAX Reagent to 24 μL Opti-MEM and incubate for 10 min.
3. Mix the two solutions above and incubate at room temperature for 5 min.
4. Add 50 μL mixture per well of a 24-well plate and gently shake the plate to mix.
5. Incubate the cells until the analysis. (We incubate the cells for 24 h.)
6. If needed (in case of cell toxicity is observed), change the medium 4 h after transfection.

3.3.4 Validation of the Switch Performance

1. Observe the cells with a fluorescence microscope. If needed, capture the cell images.
2. Aspirate the medium from the wells.
3. Wash the cells with 500 μL PBS and aspirate the PBS.
4. Add 100 μL 0.25% trypsin–EDTA and incubate at 37 °C for 5 min.
5. Add 200 μL medium.
6. Suspend the cells by pipetting and filtering them through a mesh.
7. Measure the fluorescence using a flow cytometer.
8. Evaluate the switch performance (Fig. 2 and *see* **Note** 7).

4 Notes

1. A Cas protein-responsive switch consists of a crRNA/gRNA sequence, which binds to its corresponding Cas protein as an input (trigger), and the open reading frame (ORF) of an output protein (Fig. 1a). We can construct a variety of switches by using different pairings of Cas protein and its corresponding crRNA/gRNA. In addition, we can combine existing CRISPR-based technologies such as the split-Cas system [7, 9, 10].
2. The design principle of Cas protein-responsive switches is almost the same as in the previous protocol [11]. The pAptamerCassette-EGFP plasmid contains three specific restriction sites for NheI, BamHI, and AgeI in the 5′UTR.

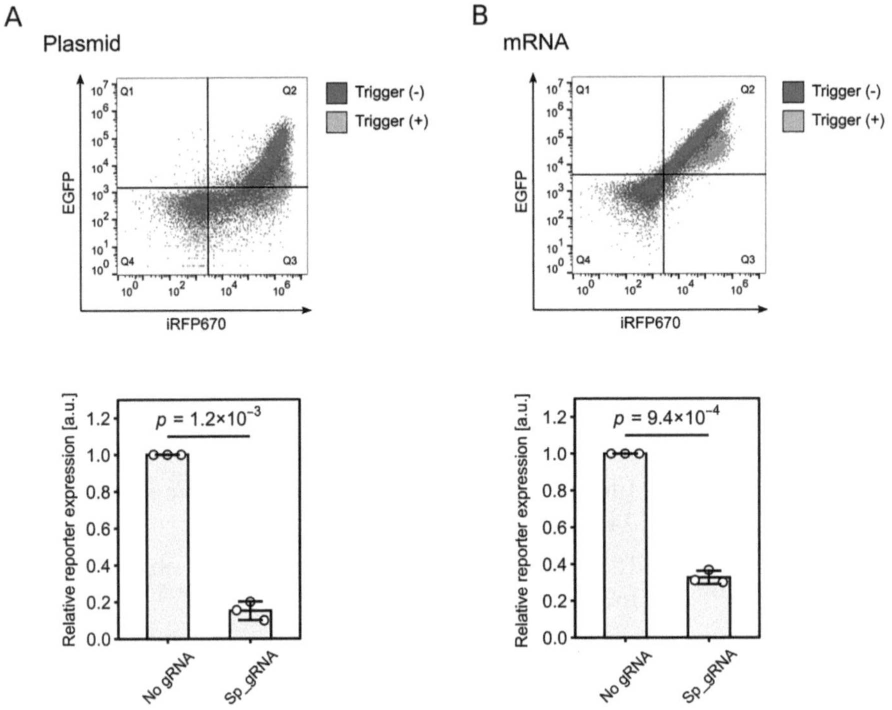

Fig. 2 Example result of SpCas9 protein-responsive switch in HEK293FT cells. (**a**) Plasmid transfection. (**b**) RNA transfection. Top: representative dot plots of expression level from switch with (blue) and without (red) SpCas9. Bottom: the relative reporter expression of a switch is shown. Data were normalized to the value of the control switch without the gRNA sequence (no gRNA). Data represent the mean ±SD ($n = 3$). Statistical analyses were performed using an unpaired two-tailed Student's t-test. The dot plots are a modified version of Ref. [7]

NheI and AgeI are located just downstream of the CMV promoter and upstream of the EGFP ORF, respectively. BamHI is located between NheI and AgeI sites. AgeI and BamHI are used for inserting the crRNA/gRNA sequence (Fig. 1b).

3. Because of the relatively short length of crRNA/gRNA sequences, we use a short dsDNA derived from two synthetic DNA oligos and insert the resulting crRNA/gRNA sequence into the switch plasmid by ligation (Fig. 1c).
4. To reduce the interferon response caused by long RNA (mRNA), pseudouridine-5′-triphosphate and 5-methylcytidine-5′-triphosphate are used instead of natural rUTP and rCTP, respectively [12]. Alternatively, 1-methylpseudouridine-5′-triphosphate can be used instead of natural rUTP. Because Cas protein-responsive switches with modified bases inhibit the interaction of Cas protein and its crRNA/gRNA, natural rNTPs are used for switch/control mRNA of the Cas protein-responsive switches.

5. For plasmid transfection, we use pcDNA3.1-myc-HisA instead of pcDNA3.1-SpCas9 as the "no-trigger control" and pAptamerCassette-EGFP instead of the Cas-protein responsive mRNA switch plasmid as the "no-crRNA/gRNA control." To validate the switch performance, we use fluorescent protein as an output protein.
6. For RNA transfection, we use EGFP mRNA instead of Gluc-Sp_gRNA-EGFP switch mRNA as the "no-crRNA/gRNA control." For the "no-trigger control," we do not transfect SpCas9 mRNA. To validate the switch performance, we use fluorescent protein as an output protein. Because switch/control mRNA consists of natural rNTPs, the interferon response may become a problem in some cell lines (e.g., HeLa, A549 cells, etc.). In that case, co-transfection of vaccinia virus-derived E3, K3, and B18R mRNAs evades this problem [7, 13].
7. To calculate Relative reporter expression, we first normalized the reporter expression level (EGFP) by the expression level of our reference protein (iRFP670) to minimize variations due to transfection efficiency (Normalized Intensity (NI)). Then, we divided the NI of the OFF state by that of the ON state to calculate Relative Intensity (RI). Finally, we normalized the effect of trigger protein induction itself by dividing the calculated RI for each experimental condition using that of the No crRNA/gRNA condition. Thus, the formulas are described as follows:

   Normalized Intensity (NI) $= 1000 \times$ median of the ratio (reporter intensity/reference intensity) of each cell.

   Relative Intensity (RI) $=$ (NI of trigger +) / (NI of trigger −).

   Relative reporter expression $=$ (RI)/(RI of No crRNA/gRNA sample).
8. Information on the switch/trigger plasmids we constructed can be obtained from [7].
9. Some switches might show weak repression efficiency by inserting crRNA/gRNA sequences into the 5′UTR only. In such cases, the efficiency may be improved by optimizing the crRNA/gRNA sequence [7, 8].

References

1. Saito H, Kobayashi T, Hara T et al (2010) Synthetic translational regulation by an L7Ae-kink-turn RNP switch. Nat Chem Biol 6(1): 71–78
2. Kawasaki S, Fujita Y, Nagaike T et al (2017) Synthetic mRNA devices that detect endogenous proteins and distinguish mammalian cells. Nucleic Acids Res 45(12):e117
3. Ausländer S, Ausländer D, Müller M et al (2012) Programmable single-cell mammalian biocomputers. Nature 487:123–127

4. Wroblewska L, Kitada T, Endo K et al (2015) Mammalian synthetic circuits with RNA binding proteins for RNA-only delivery. Nat Biotechnol 33(8):839–841
5. Jinek M, Chylinski K, Fonfara I et al (2012) Programmable dual-RNA-guided DNA endonuclease in adaptive bacterial immunity. Science 337:816–821
6. Makarova KS, Wolf YI, Iranzo J et al (2020) Evolutionary classification of CRISPR-Cas systems: a burst of class 2 and derived variants. Nat Rev Microbiol 18:67–83
7. Kawasaki S, Ono H, Hirosawa M et al (2023) Programmable mammalian translational modulators by CRISPR-associated proteins. Nat Commun 14:2243
8. Ono H, Kawasaki S, Saito H (2020) Orthogonal protein-responsive mRNA switches for mammalian synthetic biology. ACS Synth Biol 9(1):169–174
9. Zetsche B, Volz SE, Zhang F (2015) A split-Cas9 architecture for inducible genome editing and transcription modulation. Nat Biotechnol 33(2):139–142
10. Ma D, Peng S, Xie Z (2016) Integration and exchange of split dCas9 domains for transcriptional controls in mammalian cells. Nat Commun 7:13056
11. Endo K, Saito H (2014) Engineering protein-responsive mRNA switch in mammalian cells. Methods Mol Biol 1111:183–196
12. Warren L, Manos PD, Ahfeldt T et al (2010) Highly efficient reprogramming to pluripotency and directed differentiation of human cells with synthetic modified mRNA. Cell Stem Cell 7:618–630
13. Poleganov MA, Eminli S, Beissert T et al (2015) Efficient reprogramming of human fibroblasts and blood-derived endothelial progenitor cells using nonmodified RNA for reprogramming and immune evasion. Hum Gene Ther 26(11):751–766

Chapter 13

Multiplexed Transactivation of Mammalian Cells Using dFnCas12a-VPR

James W. Bryson and Susan J. Rosser

Abstract

CRISPR activation provides an invaluable tool for experimental biologists to convert correlations into causation by directly observing phenotypic changes upon targeted changes in gene expression. With few exceptions, most diseases are caused by complex polygenic interactions, with multiple genes contributing to define the output of a gene network. As such researchers are increasingly interested in tools that can offer not only control but also the capacity to simultaneously upregulate multiple genes. The adaptation of CRISPR/Cas12a has provided a system especially suited to the tightly coordinated overexpression of multiple targeted genes. Here we describe an approach to test for active targeting crRNAs for dFnCas12a-VPR, before proceeding to generate and validate longer crRNA arrays for multiplexed targeting of genes of interest.

Key words CRISPRa, Cas12a, crRNA assembly, Multiplexed activation

1 Introduction

In recent years, CRISPR/Cas9 has proven to be a highly versatile and potent tool not only to enable targeted mutagenesis [1] but further to allow the manipulation of gene expression with relative ease through CRISPR activation (CRISPRa) or CRISPR inhibition (CRISPRi) [2]. This is primarily due to the programmability of their guide RNAs (gRNAs). Each gRNA is composed of a fused crRNA (possessing the hybridizing spacer sequence) and tracrRNA, which together ensure localization of the effector Cas protein and any linked protein domains to targeted loci possessing the flanking protospacer adjacent motif (PAM). CRISPRa has been applied in a number of settings including cell reprogramming [3], optimizing genetic circuits [4], and biosensor development [5].

While *Streptococcus pyogenes* Cas9 has become widely adopted, newer variants of CRISPR have been discovered with diverse and sometimes beneficial properties. Of note Cas12a, formally Cpf1

Francesca Ceroni and Karen Polizzi (eds.), *Mammalian Synthetic Systems*, Methods in Molecular Biology, vol. 2774,
https://doi.org/10.1007/978-1-0716-3718-0_13,

[6], has been shown to be particularly amenable for multiplex targeting when utilizing plasmids for delivery of targeting gRNAs [7]. This is due to two key features of Cas12a. Firstly, targeting can be achieved only guided by a crRNA (possessing the spacer sequence responsible to loci recognition and the direct repeat sequence which is recognized by the Cas12a protein). This contrasts with Cas9 which also requires a tracrRNA for successful targeting. Secondly, Cas12a is able to process a single crRNA array, containing multiple adjacent crRNA, through endogenous RNase activity of the protein [6]. This effectively enables a single short transcript to encode for the targeting of the Cas12a effector protein to multiple unique loci. Furthermore, the short crRNA sequences enable a cheap and simple oligo-based assembly reaction to enable efficient assembly of crRNA arrays of user-defined lengths and composition. The reduced cost and relative ease of scaling crRNA numbers facilitates the expansion of experimental scope for the user.

As with Cas9, multiple variants of Cas12a have been identified with subtly different properties and diverse PAM sequences [6]. Here we provide detail around the application of dFnCas12a for CRISPRa [8], with FnCas12a being previously described to possess a simpler PAM sequence than the more commonly used As or Lb variants [9]. A simpler PAM sequence equates to higher loci availability providing more targetable sites for a given region of interest. We describe in detail the process for generating and testing the activity of individual crRNAs before subsequently designing and generating crRNA arrays for multiplexed targeting of multiple genes to enable transactivation with dFnCas12a-VPR.

The protocol here describes the design and generation of plasmids for the expression of single crRNAs. This involves key steps including gene selection, crRNA spacer design, cloning of spacer encoding oligos into an expression plasmid, and sequence verification (Fig. 1). The protocol goes on to describe the process for qRT-PCR analysis of relative gene expression after the cloned crRNA expression plasmid is delivered alongside the plasmid expressing dFnCas12a-VPR. Finally, the design and generation of plasmids expressing crRNA arrays, encoding multiple unique crRNA within a single transcript, is described alongside the subsequent screening of activity for each of the targeted genes by qRT-PCR (Fig. 2).

2 Materials

2.1 Equipment

1. Gel electrophoresis tank.
2. Thermocycler.
3. 37 °C incubator.

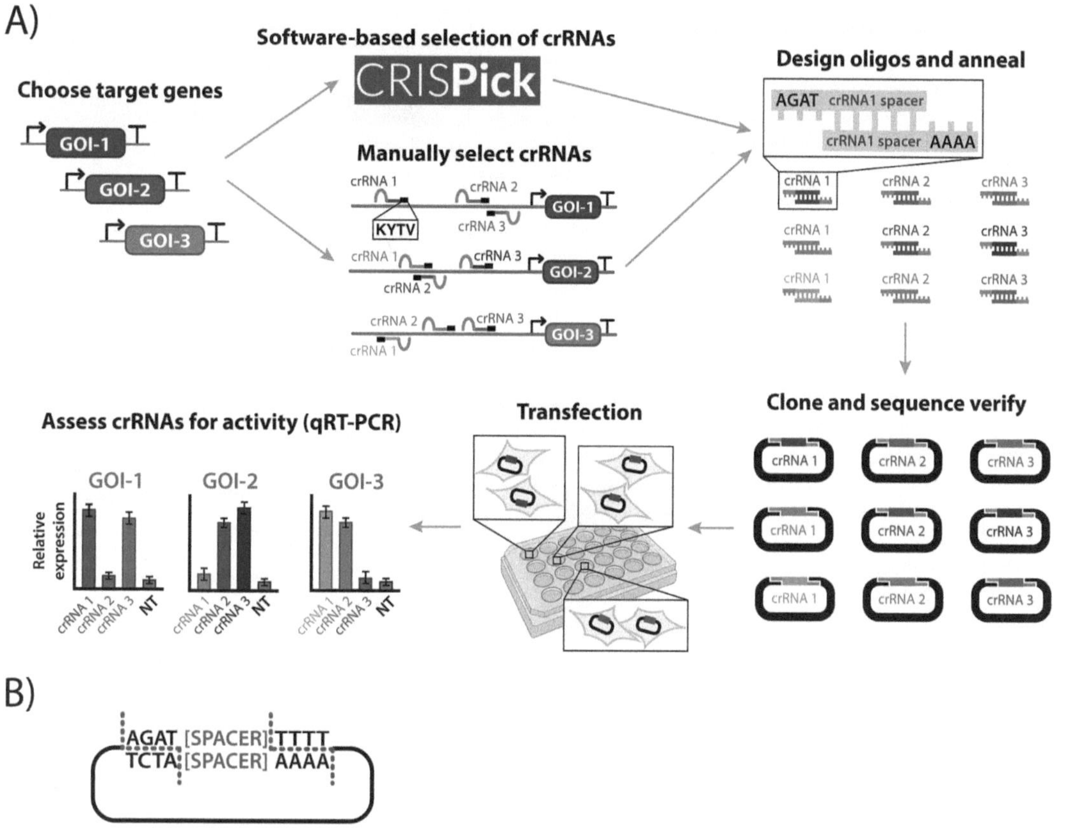

Fig. 1 Design, assembly, and validation of single crRNA constructs. (**a**) Target genes are chosen before crRNA spacers targeting their respective promoters are designed either using available software or manually. Oligos are then designed and ordered based on the chosen spacer sequences. The hybridized oligos are cloned into a digested crRNA expression plasmid before being Sanger sequence verified. Activity of the crRNA is subsequently assessed by transfecting into target cells and testing the relative expression of each targeted gene using qRT-PCR. (**b**) After linearizing the crRNA expression plasmid using Esp3I, the gel-purified plasmid backbone can be ligated with the hybridized oligo pairs

4. Dry block incubator.
5. Tissue culture flow hood.
6. Tissue culture CO_2 incubator.
7. Centrifuge.
8. qRT-PCR machine.

2.2 Consumables

1. Petri dishes.
2. LB agar.
3. Ampicillin.
4. TAE buffer.
5. Agarose.

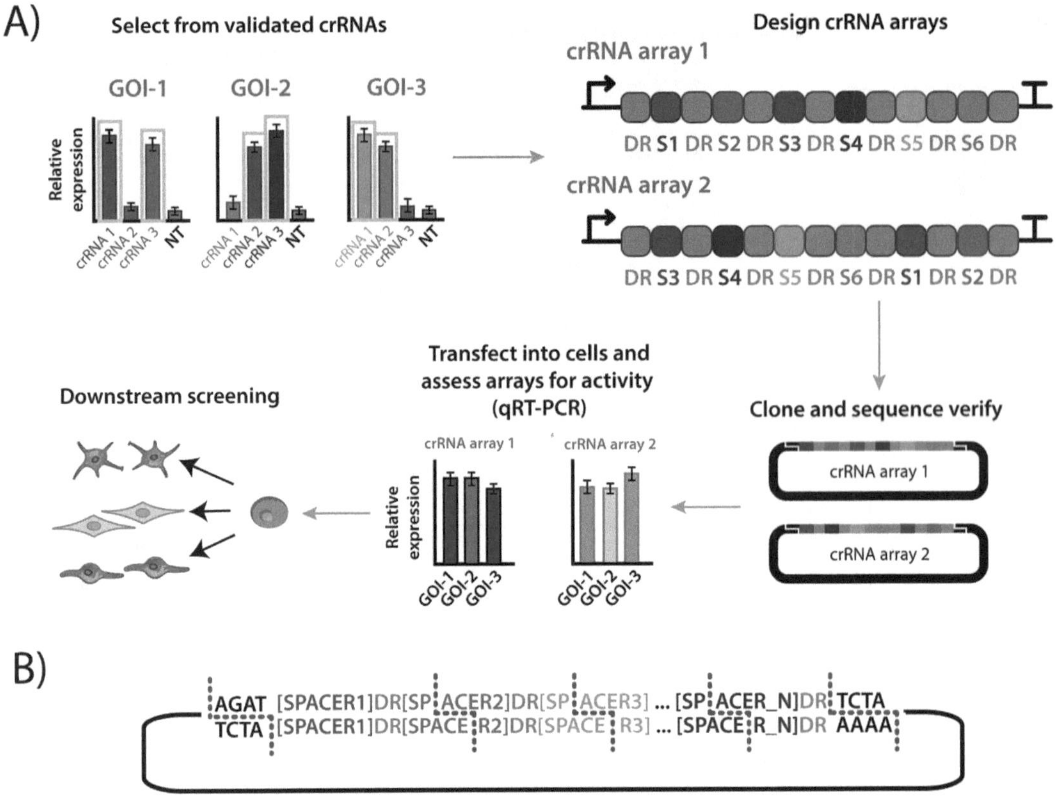

Fig. 2 Design, assembly, and validation of crRNA array constructs. (**a**) Validated crRNAs are chosen for genes of interest. Oligos are then designed and ordered based on the sequence and order of chosen crRNAs. Hybridized oligos are cloned into a digested expression plasmid before being sequence verified. Activity of the array for multiple target genes can be assessed by qRT-PCR before proceeding to functional assays. (**b**) Oligos for constructing crRNA arrays are designed as shown, where the first oligo pair will use an "AGAT" overhang and a spacer 2-dependent overhang and the final oligo pair will use a spacer N-dependent overhang (where N is the total number of spacers within the array) and an "AAAA" overhang. Intermediate oligo pairs will use an overhang such that fragment number n will have one overhang within spacer n and one overhang within spacer $n + 1$

6. LB media.
7. 50 mL falcon tubes.
8. 24-well tissue culture plate.
9. Ordered oligos.
10. Target cell line (e.g., HEK293 cells).
11. Oligo d(T)20 primer.

2.3 Commercial Reagents

1. Commercial DNA/oligo synthesis (IDT) (*see* Subheadings 3.1 and 3.3 for design of oligos).
2. The crRNA expression plasmid pU6-Fn-crRNA (Addgene plasmid #78958).

3. The dFnCas12a-VPR expression plasmid (Addgene plasmid #179520).
4. crRNA oligos (design and assembly described below, in figures and tables).
5. SYBR Safe (Thermo catalogue number—S33102).
6. Esp3I (NEB catalogue number—R0734S).
7. rCutSmart buffer (NEB catalogue number—B6004S).
8. T4 PNK (NEB catalogue number—M0201S).
9. T4 DNA ligase (NEB catalogue number—M0202S).
10. T4 DNA ligase buffer (NEB catalogue number—B0202S).
11. NEB buffer 3.1 (NEB catalogue number—B6003S).
12. DH5α chemically competent *E. coli* (Thermo catalogue number—18258012).
13. E.Z.N.A Plasmid DNA Mini kit (Omega Bio-Tek D6942-01).
14. NdeI (NEB catalogue number—R0111S).
15. XbaI (NEB catalogue number—R0145S).
16. Power SYBR Green qPCR mix (Thermo catalogue number—4368577).
17. E.Z.N.A Total RNA kit 1 (Omega Bio-Tek catalogue number—R6834-01).
18. Superscript IV (Thermo catalogue number—18090010).
19. Lipofectamine 2000 (Thermo catalogue number—11668019).

3 Methods

All the steps below should be prepared on ice unless otherwise stated.

3.1 Design and Generation of Single crRNA Plasmids

1. Choose the genes of interest to be targeted, and identify the respective promoter regions.
2. Design crRNA spacers for targeting the promoter of each gene, either using a dedicated software or by manually designing crRNAs. For each promoter region, design three to six crRNAs to maximize the chance of identifying an active crRNA (*see* **Note 1**) (Fig. 1a). For most users, we recommend crRNA design software such as CRISPick for selecting crRNA spacer sequences (*see* **Note 2**). If manually designing crRNAs, spacer sequences must be chosen such that there is a "KYTV" PAM sequence to 5′ of the displaced (unbound) DNA strand. The crRNAs should be designed within −400 to -50 nt of the annotated transcription start site [2] (*see* **Notes 3–6**).

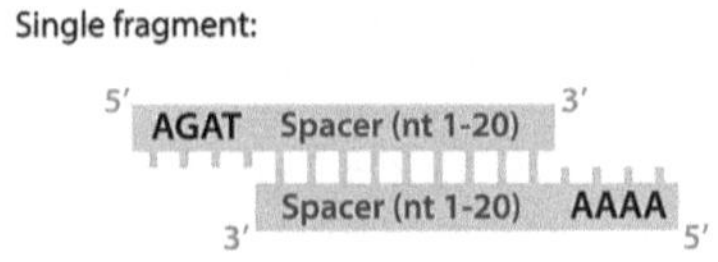

Fig. 3 Annealed oligos for the generation of single crRNA expression plasmids. The schematic is shown for the annealed oligos for insertion into the Esp3I-digested pU6-Fn-crRNA plasmid

Table 1
Example oligos for generation of single crRNA plasmid

Oligo name	Oligo sequence
ASCL1 crRNA F	agatCAAGGAGCGGGAGAAAGGAA
ASCL1 crRNA R	aaaaTTCCTTTCTCCCGCTCCTTG

The table shows an example of two oligos for the generation of a crRNA plasmid for expression of a single crRNA targeting the promoter of ASCL1. The spacer sequence is shown in red and the overhangs for cloning are shown in black

3. Commercially synthesize complementary oligo pairs encoding crRNA spacers and overhangs compatible with Esp3I digested crRNA plasmid (Fig. 3, Table 1).
4. Anneal each oligo pair in PCR tubes as follows: 42 μL water, 1.5 μL top strand oligo (100 μM), 1.5 μL bottom strand oligo (100 μM), and 5 μL NEB buffer 3.1.
5. Transfer the oligos to a thermocycler for annealing by heating to 98 °C for 10 min before lowering the temperature by 1 °C per minute to 40 °C.
6. Combine the following reagents for Esp3I digestion of the pU6-Fn-crRNA plasmid (crRNA expression plasmid): 2 μg pU6-Fn-crRNA plasmid, 2 μL Esp3I, 5 μL rCutSmart buffer, up to 50 μL water (*see* **Note 7**).
7. Incubate the digestion mixture overnight at 37 °C.
8. Run the digested plasmid on an agarose gel and cut out the linearized backbone (expecting a clear band at 2.4 Kb).
9. Simultaneously phosphorylate and ligate the annealed oligos and linearized crRNA plasmid backbone in PCR tubes as follows: 1 μL of T4 PNK, 1 μL of T4 DNA ligase, 2 μL of T4 DNA ligase buffer, 1 μL of annealed oligos, 50 ng of Esp3I linearized pU6-Fn-crRNA plasmid, and up to 20 μL with water.
10. Heat the reaction to 37 °C for 30 min.
11. In sterile conditions, transform the mixture into chemically competent *E. coli* by heat shock. Transfer 10 μL of the ligation reaction to 100 μL of chemically competent cells. Incubate on

Table 2
Sequencing primer

Oligo name	Oligo sequence
hU6-F	GAGGGCCTATTTCCCATGATT

ice for 10 min, immediately transfer to a dry block incubator set to 42 °C for 45 s, and then transfer back to ice for 5 min. Add 400 of LB media to the transformation reaction and incubate for 1 h at 37 °C.

12. Plate 100 μL of the transformation reaction on LB agar plates containing ampicillin and grow overnight at 37 °C..
13. Pick individual colonies to grow in LB media with ampicillin overnight.
14. Extract the plasmid from the resulting cultures using the E.Z. N.A Plasmid DNA Mini kit, following the manufacturer's instructions.
15. Verify cloning by Sanger sequencing, using the hU6-F primer (Table 2).
16. After sequence verification, the activity of crRNA plasmid can be assessed by qRT-PCR (*see* **Note 8**).

3.2 Screening Activity of Single crRNAs by qRT-PCR

1. To assess the activity of single crRNA, HEK293 cells can be seeded into a 24-well tissue culture plate, transferring 200,000 HEK293 cells into each well.
2. The next day, carry out the transfection using Lipofectamine 2000, following the manufacturer's protocol, delivering 250 ng of crRNA plasmid and 500 ng of dFnCas12a-VPR plasmid per well.
3. Three days post transfection, perform RNA extraction. We recommend a commercially available kit such as the E.Z.N.A Total RNA kit 1, following the manufacturer's instructions. DNase treatment of the total RNA can either be performed on the column or after RNA elution (*see* **Note 9**). RNA should be stored long term at −80 °C.
4. Carry out reverse transcription with Superscript IV or equivalent commercially available reverse transcriptase, using 1uL of oligo d(T)20 primer per cDNA reaction and otherwise following the manufacturer's instructions.
5. Before proceeding to qRT-PCR, primers targeting the gene of interest and at least one reference gene should be calibrated to ensure appropriate amplification efficiency and product size using standard and melt curves, respectively (*see* **Note 10**).

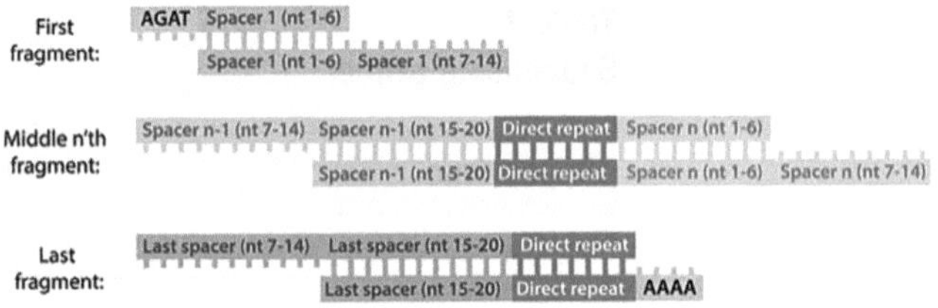

Fig. 4 Annealed oligos for the generation of crRNA arrays. The schematic shows the conceptual strategy for assembling crRNA arrays for the expression of three or more crRNAs. The designed overhangs for the beginning of the first fragment and the end of the last fragment are always consistent, matching the overhangs of the digested crRNA expression plasmid. The remaining overhangs are defined by the corresponding spacer sequences. Spacer n refers to the nth spacer within a crRNA array where n corresponds to the fragment's numeric position within the array design

6. qRT-PCR analysis can be performed using Power SYBR Green qPCR mix or an equivalent commercially available qPCR master mix, following the manufacturer's instructions.
7. Using a ΔΔcq calculation, the relative expression of the gene of interest and the reference gene can be first calculated for the samples where crRNAs have been delivered and the samples where crRNAs have not been delivered. After establishing the normalized expression of the target gene, this can then be compared between the two conditions to establish the relative (over)expression of the target gene for the samples where crRNAs were delivered (*see* **Note 11**).

3.3 Generation and Testing of Multiplexed Arrays

1. Select the crRNA spacer sequences to be employed within the multiplexed arrays, considering the ordering of crRNAs (*see* **Notes 12–14**).
2. Order complementary oligo pairs, where the overhangs required will be dependent on their relative order within an array as well as the adjacent spacer sequence. Briefly, except for the first and last overhang within the array, the 5′ overhangs are the 8 nt within the center of the 20 nt spacer sequences (Fig. 4, Table 3). For the crRNA arrays, a final direct repeat is incorporated at the end of each array within the last fragment, drawing from the designs employed by Tak and colleagues [10].
3. Anneal each oligo pair in PCR tubes as follows: 42 μL water, 1.5 μL top strand oligo (100 μM), 1.5 μL bottom strand oligo (100 μM), and 5 μL NEB buffer 3.1.
4. Transfer the oligos to a thermocycler for annealing by heating the mixtures to 98 °C for 10 min before lowering the temperature by 1 °C per minute to 40 °C.

Table 3
Example oligos for the generation of a 3-crRNA array plasmid

Oligo name	Oligo sequence
First oligo F	agatCAAGGAGCGGGAGAAAGGAATAATTTCTACTGTTGTAGAT**AAGTCC**
First oligo R	**TAGGAGTTGGACTT**ATCTACAACAGTAGAAATTATTCCTTTCTCCCGCTCCTTG
Second oligo F	**AACTCCTAAGCCAG**TAATTTCTACTGTTGTAGAT*CAGGAG*
Second oligo R	*GAGTCACCCTCCTG*ATCTACAACAGTAGAAATTA**CTGGCT**
Final oligo F	*GGTGACTCAGGCTA*TAATTTCTACTGTTGTAGAT
Final oligo R	aaaaATCTACAACAGTAGAAATTA*TAGCCT*

The table shows the oligos for assembling a 3-crRNA array, including a crRNA targeting the ASCL1 promoter (underlined), a crRNA targeting the HBB promoter (bold), and a crRNA targeting the IL1RN promoter (italicized). The Fn direct repeat sequence is shown in gray and the overhangs for cloning are shown in lower case

5. Ensure you have sufficient Esp3I-digested crRNA plasmid (either from previous digestion or otherwise carry out **steps 5** and **6** from "Generation and testing of individual crRNAs").
6. Simultaneously phosphorylate and ligate the annealed oligos and linearized crRNA plasmid backbone in PCR tubes as follows: 1 μL of T4 PNK, 1 μL of T4 DNA ligase, 2 μL of T4 DNA ligase buffer, 1 μL of each pair of annealed oligos, 50 ng of Esp3I linearized pU6-Fn-crRNA plasmid, and up to 20 μL with water.
7. Heat the reaction to 37 °C for 30 min.
8. In sterile conditions, transform the mixture into chemically competent *E. coli* by heat shock (*see* **Note 15**). Transfer 10 μL of the ligation reaction to 100 μL of chemically competent cells. Incubate on ice for 10 min, immediately transfer to a water bath set to 42 °C for 45 s, and then transfer back to ice for 5 min. Add 400 of LB media to the transformation reaction and incubate for 1 h at 37 °C.
9. Plate 100 μL of the transformation reaction on LB agar plates containing ampicillin and grow overnight at 37 °C.
10. Pick individual colonies to grow in LB media with ampicillin overnight.
11. Extract the plasmid from the resulting cultures using the E.Z.N.A Plasmid DNA Mini kit, following the manufacturer's instructions.
12. Verify cloning by Sanger sequencing, using the hU6-F primer (Table 2).

13. Assembly is somewhat less efficient for longer array assembly (*see* **Note 15**); as such an optional digestion (NdeI + XbaI) or PCR verification step can be performed at this stage. The first top strand oligo and the last bottom strand oligo can be used for PCR verification for each construct if the two oligo melting temperatures are within 5 °C of one another.
14. After sequence verification, the dFnCas12a-VPR plasmid can be delivered alongside each crRNA array before assessing activity using qRT-PCR for each of the target genes for a given array. The same approach described in the subheading "Screening Activity of Single crRNAs" can be carried out for assessing crRNA arrays.

4 Notes

1. In our hands when testing the activity of manually designed single crRNA, we found that only approximately 1/3 crRNAs show significant activity, so we recommend initially testing three to six crRNAs per gene of interest.
2. When designing crRNA sequences, for most applications, the CRISPick tool selecting As/Lb Cas12a will be appropriate as the Fn PAM sequence is included within the range selected for As/Lb PAM availability [11].
3. If manually designing crRNAs, the first step will be to identify the predicted start site for the gene of interest. For the canonical transcript of a given gene, this can be acquired using either BioMart within Ensembl [12] or through the UCSC genome browser using the "EDPnew Promoters" track [13].
4. In our hands, we found the highest activity when a C or G was present at the 3′ of the PAM sequence (e.g., NTTG or NTTC); however, we recommend checking the PAM wheel visualization of FnCas12a PAM preference to best identify preferred PAM sequences [14].
5. We have not tested for differential isoform transactivation based on crRNA localization; however, it has been previously shown that one of the key sources of isoform diversity is through differential transcription start sites [15].
6. In cases where transcription start sites are sufficiently divergent, it can be assumed that the position of crRNA will impact the relative ratio of isoform expression.
7. When digesting the crRNA expression plasmid (Addgene plasmid #78958) with Esp3I, we recommend using 5–10 ug of input plasmid and digesting for >2 h to ensure a large stock of linearized plasmid while minimizing any remaining undigested

plasmid. After observing high cloning efficiency when using this stock, it can be used with confidence repeatedly afterward.

8. It remains an open challenge to quantitatively estimate the activity of crRNAs for CRISPRa before testing them; as such any CRISPRa strategy relying on a specific expression range for a target gene will necessitate testing, calibration, and in some cases the utilization of inducible promoter systems. When testing crRNA activity, it is recommended to test the crRNAs within the context of the cell line they will be employed in.
9. While DNase treatment can be performed after RNA extraction, we found on column DNase treatment simplified the process of acquiring DNA-free total RNA.
10. We strongly recommend calibrating primers from publications for your own reagents and machines and otherwise recommend ordering three pairs of primers to expedite identifications of a suitable pair. We also recommend calibrating against a template concentration range higher than normally observed for the cell line of interest to ensure calibration is representative of the greater abundance of transcript expected after successful CRISPRa.
11. As a general trend, we observe that it is possible to induce a much higher fold increase in gene expression for lowly expressed genes compared to highly expressed genes.
12. We have previously shown that there is a modest reduction in activity of crRNAs when positioned toward the 3′ of longer crRNA arrays [8]. This should be considered for applications when high overexpression of specific genes is desired.
13. The reader should note that this property of crRNA ordering impacting relative expression of crRNAs within an array can be exploited in cases such as pathway engineering, by using the order of crRNAs as an approach to diversify relative expression of genes within a target pathway.
14. Synergistic transactivation of a gene of interest can be achieved when delivering multiple crRNA targeting the same promoter within an individual crRNA array [8]. This can be especially relevant in cases where high expression is desired or the targeting crRNAs are positioned toward the 3′ of a longer crRNA array.
15. We saw minimal evidence of recombination (with even 9-crRNA constructs showing approximately 50% of clones with perfect sequence) and as such do not see Thermo Stbl3 or similar *E. coli* strains optimized for repetitive sequences as necessary for assembly transformation steps.

References

1. Mali P, Yang L, Esvelt KM et al (2013) RNA-guided human genome engineering via Cas9. Science 339:823–826
2. Gilbert LA, Larson MH, Morsut L et al (2013) CRISPR-mediated modular RNA-guided regulation of transcription in eukaryotes. Cell 154:442–451
3. Chakraborty S, Ji H, Kabadi AM et al (2014) A CRISPR/Cas9-based system for reprogramming cell lineage specification. Stem Cell Rep 3:940–947
4. Nakamura M, Srinivasan P, Chavez M et al (2019) Anti-CRISPR-mediated control of gene editing and synthetic circuits in eukaryotic cells. Nat Commun 10:194
5. Krawczyk K, Scheller L, Kim H et al (2020) Rewiring of endogenous signaling pathways to genomic targets for therapeutic cell reprogramming. Nat Commun **11**:608
6. Zetsche B, Gootenberg JS, Abudayyeh OO et al (2015) Cpf1 is a single RNA-guided endonuclease of a class 2 CRISPR-Cas system. Cell 163:759–771
7. Campa CC, Weisbach NR, Santinha AJ et al (2019) Multiplexed genome engineering by Cas12a and CRISPR arrays encoded on single transcripts. Nat Methods 16:887–893
8. Bryson JW, Auxillos JY, Rosser SJ (2021) Multiplexed activation in mammalian cells using a split-intein CRISPR/Cas12a based synthetic transcription factor. Nucleic Acids Res 50: 549–560
9. Tu M, Lin L, Cheng Y et al (2017) A 'new lease of life': FnCpf1 possesses DNA cleavage activity for genome editing in human cells. Nucleic Acids Res 45:11295–11304
10. Tak YE, Kleinstiver BP, Nuñez JK et al (2017) Inducible and multiplex gene regulation using CRISPR–Cpf1-based transcription factors. Nat Methods 14:1163–1166
11. Kim HK, Min S, Song M et al (2018) Deep learning improves prediction of CRISPR–Cpf1 guide RNA activity. Nat Biotechnol 36:239–241
12. Martin FJ, Amode MR, Aneja A et al (2022) Ensembl 2023. Nucleic Acids Res gkac958
13. Kent WJ, Sugnet CW, Furey TS et al (2002) The human genome browser at UCSC. Genome Res 12:996–1006
14. Marshall R, Maxwell CS, Collins SP et al (2018) Rapid and scalable characterization of CRISPR technologies using an E. Coli cell-free transcription-translation. System 69:146
15. Reyes A, Huber W (2018) Alternative start and termination sites of transcription drive most transcript isoform differences across human tissues. Nucleic Acids Res 46:582–592

Chapter 14

Anti-CRISPR Proteins and Their Application to Control CRISPR Effectors in Mammalian Systems

Carolin Maja Gebhardt and Dominik Niopek

Abstract

CRISPR-Cas effectors are powerful tools for genome and transcriptome targeting and editing. Naturally, these protein–RNA complexes are part of the microbial innate immune system, which emerged from the evolutionary arms race between microbes and phages. This coevolution has also given rise to so-called anti-CRISPR (Acr) proteins that counteract the CRISPR-Cas adaptive immunity. Acrs constitutively block cognate CRISPR-Cas effectors, e.g., by interfering with guide RNA binding, target DNA/RNA recognition, or target cleavage. In addition to their important role in microbiology and evolution, Acrs have recently gained particular attention for being useful tools and switches to regulate or fine-tune the activity of CRISPR-Cas effectors. Due to their commonly small size, high inhibition potency, and structural and mechanistic versatility, Acrs offer a wide range of potential applications for controlling CRISPR effectors in heterologous systems, including mammalian cells.

Here, we review the diverse applications of Acrs in mammalian cells and organisms and discuss the underlying engineering strategies. These applications include (i) persistent blockage of CRISPR-Cas function to create write-protected cells, (ii) reduction of CRISPR-Cas off-target editing, (iii) focusing CRISPR-Cas activity to specific cell types and tissues, (iv) spatiotemporal control of CRISPR effectors based on engineered, opto-, or chemogenetic Acrs, and (v) the use of Acrs for selective binding and detection of CRISPR-Cas effectors in complex samples. We will also highlight potential future applications of Acrs in a biomedical context and point out present challenges that need to be overcome on the way.

Key words CRISPR-Cas, Anti-CRISPR, Genome editing, Mammalian cell, Optogenetics, Synthetic biology

1 Introduction

CRISPR-Cas effectors are powerful tools for precise manipulation and interrogation of the genome. In just a few years since Jennifer Doudna's and Emmanuelle Charpentier's seminal paper demonstrating sequence-specific DNA cleavage with CRISPR-Cas9 in 2012 [1], CRISPR genome editing has revolutionized the life sciences, enabled new diagnostics, and is on the verge of dramatically improving the clinical treatment of genetic disorders. CRISPR

Francesca Ceroni and Karen Polizzi (eds.), *Mammalian Synthetic Systems*, Methods in Molecular Biology, vol. 2774,
https://doi.org/10.1007/978-1-0716-3718-0_14,

effectors are derived from an adaptive immune system in prokaryotes that is capable of both memorizing and destroying invading nucleic acids by targeted generation of single- or double-strand breaks (DSBs) [1–4]. The basic principle of these systems is as follows: specialized Cas proteins recognize foreign nucleic acids and insert corresponding, short sequences called spacers into specialized loci within the microbial genome, called CRISPR arrays. This creates heritable "memories" of invading nucleic acids. These spacers are then transcribed from the CRISPR arrays to form crRNAs. Together with cognate Cas effector proteins and sometimes additional RNAs (tracrRNA in the case of Cas9), active CRISPR-Cas ribonucleoprotein (RNP) complexes are formed. Upon encountering a "memorized" invader sequence, the CRISPR-Cas RNPs can selectively target the invading nucleic acid via base-pair directed interactions and subsequently destroy it via Cas protein-mediated catalytic cleavage, thereby enabling microbial cell survival (Fig. 1a).

In general, CRISPR-Cas systems are divided into two classes. While class 1 CRISPR-Cas systems consist of multi-protein complexes, class 2 CRISPR effectors rely on single effector proteins. For this reason, class 2 systems are predominantly adapted for CRISPR-mediated applications, including genome editing, and will be focused on in the following. Due to the versatility of Cas nucleases, CRISPR-Cas systems are further divided into six types and several subtypes depending on the target nucleic acids, underlying nuclease domains and recognition sequences. The most widely used CRISPR-Cas systems are classified as class 2 type II (Cas9). These include Cas9 from *Streptococcus pyogenes* (*Spy*), which was the first Cas ortholog to be adapted for genome editing in mammalian cells a decade ago [5, 6]. Shortly thereafter, other type II effectors, including Cas9 from *Staphylococcus aureus* (*Sau*) [7], *Neisseria meningitidis* (*Nme*) [8], and several other species, were also successfully used for genome editing.

Cas12 (also known as Cpf1) is a class 2 type V effector that targets double-stranded DNA like Cas9. However, Cas12 is structurally and mechanistically profoundly different from Cas9. As a result, Cas12 DNA cleavage produces sticky overhangs, unlike Cas9, which generates blunt ends. In addition, Cas12 induces DSBs outside of the crRNA-targeted DNA sequence [9, 10]. Finally, Cas12 is able to process crRNAs from longer transcripts on its own, thus facilitating multiplexed genome editing [11].

Cas9 or Cas12 can be used to generate genomic knock-outs or knock-ins by targeted induction of DNA DSBs. In contrast, dCas9/dCas12 systems (deactivated Cas), in which the nuclease function is inactivated by mutation, are widely employed as programmable DNA-binding proteins. These can be used either to interfere with transcription (CRISPRi) or to recruit fused effector

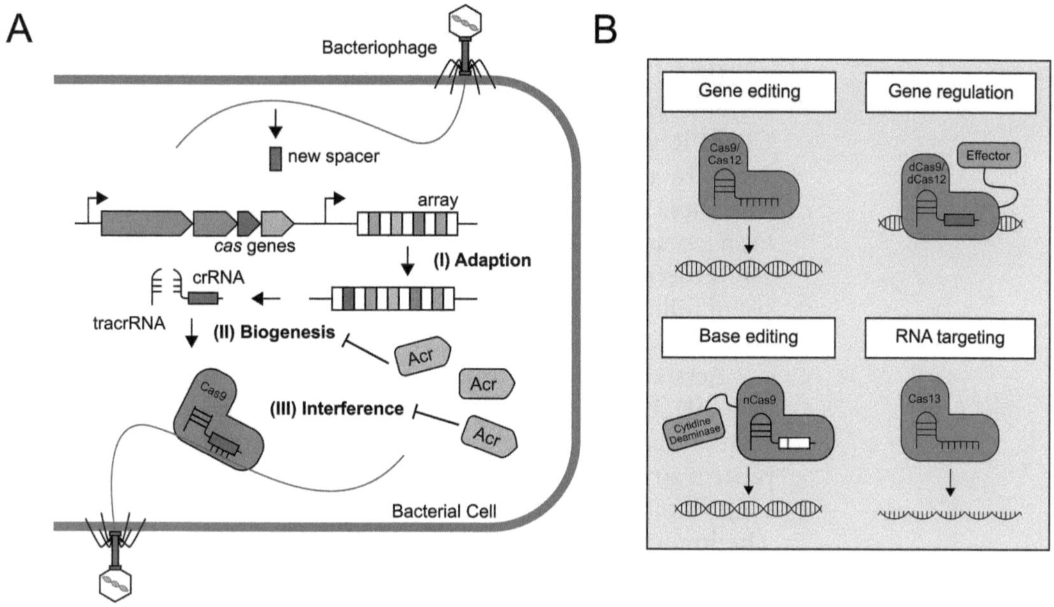

Fig. 1 Natural function and applications of CRISPR-Cas systems. (**a**) Schematic of the CRISPR type II adaptive immune system. Upon phage infection, new spacers that match the nucleic acid sequence of the invader are inserted into so-called CRISPR arrays in the bacterial genome (Step I: Adaptation). If the bacterial cell survives the infection, all daughter cells will have a genetic memory of the invader. The spacer sequence is then transcribed together with an adjacent repeat sequence (white boxes), resulting in a so-called crRNA. Together with a second RNA (tracrRNA) and Cas9, an RNP complex forms (Step II: Biogenesis). Upon encountering the same phage a second time, this RNP selectively targets the invading nucleic acid via cRNA-mediated base-pairing and then destroys it (Step 3: Interference), thus preventing secondary infections. Inhibition of CRISPR effectors can occur at several stages during the CRISPR biogenesis and interference phases. (**b**) Typical applications of CRISPR-Cas systems: Targeted induction of DNA DSBs with catalytically active Cas9/Cas12; activation or repression of endogenous genes with deactivated Cas9/Cas12 (dCas9/dCas12) fused to a transcriptional regulator or epigenetic modifier; base editing with a Cas9 nickase (nCas9) or dead Cas9 (dCas9) fused to a cytidine or adenine deaminase domain; RNA interference or RNA editing with Cas13

proteins of various types, such as transcriptional activators, repressors, epigenetic modifiers, or base-editing enzymes, to selected genomic loci. The recruitment process is accomplished by attaching effector domains to the Cas protein through genetic fusion or tethering (Fig. 1b). It is crucial to note that the activity of both Cas9 and Cas12 relies on the presence of a protospacer adjacent motif (PAM). The PAM is a clearly defined, short sequence that is found next to the crRNA-targeted DNA sequence on the DNA target. The requirement for a PAM restricts the range of sequences that can be targeted by Cas9 or Cas12.

In contrast to DNA-targeting Cas effectors, Cas13 is an RNA-directed, RNA-binding enzyme that has been used for both RNA interference (cleavage) and RNA editing (Fig. 1b) [12, 13].

For further information on the various CRISPR-Cas effector types, their manifold applications, and important structural and mechanistic considerations, we kindly redirect the readers to excellent, recent reviews [14–16].

In 2013, shortly after the adaptation of CRISPR-Cas9 for genome engineering, genetically encoded protein inhibitors of CRISPR systems were discovered in phage-derived genetic elements by the Alain Davidson lab [17]. Soon after, it became clear that these so-called anti-CRISPRs (or Acrs for short) are as common as are CRISPR-Cas systems themselves. To date, around 100 experimentally validated Acrs have been reported. These consist of inhibitors of five CRISPR-Cas subtypes, namely, class 1 types I and III and class 2 types II, V, and VI.

Acrs are just as diverse as their cognate CRISPR systems are [18–20]. In fact, different Acrs tend to share only limited similarity with one another at the sequence level. Moreover, a number of Acr protein structures have been solved by now, and they tend to adapt unique folds that are unrelated to other Acrs or proteins in general. The immense diversity of Acrs is further reflected in the multiple inhibitory mechanisms they can employ. For example, class 2 inhibitory Acrs have been shown to interfere with either crRNA processing, RNP assembly, target DNA binding/recognition, or DNA cleavage (Fig. 2a). This diversity at the sequence, structural, and functional–mechanistic levels renders the computational identification of new Acrs a particular challenge. Simple Acr sequence similarity searches are, in fact, poorly suited to identify new Acr gene or protein candidates. To circumvent this problem, Acr-associated genes (Aca) are often employed as marker genes for Arc-encoding gene clusters in "guilt-by-association" approaches [22]. Aca proteins are helix-turn-helix proteins that can function as transcriptional repressors by binding to the native Acr promoter. In nature, these genes may be necessary for the survival of invading phages, due to the putative toxicity of high levels of Acrs. This probably explains the widespread presence of Acas in Acr operons, although their biological role is still under investigation [23, 24]. In addition to guilt-by-association approaches, various machine learning strategies have recently been developed and used to identify putative Acrs [25–29]. Therefore, the main bottleneck today appears to be the experimental validation of the many thousand candidate Acrs by in vitro or in vivo functional assays rather than the identification of new Acr candidates.

The nomenclature of Acr proteins follows the nomenclature of their related CRISPR-Cas systems and the order in which the Acrs were identified (Fig. 3). For example, AcrIIA4 was the fourth Acr protein discovered to inhibit a type II-A CRISPR effector [32]. An actively maintained online spreadsheet by the Bondy-Denomy lab keeps track of experimentally verified Acrs: https://tinyurl.com/anti-CRISPR (Fig. 3) [32]. Moreover, there are several comprehensive databases, such as anti-CRISPRdb http://guolab.whu.edu.cn/anti-CRISPRdb/ and AcrHub https://pacrispr.erc.monash.

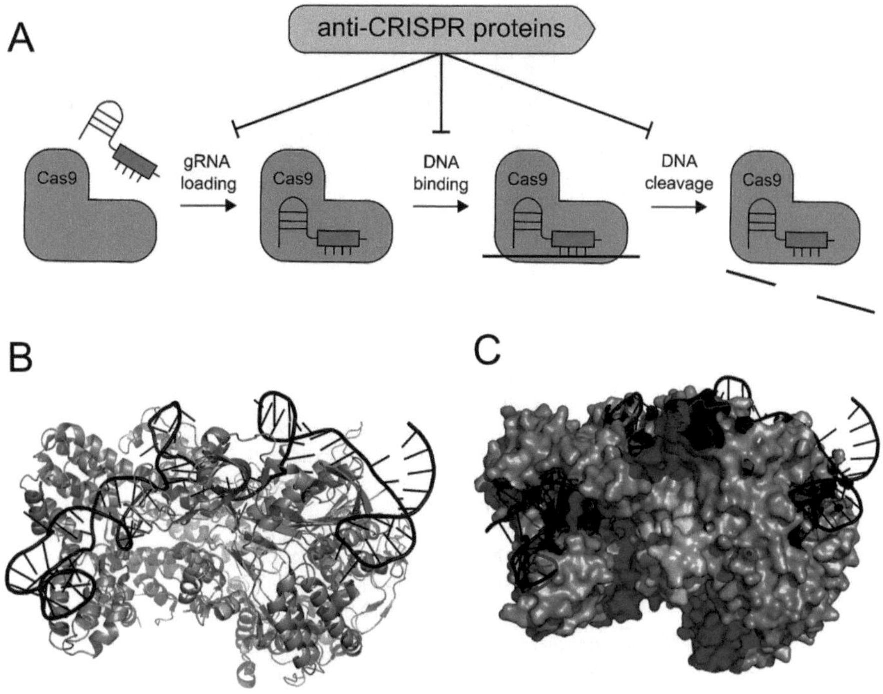

Fig. 2 Mechanisms of CRISPR-Cas inhibition by type II Acrs. (**a**) CRISPR-Cas inhibition by Acrs can occur in several ways during the RNP biogenesis and CRISPR interference phases. For example, Acrs can prevent the binding of the RNA guide to the CRISPR-Cas protein, interfere with the DNA binding of CRISPR-Cas RNPs, or selectively prevent their catalytic activity and thus DNA cleavage. (**b**, **c**) Crystal structure of *Spy*Cas9 in complex with an sgRNA in ribbon (**b**) and surface (**c**) views (PDB: 5XBL). Cas9 is colored green, the sgRNA is colored black, and AcrIIA4 is colored yellow. The region selected for LOV2 insertion to generate the CASANOVA construct [21] is shown in light blue

edu/AcrHub/, which contain both validated and putative (computationally predicted) Acrs. They further integrate mechanistic information, statistical tools to analyze Acr diversity, and Acr prediction tools [27, 33]. Together, these databases facilitate the selection of Acrs for use in a given context or application.

Despite a rather large collection of experimentally validated Acrs that exists today, the applications of Acrs in mammalian cells have thus far been driven mainly by a few well-studied members of types II, V, and VI Acr clades. In the following, we will briefly introduce a small collection of Acrs that underlie most of the applications presented in Subheading 2. These Acrs are also a good starting point for newcomers with interest in using Acrs for CRISPR-Cas control in mammalian systems.

Probably the most prominent example of a type II Acr is AcrIIA4 from *Listeria monocytogenes*, which was the first Acr identified as a potent inhibitor of the most widely used Cas9 from *S. pyogenes*. AcrIIA4 employs a dual-inhibitory mechanism: first, AcrIIA4 prevents Cas9 from target DNA binding by mimicking a

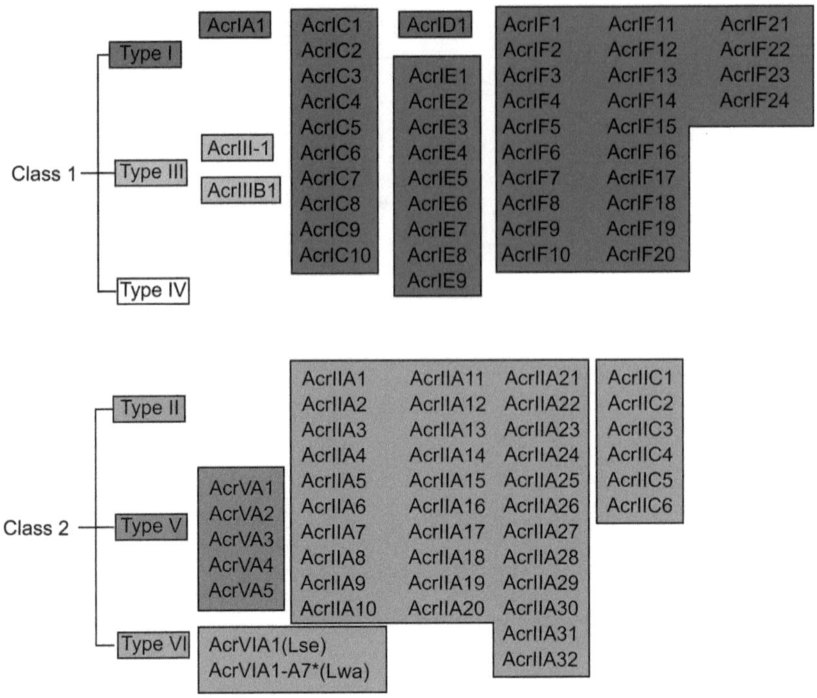

Fig. 3 Overview of experimentally validated Acrs from different Acr families. The known universe of Acrs consists of Acr types corresponding to five different CRISPR-Cas families, including class 1 types I and III and class 2 types II, V, and VI. The figure represents all known, experimentally validated Acrs according to the following database from the Bondy-Denomy lab: https:/tinyurl.com/anti-CRISPR. *AcrVIA1-A7(Lwa): Conflicting results on the functionality of these Acrs have been reported [30, 31]

PAM sequence and interacting with the respective Cas9 domains. At the same time, AcrIIA4 also binds to and blocks the RuvC domain, one of the two Cas9 catalytic domains (Fig. 2b, c) [34, 35]. AcrIIA4 is highly potent and binds Cas9 with nanomolar affinity [36].

AcrIIC1 and AcrII-C3 are important members of the type II-C Acr clade [19]. AcrIIC3 is an effective allosteric inhibitor specific for Cas9s from *Neisseria meningitides* [37]. It works by tethering two Cas9 complexes together, preventing them from binding to target DNA [38]. AcrIIC1, on the other hand, does not interfere with Cas9 DNA binding, but selectively prevents target DNA cleavage by trapping Cas9 in a DNA-bound, but catalytically inactive, state [39]. Interestingly, AcrIIC1 is a broad-spectrum inhibitor targeting a wide range of (mainly) type II-C CRISPR-Cas9 orthologs, including the *N. meningitidis*, *G. stearothermophilus*, and *C. jejuni* Cas9.

Finally, AcrIIA5 is a particularly peculiar member of the type II Acrs derived from *Streptococcus thermophilus* [40]. While the mechanism by which it blocks Cas9 is still not fully understood, AcrIIA5 is known inhibit an astonishingly broad spectrum of Cas9 orthologs, including practically all of the widely used type II-A, type II-B, and type II-C Cas9s (e.g., *Spy*Cas9, *Sau*Cas9, *Nme*Cas9) [41, 42].

In addition to type II inhibitors that target CRISPR-Cas9, type V Acrs have been identified, which effectively block Cas12a [43, 44]. Very importantly, all of the above Acrs are functional when heterologously expressed in mammalian cells.

We note that for Cas13a, protein inhibitors have been reported as well and in two independent papers, one by Meeske et al. [45] and the other by Lin et al. [30]. Meeske et al. did, however, not test their newly discovered AcrVIA1 in mammalian cells, and it thus remains to be seen if it works in mammalian species. Lin et al., on the other hand, claimed Cas13a inhibition in human cells for several of their own AcrVI candidates. However, independent reproduction of the data from Lin et al. by the Bondy-Denomy lab [31] cast doubt on the findings of this study.

For further information on the background on Acr discovery, as well as the structural diversity and inhibitory mechanisms of Acrs, we kindly refer the readers to excellent reviews [46–48].

Taken together, the diversity of Acrs and their versatility in inhibitory mechanisms suggest that the evolutionary arms race between microbes and phages has resulted in a sophisticated CRISPR counter-defense. From an application perspective, these Acrs represent tools that we can use to control and keep in check CRSISPR-Cas effectors. Due to their compact size, high inhibitory potency, and functionality across species boundaries, Acrs represent interesting building blocks to create new CRISPR-Cas regulatory tools via genetic and protein engineering for use in mammalian systems.

In the following, we will review existing applications of Acr proteins in mammalian cells and organisms and discuss the related genetic engineering and protein engineering strategies. We will divide the Acr application space into five categories, namely, (i) permanent block of CRISPR-Cas effector activity, (ii) reduction of off-target editing events by timely CRISPR-Cas inhibition, (iii) use of regulated Acr transgenes to build genetic circuits that limit CRISPR-Cas activity to specific cell types or pathologic cell states, (iv) spatiotemporal control of CRISPR effectors based on engineered, opto-, or chemogenetic Acrs, and (v) the use of Acrs for selective binding and detection of CRISPR-Cas effectors in complex samples (Fig. 4). Table 1 lists a selected panel of Acrs, which underlie the applications introduced below, as well as their mode of action (where known). Finally, we will discuss future avenues to unlock the full potential of Acrs for use in research as well as biomedicine.

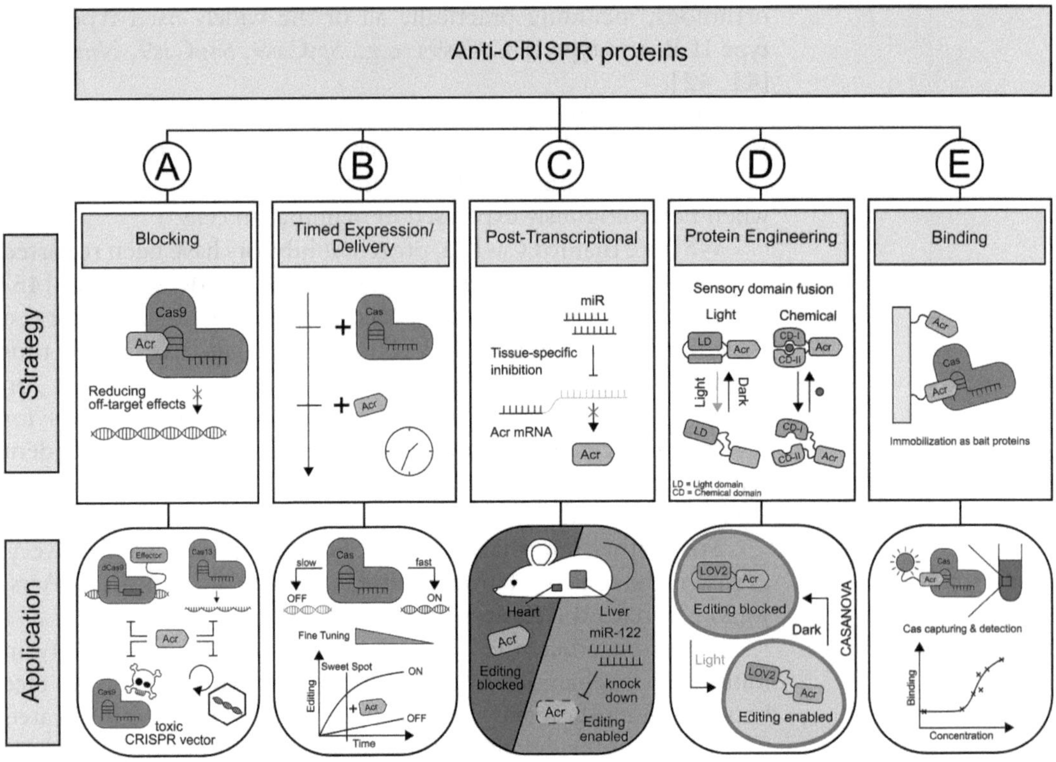

Fig. 4 Strategies for using Acrs for CRISPR-Cas control and corresponding applications in mammalian systems. (**a**) Constitutive inhibition of CRISPR-Cas activity. (**b**) Reduction of off-target editing events by timely inhibition of CRISPR-Cas. (**c**) Precise targeting of CRISPR-Cas activity to specific cell types or pathological cell states through Acr transgene control at the transcriptional and posttranscriptional levels. (**d**) Spatial and/or temporal control of CRISPR effectors using engineered, switchable Acrs that respond to light or chemical ligands. (**e**) Detection of CRISPR RNPs from complex samples using Acrs as binders. See main text for details

2 Applications of Anti-CRISPR Proteins

2.1 Constitutive Expression of Acrs to Protect Cells from CRISPR Gene Editing

Arguably the simplest application of Acrs in mammalian cells is to constitutively overexpress them from Acr-encoding transgenes, thereby rendering cells resistant to the activity of a cognate CRISPR-Cas effector. Pawluk et al. [19] and Rauch et al. [20] were the first to show that Acrs can be used to suppress CRISPR-Cas9 function in human cells. The authors identified and characterized several Acrs that target *Spy*Cas9 and *Nme*Cas9. Importantly, overexpression of these Acrs in HEK293T cells completely blocked CRISPR-Cas9-mediated DNA cleavage, preventing editing of endogenous loci or interference with reporter genes.

In subsequent work, the Lei Qi lab at Stanford University employed lentiviral vectors for genomic integration of Acr transgenes, generating "write-protected" human cell lines [49]. Subsequent delivery of either CRISPR-encoding plasmids or RNPs did not induce DNA DSBs in these engineered cells as

Table 1
Selected set of Acrs used for applications in mammalian systems

Family	Inh. type	Inhibited stage	Mechanism	PDB
AcrIIA4	II-A	DNA binding	Mimics a PAM sequence; also binds and impairs activity of the RuvC domain	5XBL 5VZL 5XN4
AcrIIA2	II-A	DNA binding	Binds to the PAM-interacting, the WED, the HNH, and the REC2 domains	6MCB6IFO
AcrIIA5	II-A	?[a]	?[a]	6LKF
AcrIIA25	II-A	DNA binding/ DNA cleavage	?	–
AcrIIA32	II-A	DNA binding/ DNA cleavage	?	–
AcrIIC1	II-C	DNA cleavage	Binds to the Cas9 HNH nuclease domain and prevents catalytic activity	5VGB 7X31
AcrIIC2	II-C	sgRNA binding	Inhibits Cas9 through interactions with the positively charged bridge helix, thereby preventing sgRNA binding	6J9M 6N05 6J9K 6J9L
AcrIIC3	II-C	DNA binding/ DNA cleavage	Interacts with the HNH domain of Cas9 and induces Cas9 dimerization	6JE9 6JHV
AcrVA1	V-A	DNA binding	Triggers cleavage of the target-recognition sequence of the Cas12a-bound guide RNA, thereby inactivating the Cas12a complex	6NMD
AcrVIA1$_{(Lse)}$	VI-A	?	Blocking RNA targeting[b]	6VRB

[a]Conflicting results [41, 42]
[b]Functionality in human cells remains to be investigated

evidenced by the absence of InDels at targeted genomic loci. This indicates that constitutive overexpression of Acrs can protect cells from the activity of selected CRISPR-Cas effectors and hence preserve their genome integrity.

This feature may be particularly relevant for the future application of gene drives. Gene drives are CRISPR-Cas-based genetic elements that bypass Mendelian inheritance by "copy-pasting" a heterozygous, CRISPR-induced mutation alongside a CRISPR-Cas9 encoding cassette to the second allele. This gene drive-mediated inheritance mechanism thus results in progeny that carry a CRISPR cassette and a related, induced mutation in their homozygous form at high frequency, which can then spread rapidly through the gene pool. Write-protected organisms could help control the spread of gene drives, as was shown by Basgall et al. in *Saccharomyces cerevisiae* [50] and by Taxiarchi et al. in the malaria vector *Anopheles gambiae* [51]. Write protection could also be an important concept for future in vivo gene therapy approaches, where CRISPR genome editing should be focused on a specific target tissue (e.g., liver, muscle, brain), but not occur throughout the rest of the body. In this case, using vectors to deliver Acrs to off-target tissues, i.e., tissues where gene editing should not occur, could help limit therapeutic genome editing to relevant tissues (*see* also Subheading 2.3).

The ability to completely silence CRISPR-Cas activity is also key to the production of self-inactivating (SIN) CRISPR-Cas vectors, e.g., based on adeno-associated virus [52, 53]. A SIN-CRISPR vector encodes a Cas9 and two sgRNAs, one of which is directed against the vector itself. Such SIN-CRISPR systems are of interest, because they allow transient CRISPR-Cas activity, which can reduce unintended off-target activity and thus potentially avoid side effects in a clinical setting. However, both cloning and production of such SIN-CRISPR vectors are challenging because SIN vectors are literally programmed for rapid "self-destruction" by CRISPR-mediated vector cleavage. Thus, even mild expression or activity of the CRISPR components interferes with both cloning and production of the vector itself. To circumvent this problem, the Sontheimer lab at the University of Massachusetts, for instance, developed an Acr-based strategy to facilitate the generation of all-in-one AAV SIN-CRISPR vectors based on a compact *Nme*2Cas9, two sgRNAs (one for locus-targeting, one for vector self-inactivation), and a homology-directed repair (HDR) template [52]. To enable both vector cloning in *E. coli* and packaging into AAV particles in HEK293T production cells, the authors overexpressed AcrIIC4 from *H. parainfluenzae* [54] in the respective hosts. Thereby, Cas9 was selectively inactivated in these hosts, resulting in successful generation of AAV SIN-CRISPR that were subsequently validated in mice.

As described above, Acrs have different mechanisms of action and interfere with their cognate CRISPR effectors at different stages of the CRISPR interference process. Depending on the Acr used, it is therefore possible to selectively interfere with the assembly, target binding, or catalytic activity of CRISPR-Cas effectors. Pawluk et al. cleverly used CRISPR labeling to demonstrate effective blockade of *Nme*Cas9 DNA binding by AcrIIC3 in human cells [19]. They co-transfected U2OS cells with constructs expressing a fluorescently labeled d*Nme*Cas9 as well as an sgRNA targeting

Cas9 to telomere repeat sequences. While in the absence of AcrIIC3, fluorescent dots in the nucleus indicated dCas9 binding to telomere repeats, these dots were absent upon co-expression of AcrIIC3.

In a complementary experiment, Nakamura et al. investigated the potential to silence a CRISPR effector comprising a d*Spy*Cas9 fusion to VPR, a potent transcriptional activator. Constitutive expression of AcrIIA4 indeed blocked the expression of a GFP reporter targeted with the Cas9-VPR construct, indicating that AcrIIA4 prevents Cas9-VPR from binding and thus activating the reporter constructs [49].

These examples indicate that Acr proteins that interfere with CRISPR-Cas DNA binding cannot only be used to prevent genome editing, but can also be employed to block the activity of dCas-based effector proteins in mammalian cells (Fig. 4a).

2.2 Timely Delivery of Acrs to Reduce CRISPR Off-Target Editing Activity and Cellular Toxicity

The above examples showcase the particular value of Acrs and Acr-encoding transgenes in completely preventing the activity of CRISPR effectors in mammalian cells. In addition to constitutive inhibition, timely blockade after a certain period of CRISPR-Cas activity is a particularly interesting approach to enhance genome editing precision. It is well-established that CRISPR effectors, including Cas9 and Cas12, can exhibit off-target editing, i.e., they bind with a certain frequency to genomic loci that are similar but not identical to the actual sgRNA-complementary target sequence. While mining of natural CRISPR-Cas enzymes as well as protein engineering of CRISPR-Cas effectors has led to a steady improvement in targeting specificity [55–57], none of the available CRISPR effectors are entirely perfect.

Off-target editing is a particular concern when working with genetic constructs that continuously express CRISPR-Cas nucleases. This is because both on-target and off-target binding and cleavage are functions of the time that active CRISPR-Cas effectors are present in cells [58]. Although off-target sites are less efficiently targeted by CRISPR-Cas effectors, they will still be bound and cleaved if given infinite time. Therefore, the longer CRISPR-Cas is active, the greater is the risk of unintended off-target mutations. Or vice versa: Limiting the time in which Cas9 is active can selectively reduce off-target editing and hence improve editing specificity without necessarily perturbing on-target editing.

This route of thinking motivated an interesting experiment from the Doudna lab at Berkeley [59]. Shin et al. studied off-target editing by *Spy*Cas9 RNPs targeting either the HBB or VEGF locus in HEK293T cells. The sgRNAs used were both known to have prominent off-target sites. Indeed, when the authors nucleofected the Cas9 RNPs and measured InDels 4 days later, high levels of both on- and off-target editing were observed.

However, when the authors supplied AcrIIA4 either as protein or plasmid 6 h after the initial nucleofection of the Cas9 RNP, off-target editing could be selectively reduced. These findings indicated that timely supply of Acrs can improve genome editing precision in mammalian cells.

Importantly, off-target editing is an issue not only in the context of DSB-dependent genome editing but also in the context of base editing. Base editing is an approach to achieve targeted nucleotide substitutions without inducing DNA DSBs. This strategy is particularly useful because most known human pathogenic mutations are single point mutations, and efficient reversal of these mutations hence provides a targeted, curative strategy [60]. Base editors most commonly consist of a Cas9 nickase (nCas9) fused to a DNA base-altering enzyme. There are two classes of base editors: cytosine base editors (CBEs), which convert C to T, and adenine base editors (ABEs), which convert A to G [61, 62]. Both ABEs and CBEs can induce significant off-target editing of RNAs. Moreover, CBEs in particular have been shown to exhibit DNA off-target editing activity, which raises safety concerns regarding their clinical use [63, 64].

While off-target base editing can be reduced by directly modifying the Cas nuclease [65], Acrs can again provide an elegant means to enhance base-editing specificity [66]. Mingming Liang and co-authors studied AcrIIA5 and demonstrated that it can impair the activity of CRISPR base editors (upon co-delivery with BE3 and ABE7.10) [66]. Building on the strategy of Shin et al., the co-authors then separated the delivery of plasmids encoding the base editor components from the supply of plasmids encoding AcrIIA5 by 3 or 6 h. In analogy to the findings of Shin et al., this timed inhibition strategy resulted in a strong reduction of off-target base editing [66].

Although sequential delivery of CRISPR-Cas effectors and Acrs provides a generalizable strategy for off-target reduction, it can be challenging to implement two individual delivery steps, one for Cas9 and one for an Acr, in numerous relevant application settings. Our group therefore investigated alternative ways to use Acrs for off-target reduction by generally taming CRISPR-Cas9 enzymes, i.e., fine-tuning their activity to just to the level needed to maintain on-target editing while selectively suppressing off-target editing. Using a combination of mathematical modeling and experiments, we studied fusion proteins between *Spy*Cas9 and "artificial inhibitory domains." The latter consisted of AcrIIA4 attenuated by point mutation or insertional mutagenesis, i.e., an artificially weakened Acr only partially inhibiting Cas9. Interestingly, the resulting Cas-Acr fusions exhibited highly favorable on- and off-target editing patterns for several sgRNAs tested and in some cases outperformed previously engineered, high-fidelity variants of Cas9. Mathematical modeling suggested that this gain in target specificity observed at the experimental endpoints was due to what we termed "kinetic insulation" [67]. This term refers to the observation on-target editing usually follows a considerably faster kinetics than off-target editing. Thus, by fine-tuning the activity of Cas9 to selected levels, it is possible to limit genome editing mainly to on-target sites, given a selected experimental observation period. However, this kinetic insulation is only effective if, indeed, the editing kinetics between on-target and corresponding off-target sites are very different, which is not always the case.

Apart from generally increasing chances for off-target editing, the constitutive expression of CRISPR-Cas9 has also been associated with p53-mediated cellular toxicity, particularly in human pluripotent stem cells (hPSCs) [68, 69]. This is a major obstacle for both CRISPR screens and therapeutic editing of stem cells. André Lieber and co-authors sought to address this issue in the context of stem cell engineering for β-thalassemia and sickle cell disease [70]. The authors developed an adenovirus vector system to reactivate the expression of fetal γ-globin. While their CRISPR vector was both well tolerated and effective in human CD34+ cells, and did not affect their in vitro expansion or erythroid differentiation, significant toxicity was observed in primitive hematopoietic stem cells. This was due to prolonged expression and activity of CRISPR-Cas9, which resulted in p53-mediated cellular toxicity. To overcome this obstacle, the authors generated a second adenoviral vector expressing two anti-CRISPR peptides, AcrIIA4 and AcrIIA2. Sequential infection with the CRISPR- and Acr-encoding vectors eventually resulted in a marked improvement in CD34+ cell engraftment in irradiated NOD/Shi-scid/interleukin-2 receptor γ null mice.

Taken together, these data suggest that the timely delivery and/or expression of Acrs can be used to prevent off-target editing and even reduce the toxicity resulting from prolonged CRISPR-Cas9 activity in hematopoietic stem cells (Fig. 4b).

2.3 Transcriptional and Posttranscriptional Control of Acr Transgenes Facilitates Complex CRISPR Circuits As Well As Cell-Type-Specific Genome Editing

So far, we have mostly considered use cases for constitutively overexpressed Acrs. In the next two sections, we will now take a synthetic biology perspective and explore how Acrs or Acr transgenes can be engineered for conditional control of CRISPR-Cas effectors using transcriptional and posttranscriptional (this section) as well as posttranslational (Subheading 2.4) control mechanisms.

The development of conditional Acr transgenes and their combination with conditional CRISPR-Cas-encoding transgenes provides an exciting avenue for the construction of synthetic gene circuits and networks that sense one or more defined inputs, perform a logical operation, and have a defined CRISPR-Cas effector activity as an output.

This concept was nicely exemplified in a recent study by Jianghuai Liu et al. [71]. The researchers created multiple CRISPRa

AND–NOT circuits that link the presence and absence of tumor-promoting or tumor-suppressing transcription factors (TFs) to an immunoregulatory output. The circuit is composed of two modules.

The first module encodes a d*Spy*Cas9 transcriptional activator that targets the INFG gene encoding INFγ, a cytokine involved in tumor immune surveillance. Activation of the circuit would thus generate an immunoregulatory output. The dCas9 effector hereby is driven by a promoter dependent on a selected oncogenic TF and thus only expressed when the TF is present at high abundance.

The second module takes advantage of the fact that loss or impairment of the p53 tumor suppressor is a hallmark of many cancers. The module encodes a NOT gate consisting of AcrIIA4 driven by a p53-dependent promoter. In tumor cells lacking active p53, no AcrIIA4 is made. In healthy cells with normal levels of functional p53, however, AcrIIA4 will be expressed and hence impair *Spy*Cas9 DNA binding.

The combination of both modules therefore results in a logic circuit that selectively promotes an immune response only when p53 is absent and a selected oncogenic TF is present in a given cell. Tumor engraftment experiments in mice with p53-positive and p53-negative cancer cells showed that the CRISPRa gene circuit selectively limits the growth of p53-negative tumors, but not p53-positive tumors. In addition, qPCR analysis of tumor samples at the endpoint of the experiment showed that p53-negative tumors expressed marker genes of the IFNγ axis as well as markers of cytotoxic T cell activity.

Kempton et al. also used Acrs as basic parts of NOT logic, but in a completely different context [72]. They created a split Cas12a-based multi-input, multi-output logic platform that was initially limited to AND operations depending on up to four inputs. By using the Cas12a targeting AcrVA1 under the control of a single input, the researchers were able to combine their multi-AND logic with NOT logic.

The aforementioned study by Jianghuai Liu et al. rendered CRISPR-Cas and Acr expression dependent on cell-type-specific transcription factors to discriminate between healthy and cancer cells. Our group has developed a complementary approach to discriminate between cell types based not on transcriptional control but on posttranscriptional control via mi(cro)RNAs. These are small, regulatory, and noncoding RNAs that play an essential role in the control of gene expression in eukaryotes. As part of the RNA-induced silencing complex (RISC), miRNAs recognize complementary sequence stretches on mRNAs, which are then (usually) silenced either by RISC-mediated mRNA degradation or by transcriptional repression mechanisms [73, 74]. A number of miRNAs have been identified that are present exclusively in specific cell types, most famously miR-122, which is expressed only in hepatocytes.

To make the abundance of such cell-type-specific miRNAs dependent on CRISPR-Cas activity, we generated Acr-encoding transgenes bearing miRNA binding sites in the 3′ untranslated region (UTR). Upon delivery into cells expressing the corresponding miRNA, the Acr-encoding mRNA is knocked down, allowing CRISPR-Cas activity. Conversely, in all cells that do not contain the miRNA trigger, the Acr is expressed at high levels and thus blocks CRISPR-Cas activity. In a study in 2019, we presented this concept for the first time and demonstrated cell-type-specific editing in hepatocyte- and cardiomyocyte-derived cell lines using miR-122- and miR-1-dependent Acr transgenes, respectively [75]. In the same study, we also showed that our approach is compatible with plasmids and viral vectors as well as with different CRISPR-Cas9 orthologs (*Spy*Cas9 and *Nme*Cas9) and cognate Acrs (AcrIIA4, AcrII-C1, and AcrII-C3).

Shortly after our initial report, two additional studies appeared, both proposing very similar concepts [76, 77]. Importantly, the study by Lee et al. included an in vivo validation in mice and was the first to deliver an Acr into a mammalian organism. In this particular experiment, the authors used a dual AAV vector system encoding (i) *Nme2*Cas9 and (ii) a miRNA-122-dependent AcrIIC3 transgene as well as sgRNA targeting the Rosa26 locus. The AAV serotype used was AAV9, which is known to target a broad spectrum of cell types and tissues throughout the mouse body. When the AAVs were co-injected via the mouse tail vein, genome editing occurred exclusively in the liver and not in the heart, the latter of which served as a prominent off-target tissue.

As controls, the authors tested additional vector combinations in which the miRNA-122-dependent AcrIIC3 encoded on the second AAV vector was replaced by a miRNA-122-independent AcrIIC3 cassette (lacking the miR-122 binding sites) or AcrIIA4 (unable to inhibit *Nme2*Cas9 at all) [77]. As expected, these resulted in complete suppression of genome editing (miR-122-independent AcrIIC3 control) or strong editing (AcrIIA4 control) in both the liver and heart, respectively. Together, these findings indicate that the transcriptional and posttranscriptional control of Acr transgenes can be employed to create logic gates or circuits that permit the activity of CRISPR effectors only in selected cell types or pathogenetic cell states.

In addition to the aforementioned examples exploring the use of Acrs to create different types of CRISPR NOT gates, Acrs transgenes can also be used to establish dynamic negative feedback. In an interesting experiment, Nakamura et al. created a negative feedback loop from the activity of a CRISPR effector to its own expression by placing an AcrIIA4-encoding transgene under the control of the Cas9-VPR transcriptional activator in HEK293T cells [49]. Upon induction of Cas9-VPR expression, Cas9-VPR activity was observed as evidenced by activation of a corresponding

GFP reporter. However, as AcrIIA4 levels increased with reporter levels, Cas9-VPR activity was increasingly inhibited, resulting in a pulsatile dynamic of Cas9-VPR.

Collectively, these studies demonstrate the potential of both transcriptional and posttranscriptional regulation of Acr transgenes to target the activity of CRISPR-Cas effectors to specific cells and pathological states, as well as to control the dynamic behavior of CRISPR-Cas effectors (Fig. 4c).

2.4 Engineered Acr-Based Switches for Light- or Ligand-Controlled CRISPR-Cas Inhibition

Complementary to controlling Acrs at the DNA or RNA level, there are several strategies to control Acr activity at the protein level, which arguably is a more direct and hence immediate control modality.

One approach to regulating Acrs is to fuse them to chemically dependent destabilization domains (DDs). In the absence of a specific small molecule, these domains are misfolded and rapidly degraded by the proteasome. The addition of the corresponding ligand, however, restores the proper protein fold, thereby avoiding degradation and facilitating protein expression and accumulation.

In 2019, Nakamura et al. demonstrated the feasibility of controlling CRISPR-based gene regulation using a DD based on a mutant FKBP, which the authors fused to the N-terminus of AcrIIA4. In the absence of the Shield-1 ligand, the DD-AcrIIA4 construct is degraded, whereas Shield-1 stabilized DD-AcrIIA4 in mammalian cells, preventing Cas9 activity [49]. Using a dCas9-VPR as an effector to drive a GFP reporter, the authors demonstrated titratable GFP expression dependent on the concentration of Shield-1 delivered.

In 2021, Huimin Zhao's group established an alternative to the above system consisting of AcrIIA4 fused to a DHFR DD domain, which is stabilized by the FDA-approved drug trimethoprim (TMP) [78]. After optimizing the linkers between AcrIIA4 and the DD, the authors generated a stable HEK293T cell line expressing their construct (called "CRISPR controller"). The researchers then performed editing experiments with different sgRNAs and measured InDel frequencies at on- and off-target loci using GUIDE-seq [79]. Interestingly, the CRISPR controller construct reduced Cas9-mediated off-target editing when low doses of TMP (1 μM) were added to samples.

Apart from the use of exogenous ligands, conditional destabilization of Acr proteins is also an interesting approach to link CRISPR-Cas inhibition to endogenous cell states. It is well known that the efficiency of HDR compared to NHEJ differs across cell cycle stages. HDR is dominant in the S and G2 phases of the cell cycle, whereas NHEJ is more likely to occur in G1. To direct the repair outcome of DNA DSB induction toward HDR, Matsumoto et al. developed a strategy based on cell cycle-dependent expression of an Acr [80]. Specifically, the authors fused AcrIIA4 to the

N-terminal region of human chromatin licensing and DNA replication factor 1 (Cdt1), which is involved in the formation of the pre-replication complex. The AcrIIA4-Cdt1 fusion was then co-expressed with *Spy*Cas9 in mammalian cells from a single episomal vector using a 2A peptide strategy [81]. Importantly, Cdt1 is expressed in G1, but selectively degraded in S and G2. As a consequence, the Cdt1-AcrIIA4 fusion is active in G1, but selectively inactivated in S/G2, thereby facilitating Cas9 activity and the formation of targeted DSBs only in the latter cell cycle stages, which are then frequently repaired via the HDR pathway. Experimental validation in human cells showed that this strategy indeed shifts the balance of repair outcomes after CRISPR-Cas DSB induction strongly toward HDR. In addition, editing at off-target sites was likewise reduced.

An alternative to the use of destabilization domains is the control of Acrs by chemical induction of protein spicing. The group of Yong Tian presented a strategy to construct chemically inducible variants via fusion to a 4-hydroxytamoxifen (4-HT)-responsive intein (37R3-2) [82, 83]. In the presence of the 4-HT ligand, the intein part of the fusion protein undergoes self-splicing, thereby restoring a functional Acr protein capable of inhibiting Cas9. The researchers demonstrated their strategy on AcrIIA25.1 and AcrIIA32.1, two Acrs they identified in this study [83].

While chemical triggers can be easily delivered to cells or organisms, it is difficult to control their activity with high precision in time and space. This is because chemical manipulation is often irreversible and spatial precision is limited by both precision of ligand supply and diffusion.

Light is a powerful alternative to chemical stimulation because it can be applied noninvasively and with spatiotemporal precision. Optogenetics aims to harness light-sensitive proteins and protein domains from nature to control cellular processes. In 2018, our group reported the design of an optogenetic AcrIIA4 whose inhibitory activity can be switched off by blue light. To this end, we generated hybrids between AcrIIA4 and the light-oxygen-voltage 2 (LOV2) domain of *Avena sativa* phototropin-1 by inserting LOV2 into an allosteric surface site of the Acr [21]. The resulting Acr-LOV2 fusion protein, named CASANOVA (for CRISPR-Cas9 activity switching via a novel optogenetic variant of AcrIIA4), blocks *Spy*Cas9 activity in the dark similarly to wild-type AcrIIA4. However, in the presence of blue light, the LOV2 photosensory domain gains flexibility by undocking and unfolding of its terminal helices. This causes a structural change in the AcrIIA4 part of the fusion protein, rendering the inhibitor inactive and hence releasing Cas9 activity. Importantly, since CASANOVA was constructed from an Acr that functions by blocking Cas9 DNA binding, the CASANOVA tool can be used to control both genome editing by catalytically active Cas9 and the activity of dCas9-based effectors.

The latter was demonstrated in two experiments, one using an epigenetic CRISPR construct (Cas9-p300) to induce gene expression from an endogenous locus and the other using light for timed induction of CRISPR labeling of telomeres [21].

The CASANOVA approach can also be adapted to other Acrs and Cas orthologs, as we recently demonstrated for AcrIIC3 and *Nme*Cas9 [84], demonstrating its generalizability as a strategy for engineering light-switchable CRISPR inhibitors.

As sidenote, in collaboration with Bruno E. Correia's lab at EPFL, our group has also applied protein engineering strategies to generally improve natural Acrs. Using the broad-spectrum inhibitor AcrIIC1 as a scaffold, we employed protein domain insertion as well as structure-based protein surface design approaches to increase the inhibitory potency of AcrIIC1 for two of its Cas9 target orthologs, namely, *Nme*Cas9 and *Sau*Cas9 [85]. Thus, in addition to creating chemically or light-controlled switchable Acrs, protein engineering can also be used to generally improve the inhibition potency of Acrs if needed.

Together, the above examples illustrate the power of engineering and regulating Acrs at the protein level (Fig. 4d). Fusing Acrs to receptors or inducible inteins responding to exogenous triggers such as light or chemicals enables the temporal and/or spatial control of CRISPR-Cas effector activity. In turn, pairing Acrs with protein domains that are part of the endogenous cell regulation can limit CRISPR-Cas activity to defined cellular conditions, such as selected cell cycle stages.

2.5 Employing Acrs for CRISPR-Cas Detection

In all of the above examples, Acrs were delivered to cells to either constitutively or conditionally block CRISPR-Cas activity. However, the fact that many natural Acrs tightly bind their cognate Cas effectors also makes Acrs interesting agents for affinity-based CRISPR-Cas detection, similar to antibodies or nanobodies. Such platforms to detect CRISPR-Cas effectors and determine their concentrations even from complex samples such as body fluids are of great clinical relevance, e.g., to study and optimize the pharmacokinetics and pharmacodynamics as well as the distribution of CRISPR-Cas-based in vivo therapeutics.

Due to their commonly high binding affinity to their cognate Cas orthologs, Acrs are ideal candidates for the detection of CRISPR RNPs. Both broad-spectrum and single-ortholog-specific Acrs are of interest in this context. Broad-spectrum Acrs can be used to bind and detect a wide range of CRISPR-Cas orthologs in a single assay reaction. In turn, highly specific Acrs that bind only a single Cas ortholog can potentially be used to discriminate between different Cas effectors present in the same sample.

In 2019, two studies reported the use of AcrIIA4 and AcrIIC1, respectively, for the implementation of CRISPR-Cas9 detection platforms [86, 87]. Acrs were therefore immobilized as bait proteins on surfaces. After incubation with CRISPR-RNA-containing samples, Cas9 was detected and quantified using electrochemical, fluorescence, or colorimetric methods. For AcrIIC1, the authors demonstrated the detection of several type II-C Cas9 orthologs, including *N. meningitidis*, *G. stearothermophilus*, and *C. jejuni* Cas9. In addition, the platform was compatible with complex samples, including whole blood.

Similar detection platforms were also built on the basis of type V Acrs to detect Cas12 RNPs [88].

These studies highlight the potential of Acrs for the development of biosensor platforms for the specific detection of Cas proteins and RNPs in complex samples, including clinical samples (Fig. 4e).

3 Conclusions and Future Perspectives

The discovery of Acrs has started a rapidly growing line of research and provided a highly valuable addition to the growing CRISPR-Cas toolbox. The versatility of Acr proteins is demonstrated by the wide range of applications they already offer today, including (i) sustained blockade of CRISPR-Cas function; (ii) reduction of off-target editing events through timely Cas inhibition; (iii) targeting of CRISPR-Cas activity to specific cell types or pathologic cell states via transcriptional and posttranscriptional Acr transgene control; (iv) spatiotemporal control of CRISPR effectors using engineered, switchable Acrs; and (v) detection of CRISPR RNPs from complex (human) samples. Since Acrs can be delivered in trans, the above strategies are in principle compatible with many existing CRISPR-Cas vectors and cellular and organismal systems.

It is important to note that Acr-based regulatory strategies are not intended to (and probably cannot) replace alternative strategies for controlling and/or safeguarding CRISPR technologies. Instead, it is expected that Acr-mediated CRISPR-Cas control will in the future be combined with complementary approaches, e.g., based on conditional CRISPR-Cas transgenes or engineered Cas nucleases or sgRNAs [1, 57, 89, 90]. Such synergistic combination could provide an unprecedented level of control over CRISPR systems, enabling highly targeted and precise genome and transcriptome perturbations in mammalian cells and organisms.

Recent advances in protein engineering, e.g., based on Alphafold-2 [91–93], will likely have an impact on our ability to tinker with Acrs and enable the creation of new types of switchable or otherwise improved Acr variants, potentially even de novo inhibitors of CRISPR-Cas systems. Thus far, switchable Acrs have been engineered, with few exceptions, on the basis of AcrIIA4, and the corresponding variants respond either to blue light or to a set of chemical ligands of limited clinical utility. Also, considerable leakiness and a suboptimal dynamic range of control are factors thus far limiting the broader applicability of Acr-based switches. In our view, it would be of great value to equip Acrs targeting a wide range of Cas9, Cas12, or Cas13 orthologs with tight control via a set of clinically approved drugs [94, 95]. The resulting Acr switches could help "dose" the activity of CRISPR effectors in in vivo application settings to maintain efficiency and likewise avoid potential adverse effects.

Moreover, the control of Acrs with optogenetic strategies is also still at the beginning. In addition to the *As*LOV2 domain, there are many other LOV domains that have been found in plants, fungi, and bacteria [96]. These could potentially be used to create powerful optogenetic Acrs with improved dynamic range of light control and functioning in different modes (light-on and light-off). Furthermore, since blue light cannot penetrate deeply into tissues, it would be advantageous to develop optogenetic Acrs based on red or even far-red light sensing photoreceptors, such as BphP1 from *Rhodopseudomonas palustris*, a well-studied bacterial phytochrome [97, 98].

On top of engineering new types of switchable Acrs, nature itself has evolved Acrs that exhibit switchable behavior depending on environmental conditions. For example, AcrIIA2 is a highly potent *Spy*Cas9 inhibitor at room temperature. At 37 °C, however, AcrIIA2 barely functions and is therefore unable to efficiently block genome editing in human cells (which are cultured at 37 °C), even when supplied in high amounts [99]. In addition, *Nme*AcrIIC1 and several of its orthologs were recently found to undergo a redox dependent monomer–dimer interconversion mediated by a pair of cysteines [100]. Oxidizing conditions lead to the formation of impaired AcrIIC1 dimers, while reducing conditions result in active AcrIIC1 monomers. Considering that cancer cells typically show high levels of reactive oxygen species (ROS) compared to healthy cells, sensitive Acr redox switches may provide interesting vehicles to target CRISPR-Cas activity to malignant tissue. The exploration of such natural Acr switches for CRISPR-Cas control could therefore greatly complement the aforementioned Acr protein engineering efforts.

Regarding clinical applications, in the nearer future, Acrs could become very useful in phage therapy protocols to treat infections with multidrug-resistant (MDR) bacteria. Many human pathogens have CRISPR-Cas systems that render them resistant to certain types of phages. Therefore, phages used for phage therapy should be equipped with a battery of broad-spectrum Acrs to broaden their host range and prevent the bacterial CRISPR immune system from adapting to the phage therapy. The research group led by Min Wu recently demonstrated this concept. They engineered phages with three different type I Acrs, enabling them to block CRISPR-Cas immunity and thus infect and replicate efficiently in MDR *Pseudomonas aeruginosa* [101].

Further into the future, complex CRISPR-Cas gene circuits employing Acrs as mediators hold great potential to limit CRISPR-mediated genome or transcriptome perturbations to specific cell types or diseased tissues, thereby improving the specificity and safety of CRISPR-based therapeutics. To achieve this goal, however, it will be crucial to conduct in vivo studies of Acrs in mammalian model organisms. The Sontheimer lab has already taken an important first step in this direction, finding no evidence of toxicity or adverse effects following viral delivery of AcrIIC3 and AcrIIA4 to mice [77]. Additional research dedicated at studying Acrs in mammalian organisms, including their expression, activity, immunogenicity, and potential toxicity, will undoubtedly be key for the future development of CRISPR therapeutics that benefit from Acr control.

Taken together, the discovery, study, and engineering of Acrs has provided the CRISPR-Cas community with a powerful regulatory layer to keep CRISPR-Cas effectors in check. This regulatory layer enhances the precision with which we can make CRISPR genome and transcriptome perturbations in mammalian cells, thereby paving the way for a broad range of applications in basic research, synthetic biology, and biomedicine.

Acknowledgments

We thank the research group of Dominik Niopek for feedback on this manuscript. C.M.G. is a member of the Graduate School Life Science Engineering at TU Darmstadt. D.N. acknowledges funding by the German Research Foundation (DFG) [project no. 453202693], the Schwiete Stiftung, and the Aventis Foundation.

References

1. Jinek M, Chylinski K, Fonfara I, Hauer M, Doudna JA, Charpentier E (2012) A programmable dual-RNA–guided DNA endonuclease in adaptive bacterial immunity. Science 337:816–821. https://doi.org/10.1126/science.1225829
2. Barrangou R, Fremaux C, Deveau H, Richards M, Boyaval P, Moineau S, Romero DA, Horvath P (2007) CRISPR provides acquired resistance against viruses in prokaryotes. Science 315:1709–1712. https://doi.org/10.1126/science.1138140
3. Brouns SJJ, Jore MM, Lundgren M, Westra ER, Slijkhuis RJH, Snijders APL, Dickman MJ, Makarova KS, Koonin EV, van der Oost J (2008) Small CRISPR RNAs guide antiviral defense in prokaryotes. Science 321:960–964. https://doi.org/10.1126/science.1159689
4. Mojica FJM, Díez-Villaseñor C, García-Martínez J, Soria E (2005) Intervening sequences of regularly spaced prokaryotic repeats derive from foreign genetic elements. J Mol Evol 60:174–182. https://doi.org/10.1007/s00239-004-0046-3
5. Cong L, Ran FA, Cox D, Lin S, Barretto R, Habib N, Hsu PD, Wu X, Jiang W, Marraffini LA, Zhang F (2013) Multiplex genome

engineering using CRISPR/Cas systems. Science 339:819–823. https://doi.org/10.1126/science.1231143

6. Mali P, Yang L, Esvelt KM, Aach J, Guell M, DiCarlo JE, Norville JE, Church GM (2013) RNA-guided human genome engineering via Cas9. Science 339:823–826. https://doi.org/10.1126/science.1232033
7. Ran FA, Cong L, Yan WX, Scott DA, Gootenberg JS, Kriz AJ, Zetsche B, Shalem O, Wu X, Makarova KS, Koonin EV, Sharp PA, Zhang F (2015) In vivo genome editing using Staphylococcus aureus Cas9. Nature 520:186–191. https://doi.org/10.1038/nature14299
8. Hou Z, Zhang Y, Propson NE, Howden SE, Chu L-F, Sontheimer EJ, Thomson JA (2013) Efficient genome engineering in human pluripotent stem cells using Cas9 from Neisseria meningitidis. Proc Natl Acad Sci U S A 110:15644–15649. https://doi.org/10.1073/pnas.1313587110
9. Shmakov S, Abudayyeh OO, Makarova KS, Wolf YI, Gootenberg JS, Semenova E, Minakhin L, Joung J, Konermann S, Severinov K, Zhang F, Koonin EV (2015) Discovery and functional characterization of diverse class 2 CRISPR-Cas systems. Mol Cell 60:385–397. https://doi.org/10.1016/j.molcel.2015.10.008
10. Zetsche B, Gootenberg JS, Abudayyeh OO, Slaymaker IM, Makarova KS, Essletzbichler P, Volz SE, Joung J, van der Oost J, Regev A, Koonin EV, Zhang F (2015) Cpf1 is a single RNA-guided endonuclease of a class 2 CRISPR-Cas system. Cell 163:759–771. https://doi.org/10.1016/j.cell.2015.09.038
11. Zetsche B, Heidenreich M, Mohanraju P, Fedorova I, Kneppers J, DeGennaro EM, Winblad N, Choudhury SR, Abudayyeh OO, Gootenberg JS, Wu WY, Scott DA, Severinov K, van der Oost J, Zhang F (2017) Multiplex gene editing by CRISPR–Cpf1 using a single crRNA array. Nat Biotechnol 35:31–34. https://doi.org/10.1038/nbt.3737
12. Abudayyeh OO, Gootenberg JS, Essletzbichler P, Han S, Joung J, Belanto JJ, Verdine V, Cox DBT, Kellner MJ, Regev A, Lander ES, Voytas DF, Ting AY, Zhang F (2017) RNA targeting with CRISPR–Cas13. Nature 550:280–284. https://doi.org/10.1038/nature24049
13. Cox DBT, Gootenberg JS, Abudayyeh OO, Franklin B, Kellner MJ, Joung J, Zhang F (2017) RNA editing with CRISPR-Cas13. Science 358:1019–1027. https://doi.org/10.1126/science.aaq0180
14. Anzalone AV, Koblan LW, Liu DR (2020) Genome editing with CRISPR–Cas nucleases, base editors, transposases and prime editors. Nat Biotechnol 38:824–844. https://doi.org/10.1038/s41587-020-0561-9
15. Wang JY, Pausch P, Doudna JA (2022) Structural biology of CRISPR–Cas immunity and genome editing enzymes. Nat Rev Microbiol 20:641–656. https://doi.org/10.1038/s41579-022-00739-4
16. Zhang F (2019) Development of CRISPR-Cas systems for genome editing and beyond. Q Rev Biophys 52:e6. https://doi.org/10.1017/S0033583519000052
17. Bondy-Denomy J, Pawluk A, Maxwell KL, Davidson AR (2013) Bacteriophage genes that inactivate the CRISPR/Cas bacterial immune system. Nature 493:429–432. https://doi.org/10.1038/nature11723
18. Bondy-Denomy J, Garcia B, Strum S, Du M, Rollins MF, Hidalgo-Reyes Y, Wiedenheft B, Maxwell KL, Davidson AR (2015) Multiple mechanisms for CRISPR–Cas inhibition by anti-CRISPR proteins. Nature 526:136–139. https://doi.org/10.1038/nature15254
19. Pawluk A, Amrani N, Zhang Y, Garcia B, Hidalgo-Reyes Y, Lee J, Edraki A, Shah M, Sontheimer EJ, Maxwell KL, Davidson AR (2016) Naturally occurring off-switches for CRISPR-Cas9. Cell 167:1829–1838.e9. https://doi.org/10.1016/j.cell.2016.11.017
20. Rauch BJ, Silvis MR, Hultquist JF, Waters CS, McGregor MJ, Krogan NJ, Bondy-Denomy J (2017) Inhibition of CRISPR-Cas9 with bacteriophage proteins. Cell 168:150–158. https://doi.org/10.1016/j.cell.2016.12.009
21. Bubeck F, Hoffmann MD, Harteveld Z, Aschenbrenner S, Bietz A, Waldhauer MC, Börner K, Fakhiri J, Schmelas C, Dietz L, Grimm D, Correia BE, Eils R, Niopek D (2018) Engineered anti-CRISPR proteins for optogenetic control of CRISPR–Cas9. Nat Methods 15:924–927. https://doi.org/10.1038/s41592-018-0178-9
22. Forsberg KJ (2023) Anti-CRISPR discovery: using magnets to find needles in haystacks. J Mol Biol 167952. https://doi.org/10.1016/j.jmb.2023.167952
23. Shehreen S, Birkholz N, Fineran PC, Brown CM (2022) Widespread repression of anti-CRISPR production by anti-CRISPR-associated proteins. Nucleic Acids Res 50:8615–8625. https://doi.org/10.1093/nar/gkac674

24. Stanley SY, Borges AL, Chen K-H, Swaney DL, Krogan NJ, Bondy-Denomy J, Davidson AR (2019) Anti-CRISPR-associated proteins are crucial repressors of anti-CRISPR transcription. Cell 178:1452–1464.e13. https://doi.org/10.1016/j.cell.2019.07.046
25. Eitzinger S, Asif A, Watters KE, Iavarone AT, Knott GJ, Doudna JA, Minhas F u AA (2020) Machine learning predicts new anti-CRISPR proteins. Nucleic Acids Res 48:4698–4708. https://doi.org/10.1093/nar/gkaa219
26. Gussow AB, Park AE, Borges AL, Shmakov SA, Makarova KS, Wolf YI, Bondy-Denomy J, Koonin EV (2020) Machine-learning approach expands the repertoire of anti-CRISPR protein families. Nat Commun 11: 3784. https://doi.org/10.1038/s41467-020-17652-0
27. Wang J, Dai W, Li J, Xie R, Dunstan RA, Stubenrauch C, Zhang Y, Lithgow T (2020) PaCRISPR: a server for predicting and visualizing anti-CRISPR proteins. Nucleic Acids Res 48:W348–W357. https://doi.org/10.1093/nar/gkaa432
28. Yi H, Huang L, Yang B, Gomez J, Zhang H, Yin Y (2020) AcrFinder: genome mining anti-CRISPR operons in prokaryotes and their viruses. Nucleic Acids Res 48:W358–W365. https://doi.org/10.1093/nar/gkaa351
29. Zhu L, Wang X, Li F, Song J (2022) PreAcrs: a machine learning framework for identifying anti-CRISPR proteins. BMC Bioinform 23: 444. https://doi.org/10.1186/s12859-022-04986-3
30. Lin P, Qin S, Pu Q, Wang Z, Wu Q, Gao P, Schettler J, Guo K, Li R, Li G, Huang C, Wei Y, Gao GF, Jiang J, Wu M (2020) CRISPR-Cas13 inhibitors block RNA editing in bacteria and mammalian cells. Mol Cell 78: 850–861.e5. https://doi.org/10.1016/j.molcel.2020.03.033
31. Johnson MC, Hille LT, Kleinstiver BP, Meeske AJ, Bondy-Denomy J (2022) Lack of Cas13a inhibition by anti-CRISPR proteins from Leptotrichia prophages. Mol Cell 82: 2161–2166.e3. https://doi.org/10.1016/j.molcel.2022.05.002
32. Bondy-Denomy J, Davidson AR, Doudna JA, Fineran PC, Maxwell KL, Moineau S, Peng X, Sontheimer EJ, Wiedenheft B (2018) A unified resource for tracking anti-CRISPR names. CRISPR J 1:304–305. https://doi.org/10.1089/crispr.2018.0043
33. Dong C, Wang X, Ma C, Zeng Z, Pu D-K, Liu S, Wu C-S, Chen S, Deng Z, Guo F-B (2022) Anti-CRISPRdb v2.2: an online repository of anti-CRISPR proteins including information on inhibitory mechanisms, activities and neighbors of curated anti-CRISPR proteins. Database 2022:baac010. https://doi.org/10.1093/database/baac010
34. Dong D, Guo M, Wang S, Zhu Y, Wang S, Xiong Z, Yang J, Xu Z, Huang Z (2017) Structural basis of CRISPR–SpyCas9 inhibition by an anti-CRISPR protein. Nature 546: 436–439. https://doi.org/10.1038/nature22377
35. Yang H, Patel DJ (2017) Inhibition mechanism of an anti-CRISPR suppressor AcrIIA4 targeting SpyCas9. Mol Cell 67:117–127.e5. https://doi.org/10.1016/j.molcel.2017.05.024
36. Kim I, Jeong M, Ka D, Han M, Kim N-K, Bae E, Suh J-Y (2018) Solution structure and dynamics of anti-CRISPR AcrIIA4, the Cas9 inhibitor. Sci Rep 8:3883. https://doi.org/10.1038/s41598-018-22177-0
37. Kim Y, Lee SJ, Yoon H, Kim N, Lee B, Suh J (2019) Anti-CRISPR AcrIIC3 discriminates between Cas9 orthologs via targeting the variable surface of the HNH nuclease domain. FEBS J 286:4661–4674. https://doi.org/10.1111/febs.15037
38. Sun W, Yang J, Cheng Z, Amrani N, Liu C, Wang K, Ibraheim R, Edraki A, Huang X, Wang M, Wang J, Liu L, Sheng G, Yang Y, Lou J, Sontheimer EJ, Wang Y (2019) Structures of Neisseria meningitidis Cas9 complexes in catalytically poised and anti-CRISPR-inhibited states. Mol Cell 76:938–952.e5. https://doi.org/10.1016/j.molcel.2019.09.025
39. Harrington LB, Doxzen KW, Ma E, Liu J-J, Knott GJ, Edraki A, Garcia B, Amrani N, Chen JS, Cofsky JC, Kranzusch PJ, Sontheimer EJ, Davidson AR, Maxwell KL, Doudna JA (2017) A broad-Spectrum inhibitor of CRISPR-Cas9. Cell 170:1224–1233.e15. https://doi.org/10.1016/j.cell.2017.07.037
40. Hynes AP, Rousseau GM, Lemay M-L, Horvath P, Romero DA, Fremaux C, Moineau S (2017) An anti-CRISPR from a virulent streptococcal phage inhibits streptococcus pyogenes Cas9. Nat Microbiol 2:1374–1380. https://doi.org/10.1038/s41564-017-0004-7
41. Garcia B, Lee J, Edraki A, Hidalgo-Reyes Y, Erwood S, Mir A, Trost CN, Seroussi U, Stanley SY, Cohn RD, Claycomb JM, Sontheimer EJ, Maxwell KL, Davidson AR (2019) Anti-CRISPR AcrIIA5 potently inhibits all Cas9 homologs used for genome editing. Cell Rep 29:1739–1746.e5. https://doi.org/10.1016/j.celrep.2019.10.017

42. Song G, Zhang F, Zhang X, Gao X, Zhu X, Fan D, Tian Y (2019) AcrIIA5 inhibits a broad range of Cas9 orthologs by preventing DNA target cleavage. Cell Rep 29:2579–2589. https://doi.org/10.1016/j.celrep.2019.10.078
43. Marino ND, Zhang JY, Borges AL, Sousa AA, Leon LM, Rauch BJ, Walton RT, Berry JD, Joung JK, Kleinstiver BP, Bondy-Denomy J (2018) Discovery of widespread type I and type V CRISPR-Cas inhibitors. Science 362: 240–242. https://doi.org/10.1126/science.aau5174
44. Watters KE, Fellmann C, Bai HB, Ren SM, Doudna JA (2018) Systematic discovery of natural CRISPR-Cas12a inhibitors. Science 362:236–239. https://doi.org/10.1126/science.aau5138
45. Meeske AJ, Jia N, Cassel AK, Kozlova A, Liao J, Wiedmann M, Patel DJ, Marraffini LA (2020) A phage-encoded anti-CRISPR enables complete evasion of type VI-A CRISPR-Cas immunity. Science 369:54–59. https://doi.org/10.1126/science.abb6151
46. Jia N, Patel DJ (2021) Structure-based functional mechanisms and biotechnology applications of anti-CRISPR proteins. Nat Rev Mol Cell Biol 22:563–579. https://doi.org/10.1038/s41580-021-00371-9
47. Davidson AR, Lu W-T, Stanley SY, Wang J, Mejdani M, Trost CN, Hicks BT, Lee J, Sontheimer EJ (2020) Anti-CRISPRs: protein inhibitors of CRISPR-Cas systems. Annu Rev Biochem 89:309–332. https://doi.org/10.1146/annurev-biochem-011420-111224
48. Marino ND, Pinilla-Redondo R, Csörgő B, Bondy-Denomy J (2020) Anti-CRISPR protein applications: natural brakes for CRISPR-Cas technologies. Nat Methods 17:471–479. https://doi.org/10.1038/s41592-020-0771-6
49. Nakamura M, Srinivasan P, Chavez M, Carter MA, Dominguez AA, La Russa M, Lau MB, Abbott TR, Xu X, Zhao D, Gao Y, Kipniss NH, Smolke CD, Bondy-Denomy J, Qi LS (2019) Anti-CRISPR-mediated control of gene editing and synthetic circuits in eukaryotic cells. Nat Commun 10:194. https://doi.org/10.1038/s41467-018-08158-x
50. Basgall EM, Goetting SC, Goeckel ME, Giersch RM, Roggenkamp E, Schrock MN, Halloran M, Finnigan GC (2018) Gene drive inhibition by the anti-CRISPR proteins AcrIIA2 and AcrIIA4 in Saccharomyces cerevisiae. Microbiology 164:464–474. https://doi.org/10.1099/mic.0.000635
51. Taxiarchi C, Beaghton A, Don NI, Kyrou K, Gribble M, Shittu D, Collins SP, Beisel CL, Galizi R, Crisanti A (2021) A genetically encoded anti-CRISPR protein constrains gene drive spread and prevents population suppression. Nat Commun 12:3977. https://doi.org/10.1038/s41467-021-24214-5
52. Ibraheim R, Tai PWL, Mir A, Javeed N, Wang J, Rodríguez TC, Namkung S, Nelson S, Khokhar ES, Mintzer E, Maitland S, Chen Z, Cao Y, Tsagkaraki E, Wolfe SA, Wang D, Pai AA, Xue W, Gao G, Sontheimer EJ (2021) Self-inactivating, all-in-one AAV vectors for precision Cas9 genome editing via homology-directed repair in vivo. Nat Commun 12:6267. https://doi.org/10.1038/s41467-021-26518-y
53. Li A, Lee CM, Hurley AE, Jarrett KE, De Giorgi M, Lu W, Balderrama KS, Doerfler AM, Deshmukh H, Ray A, Bao G, Lagor WR (2019) A self-deleting AAV-CRISPR system for in vivo genome editing. Mol Ther Methods Clin Dev 12:111–122. https://doi.org/10.1016/j.omtm.2018.11.009
54. Lee J, Mir A, Edraki A, Garcia B, Amrani N, Lou HE, Gainetdinov I, Pawluk A, Ibraheim R, Gao XD, Liu P, Davidson AR, Maxwell KL, Sontheimer EJ (2018) Potent Cas9 inhibition in bacterial and human cells by AcrIIC4 and AcrIIC5 anti-CRISPR proteins. mBio 9:e02321–18. https://doi.org/10.1128/mBio.02321-18
55. Amrani N, Gao XD, Liu P, Edraki A, Mir A, Ibraheim R, Gupta A, Sasaki KE, Wu T, Donohoue PD, Settle AH, Lied AM, McGovern K, Fuller CK, Cameron P, Fazzio TG, Zhu LJ, Wolfe SA, Sontheimer EJ (2018) NmeCas9 is an intrinsically high-fidelity genome-editing platform. Genome Biol 19: 214. https://doi.org/10.1186/s13059-018-1591-1
56. Hu JH, Miller SM, Geurts MH, Tang W, Chen L, Sun N, Zeina CM, Gao X, Rees HA, Lin Z, Liu DR (2018) Evolved Cas9 variants with broad PAM compatibility and high DNA specificity. Nature 556:57–63. https://doi.org/10.1038/nature26155
57. Slaymaker IM, Gao L, Zetsche B, Scott DA, Yan WX, Zhang F (2016) Rationally engineered Cas9 nucleases with improved specificity. Science 351:84–88. https://doi.org/10.1126/science.aad5227
58. Eslami-Mossallam B, Klein M, Smagt CVD, Sanden KVD, Jones SK, Hawkins JA, Finkelstein IJ, Depken M (2022) A kinetic model predicts SpCas9 activity, improves off-target classification, and reveals the physical basis of

targeting fidelity. Nat Commun 13:1367. https://doi.org/10.1038/s41467-022-28994-2

59. Shin J, Jiang F, Liu J-J, Bray NL, Rauch BJ, Baik SH, Nogales E, Bondy-Denomy J, Corn JE, Doudna JA (2017) Disabling Cas9 by an anti-CRISPR DNA mimic. Sci Adv 3: e1701620. https://doi.org/10.1126/sciadv.1701620
60. Rees HA, Liu DR (2018) Base editing: precision chemistry on the genome and transcriptome of living cells. Nat Rev Genet 19:770–788. https://doi.org/10.1038/s41576-018-0059-1
61. Komor AC, Kim YB, Packer MS, Zuris JA, Liu DR (2016) Programmable editing of a target base in genomic DNA without double-stranded DNA cleavage. Nature 533:420–424. https://doi.org/10.1038/nature17946
62. Gaudelli NM, Komor AC, Rees HA, Packer MS, Badran AH, Bryson DI, Liu DR (2017) Programmable base editing of A•T to G•C in genomic DNA without DNA cleavage. Nature 551:464–471. https://doi.org/10.1038/nature24644
63. Grünewald J, Zhou R, Garcia SP, Iyer S, Lareau CA, Aryee MJ, Joung JK (2019) Transcriptome-wide off-target RNA editing induced by CRISPR-guided DNA base editors. Nature 569:433–437. https://doi.org/10.1038/s41586-019-1161-z
64. Zuo E, Sun Y, Wei W, Yuan T, Ying W, Sun H, Yuan L, Steinmetz LM, Li Y, Yang H (2019) Cytosine base editor generates substantial off-target single-nucleotide variants in mouse embryos. Science 364:289–292. https://doi.org/10.1126/science.aav9973
65. Liu Y, Zhou C, Huang S, Dang L, Wei Y, He J, Zhou Y, Mao S, Tao W, Zhang Y, Yang H, Huang X, Chi T (2020) A Cas-embedding strategy for minimizing off-target effects of DNA base editors. Nat Commun 11:6073. https://doi.org/10.1038/s41467-020-19690-0
66. Liang M, Sui T, Liu Z, Chen M, Liu H, Shan H, Lai L, Li Z (2020) AcrIIA5 suppresses Base editors and reduces their off-target effects. Cell 9:1786. https://doi.org/10.3390/cells9081786
67. Aschenbrenner S, Kallenberger SM, Hoffmann MD, Huck A, Eils R, Niopek D (2020) Coupling Cas9 to artificial inhibitory domains enhances CRISPR-Cas9 target specificity. Sci Adv 6:eaay0187. https://doi.org/10.1126/sciadv.aay0187
68. Álvarez MM, Biayna J, Supek F (2022) TP53-dependent toxicity of CRISPR/Cas9 cuts is differential across genomic loci and can confound genetic screening. Nat Commun 13: 4520. https://doi.org/10.1038/s41467-022-32285-1
69. Ihry RJ, Worringer KA, Salick MR, Frias E, Ho D, Theriault K, Kommineni S, Chen J, Sondey M, Ye C, Randhawa R, Kulkarni T, Yang Z, McAllister G, Russ C, Reece-Hoyes J, Forrester W, Hoffman GR, Dolmetsch R, Kaykas A (2018) p53 inhibits CRISPR–Cas9 engineering in human pluripotent stem cells. Nat Med 24:939–946. https://doi.org/10.1038/s41591-018-0050-6
70. Li C, Psatha N, Gil S, Wang H, Papayannopoulou T, Lieber A (2018) HDAd5/35++ adenovirus vector expressing anti-CRISPR peptides decreases CRISPR/Cas9 toxicity in human hematopoietic stem cells. Mol Ther Methods Clin Dev 9:390–401. https://doi.org/10.1016/j.omtm.2018.04.008
71. Wang Y, Zhang G, Meng Q, Huang S, Guo P, Leng Q, Sun L, Liu G, Huang X, Liu J (2022) Precise tumor immune rewiring via synthetic CRISPRa circuits gated by concurrent gain/loss of transcription factors. Nat Commun 13: 1454. https://doi.org/10.1038/s41467-022-29120-y
72. Kempton HR, Goudy LE, Love KS, Qi LS (2020) Multiple input sensing and signal integration using a Split Cas12a system. Mol Cell 78:184–191.e3. https://doi.org/10.1016/j.molcel.2020.01.016
73. He L, Hannon GJ (2004) MicroRNAs: small RNAs with a big role in gene regulation. Nat Rev Genet 5:522–531. https://doi.org/10.1038/nrg1379
74. Schmiedel JM, Klemm SL, Zheng Y, Sahay A, Blüthgen N, Marks DS, van Oudenaarden A (2015) MicroRNA control of protein expression noise. Science 348:128–132. https://doi.org/10.1126/science.aaa1738
75. Hoffmann MD, Aschenbrenner S, Grosse S, Rapti K, Domenger C, Fakhiri J, Mastel M, Börner K, Eils R, Grimm D, Niopek D (2019) Cell-specific CRISPR–Cas9 activation by microRNA-dependent expression of anti-CRISPR proteins. Nucleic Acids Res 47: e75–e75. https://doi.org/10.1093/nar/gkz271
76. Hirosawa M, Fujita Y, Saito H (2019) Cell-type-specific CRISPR activation with MicroRNA-responsive AcrIlA4 switch. ACS Synth Biol 8:1575–1582. https://doi.org/10.1021/acssynbio.9b00073
77. Lee J, Mou H, Ibraheim R, Liang S-Q, Liu P, Xue W, Sontheimer EJ (2019) Tissue-restricted genome editing in vivo specified by

microRNA-repressible anti-CRISPR proteins. RNA 25:1421–1431. https://doi.org/10.1261/rna.071704.119

78. Jain S, Xun G, Abesteh S, Ho S, Lingamaneni M, Martin TA, Tasan I, Yang C, Zhao H (2021) Precise regulation of Cas9-mediated genome engineering by anti-CRISPR-based inducible CRISPR controllers. ACS Synth Biol 10:1320–1327. https://doi.org/10.1021/acssynbio.0c00548

79. Tsai SQ, Zheng Z, Nguyen NT, Liebers M, Topkar VV, Thapar V, Wyvekens N, Khayter C, Iafrate AJ, Le LP, Aryee MJ, Joung JK (2015) GUIDE-seq enables genome-wide profiling of off-target cleavage by CRISPR-Cas nucleases. Nat Biotechnol 33:187–197. https://doi.org/10.1038/nbt.3117

80. Matsumoto D, Tamamura H, Nomura W (2020) A cell cycle-dependent CRISPR-Cas9 activation system based on an anti-CRISPR protein shows improved genome editing accuracy. Commun Biol 3:601. https://doi.org/10.1038/s42003-020-01340-2

81. Liu Z, Chen O, Wall JBJ, Zheng M, Zhou Y, Wang L, Ruth Vaseghi H, Qian L, Liu J (2017) Systematic comparison of 2A peptides for cloning multi-genes in a polycistronic vector. Sci Rep 7:2193. https://doi.org/10.1038/s41598-017-02460-2

82. Peck SH, Chen I, Liu DR (2011) Directed evolution of a small-molecule-triggered Intein with improved splicing properties in mammalian cells. Chem Biol 18:619–630. https://doi.org/10.1016/j.chembiol.2011.02.014

83. Song G, Zhang F, Tian C, Gao X, Zhu X, Fan D, Tian Y (2022) Discovery of potent and versatile CRISPR–Cas9 inhibitors engineered for chemically controllable genome editing. Nucleic Acids Res 50:2836–2853. https://doi.org/10.1093/nar/gkac099

84. Hoffmann MD, Mathony J, Upmeier zu Belzen J, Harteveld Z, Aschenbrenner S, Stengl C, Grimm D, Correia BE, Eils R, Niopek D (2021) Optogenetic control of Neisseria meningitidis Cas9 genome editing using an engineered, light-switchable anti-CRISPR protein. Nucleic Acids Res 49:e29–e29. https://doi.org/10.1093/nar/gkaa1198

85. Mathony J, Harteveld Z, Schmelas C, Upmeier zu Belzen J, Aschenbrenner S, Sun W, Hoffmann MD, Stengl C, Scheck A, Georgeon S, Rosset S, Wang Y, Grimm D, Eils R, Correia BE, Niopek D (2020) Computational design of anti-CRISPR proteins with improved inhibition potency. Nat Chem Biol 16:725–730. https://doi.org/10.1038/s41589-020-0518-9

86. Johnston RK, Seamon KJ, Saada EA, Podlevsky JD, Branda SS, Timlin JA, Harper JC (2019) Use of anti-CRISPR protein AcrIIA4 as a capture ligand for CRISPR/Cas9 detection. Biosens Bioelectron 141:111361. https://doi.org/10.1016/j.bios.2019.111361

87. Phaneuf CR, Seamon KJ, Eckles TP, Sinha A, Schoeniger JS, Harmon B, Meagher RJ, Abhyankar VV, Koh C-Y (2019) Ultrasensitive multi-species detection of CRISPR-Cas9 by a portable centrifugal microfluidic platform. Anal Methods 11:559–565. https://doi.org/10.1039/C8AY02726A

88. Qiao J, Lin S, Sun W, Ma L, Liu Y (2020) A method for the quantitative detection of Cas12a ribonucleoproteins. Chem Commun 56:12616–12619. https://doi.org/10.1039/D0CC04019C

89. Pattanayak V, Lin S, Guilinger JP, Ma E, Doudna JA, Liu DR (2013) High-throughput profiling of off-target DNA cleavage reveals RNA-programmed Cas9 nuclease specificity. Nat Biotechnol 31:839–843. https://doi.org/10.1038/nbt.2673

90. Mir A, Alterman JF, Hassler MR, Debacker AJ, Hudgens E, Echeverria D, Brodsky MH, Khvorova A, Watts JK, Sontheimer EJ (2018) Heavily and fully modified RNAs guide efficient SpyCas9-mediated genome editing. Nat Commun 9:2641. https://doi.org/10.1038/s41467-018-05073-z

91. Anishchenko I, Pellock SJ, Chidyausiku TM, Ramelot TA, Ovchinnikov S, Hao J, Bafna K, Norn C, Kang A, Bera AK, DiMaio F, Carter L, Chow CM, Montelione GT, Baker D (2021) De novo protein design by deep network hallucination. Nature 600:547–552. https://doi.org/10.1038/s41586-021-04184-w

92. Jendrusch M, Korbel JO, Sadiq SK (2021) AlphaDesign: a de novo protein design framework based on AlphaFold. bioRxiv. https://doi.org/10.1101/2021.10.11.463937

93. Jumper J, Evans R, Pritzel A, Green T, Figurnov M, Ronneberger O, Tunyasuvunakool K, Bates R, Žídek A, Potapenko A, Bridgland A, Meyer C, Kohl SAA, Ballard AJ, Cowie A, Romera-Paredes B, Nikolov S, Jain R, Adler J, Back T, Petersen S, Reiman D, Clancy E, Zielinski M, Steinegger M, Pacholska M, Berghammer T, Bodenstein S, Silver D, Vinyals O, Senior AW, Kavukcuoglu K, Kohli P, Hassabis D (2021) Highly accurate protein structure prediction with AlphaFold. Nature 596:583–589.

https://doi.org/10.1038/s41586-021-03819-2
94. Rihtar E, Lebar T, Lainšček D, Kores K, Lešnik S, Bren U, Jerala R (2023) Chemically inducible split protein regulators for mammalian cells. Nat Chem Biol 19:64–71. https://doi.org/10.1038/s41589-022-01136-x
95. Shui S, Gainza P, Scheller L, Yang C, Kurumida Y, Rosset S, Georgeon S, Di Roberto RB, Castellanos-Rueda R, Reddy ST, Correia BE (2021) A rational blueprint for the design of chemically-controlled protein switches. Nat Commun 12:5754. https://doi.org/10.1038/s41467-021-25735-9
96. Hoffmann MD, Bubeck F, Eils R, Niopek D (2018) Controlling cells with light and LOV. Adv Biosys 2:1800098. https://doi.org/10.1002/adbi.201800098
97. Shcherbakova DM, Shemetov AA, Kaberniuk AA, Verkhusha VV (2015) Natural photoreceptors as a source of fluorescent proteins, biosensors, and optogenetic tools. Annu Rev Biochem 84:519–550. https://doi.org/10.1146/annurev-biochem-060614-034411
98. Shcherbakova DM, Baloban M, Emelyanov AV, Brenowitz M, Guo P, Verkhusha VV (2016) Bright monomeric near-infrared fluorescent proteins as tags and biosensors for multiscale imaging. Nat Commun 7:12405. https://doi.org/10.1038/ncomms12405
99. Jiang F, Liu J-J, Osuna BA, Xu M, Berry JD, Rauch BJ, Nogales E, Bondy-Denomy J, Doudna JA (2019) Temperature-responsive competitive inhibition of CRISPR-Cas9. Mol Cell 73:601–610.e5. https://doi.org/10.1016/j.molcel.2018.11.016
100. Zhao Y, Hu J, Yang S-S, Zhong J, Liu J, Wang S, Jiao Y, Jiang F, Zhai R, Ren B, Cong H, Zhu Y, Han F, Zhang J, Xu Y, Huang Z, Zhang S, Yang F (2022) A redox switch regulates the assembly and anti-CRISPR activity of AcrIIC1. Nat Commun 13:7071. https://doi.org/10.1038/s41467-022-34551-8
101. Qin S, Liu Y, Chen Y, Hu J, Xiao W, Tang X, Li G, Lin P, Pu Q, Wu Q, Zhou C, Wang B, Gao P, Wang Z, Yan A, Nadeem K, Xia Z, Wu M (2022) Engineered bacteriophages containing anti-CRISPR suppress infection of antibiotic-Resistant P. aeruginosa. Microbiol Spectr 10:e01602–e01622. https://doi.org/10.1128/spectrum.01602-22

Chapter 15

Posttranslational Remote Control Mediated by Physical Inducers for Rapid Protein Release in Engineered Mammalian Cells

Maysam Mansouri and Martin Fussenegger

Abstract

Physical cues such as light, heat, or an electrical field can be utilized for traceless, on-demand activation of the expression of a desired therapeutic gene in appropriately engineered cells with excellent spatiotemporal resolution, good inducibility, and simple reversibility. A similar approach can be applied to build a depolarization-based protein secretion system that enables rapid release of a therapeutic protein pre-stored in intracellular vesicles in mammalian cells. Here, we present a protocol to create designer β-cells that exhibit light-controllable rapid release (within 15 min) of a pre-synthesized proinsulin–nanoluciferase construct from vesicular stores. The construct is cleaved extracellularly to afford secreted insulin as a therapeutic protein and nanoluciferase as a reporter molecule. Such posttranslational remote control offers a much faster response than expression-based systems.

Key words Synthetic biology, Mammalian cell engineering, Gene regulation, Physical inducers, Rapid protein secretion

1 Introduction

Synthetic biology is a multidisciplinary field of science that involves redesigning organisms to provide new capabilities by introducing genetic circuits in the form of a "sensing–processing–production" platform that can sense a user-defined input, process the biological information, and generate a customized output in response [1, 2]. Regulation of such genetic circuits can take place at various levels, such as transcription (RNA synthesis), posttranscription (RNA modification or editing), translation (protein synthesis), and posttranslation (protein modification). Although the programming of cell behavior in the earlier steps often leads to an efficient and robust functional activity (gene expression), there is a significant lag time in the production of the desired output [3]. In

Francesca Ceroni and Karen Polizzi (eds.), *Mammalian Synthetic Systems*, Methods in Molecular Biology, vol. 2774,
https://doi.org/10.1007/978-1-0716-3718-0_15,

contrast, posttranslational control systems can provide a customized response in a short period of time [4].

The input (inducer) that triggers the genetic circuit may be either chemical or physical [2, 5]. Chemical inducers, including organic and inorganic compounds, stimulatory peptides, and odorants, offer high induction potential and ease of use, but may suffer from poor biodistribution, insufficient bioavailability, or inappropriate pharmacodynamics [1]. Conversely, physical stimuli provide a robust, safe, and efficient way to control desired target tissues with a high level of spatiotemporal resolution. Physical cues also reduce the likelihood of cross-reactivity and off-target effects, as well as the need for invasive access to the targeted tissue or organ [6, 7].

We have used three well-known physical stimuli, light, heat, and an electrical field, to remotely program therapeutic cell behaviors. The therapeutic designer cell is an engineered pancreatic β-cell line, 1.1E7, expressing a synthetic proinsulin–nanoluciferase (nLuc) construct that leads to the co-secretion of insulin and nLuc in equimolar amounts [8–10]. Insulin and nLuc are produced as interconnected prohormone that traverses the endoplasmic reticulum and the Golgi apparatus before reaching the secretory vesicles where the prohormone convertases 2 and 1/3 process the prohormone to native insulin and nLuc [11]. In the unstimulated condition, these designer β-cells produce and store proinsulin–nLuc within granular vesicles. The vesicles release the stored protein only upon user-defined physical stimulation, irrespective of the glucose concentration in the vicinity. In order to create light-, heat-, and electro-responsive designer β-cells, appropriate sensors were engineered into the cells. When these sensors detect the trigger signal, they activate an intracellular calcium pathway within the cells, depolarizing the cell membrane and releasing the pre-synthesized proinsulin–nLuc (Fig. 1a). To create light-responsive β-cells, which release the stored protein in response to blue light (≈470 nm), we employed ectopically expressed melanopsin, a blue light-sensitive G-protein-coupled receptor, as a sensor (Fig. 1b) [12]. To engineer heat-responsive cells, we constructed β-cells with constitutive expression of the TRPV1 ion channel, which mediates Ca^{2+} influx, leading to release of the stored protein, in response to high temperature (42 °C) (Fig. 1c) [13]. Finally, to obtain electro-responsive β-cells, we utilized a voltage-gated channel circuit consisting of an L-type voltage-gated calcium channel ($Ca_V1.2$) and an inwardly rectifying potassium channel ($K_{ir}2.1$) (Fig. 1d) [8]. All three types of engineered β-cells showed peak secretion of insulin and nLuc within 10–15 min after stimulation.

The following protocol describes in detail how to create a depolarization-based fast-release protein secretion system in mammalian cells, focusing on light-controllable β-cells as an example. A similar protocol would be applicable for different physical stimuli and for other therapeutic proteins and/or reporter molecules.

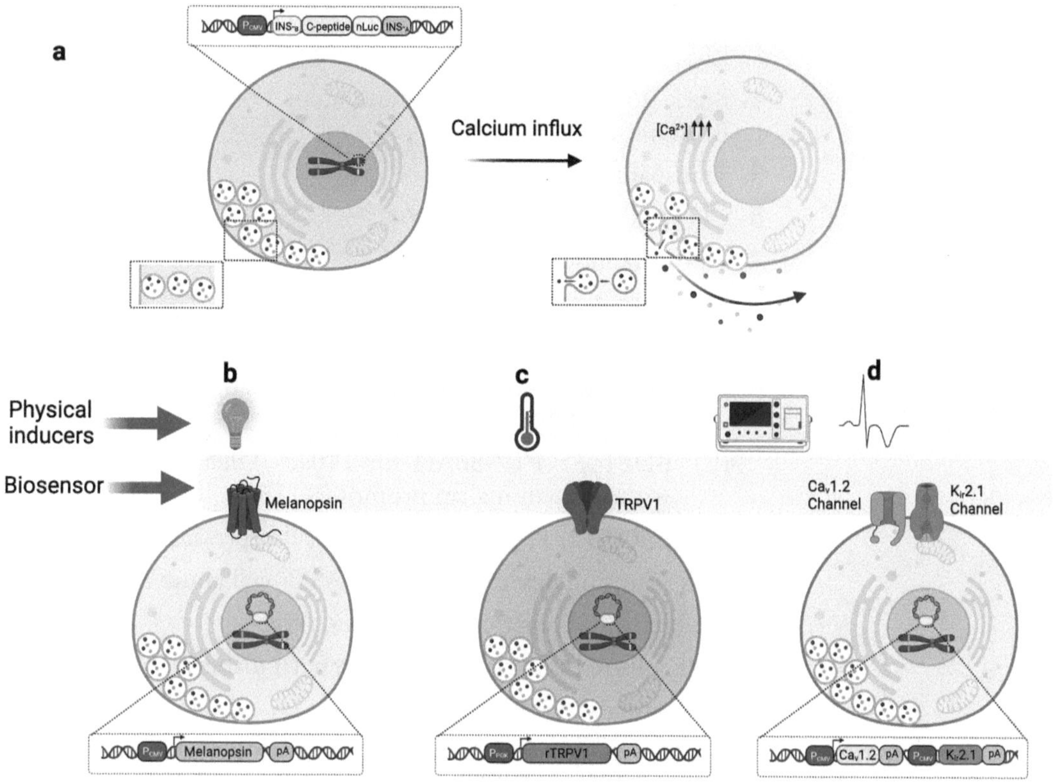

Fig. 1 Remote control of rapid therapeutic-reporter protein secretion by engineered β-cells. (**a**) 1. 1E7-derived cell clone deficient in glucose-sensitive insulin secretion was engineered to co-express proinsulin–nLuc from a synthetic expression unit. In the unstimulated condition, engineered β-cells produce and store proinsulin–nLuc within granular vesicles. These granules release stored protein only in response to user-defined physical stimulation. (**b–d**) Physical inducers. Light (**b**), heat (**c**), and an electrical field (**d**) can be used to remotely induce release of proinsulin–nLuc from granular vesicles from engineered cells equipped with the appropriate genetically encoded biosensor to convert the physical trigger signal into a biological signal (calcium influx)

2 Materials

2.1 Construction of a Stable β-Cell Line Co-expressing Insulin–nLuc

1. 1.1E7 β cells.
2. An 1.1E7-derived cell clone deficient in glucose-sensitive insulin secretion was constructed as described in detail elsewhere [8]).
3. A monoclonal stable cell (called INS_{vesc}-A12-A15) was selected and maintained in culture medium containing 5 μg/mL blasticidin for 14 days, and cloned by limiting dilution.

2.2 Cell Culture

1. Roswell Park Memorial Institute 1640 medium supplemented with 10% fetal bovine serum, 100 U/mL penicillin, and 100 μg/mL streptomycin (penicillin–streptomycin solution 100x).

2. Trypsin–EDTA solution (0.05% (v/v)).
3. PBS.
4. Black 96-well plates with transparent bottoms (e.g., Greiner Bio-One).
5. Transparent 96-well plates.
6. CO_2 tissue culture incubator.
7. Lipofectamine 3000.
8. 1 mM stock of all-trans-retinal (ATR).
9. Plasmids (*see* **Note 1**):
   (a) pHY42; P_{CMV}-OPN4-pA; Melanopsin expression vector [14].
   (b) pDF145; PT7-SpAH-Env140ac; Filler DNA (plasmid without mammalian promoter) [15].
   (c) pFOX12; P_{CMV}-eGFP-pA; eGFP expression vector.
10. Cell counting device (e.g., the CASY® cell counter).
11. Multichannel pipette and tray (reservoir).
12. Cell culture plasticware (100 mm petri dishes).
13. Centrifuge.
14. Multi-well light induction platform with blue LEDs (475 nm, 20–30 cd, 50 mA, 5 mm) [16].
15. Arduino™ microcontroller and software and Arduino drivers.
16. Opti-MEM™.

2.3 nLuc Measurement

1. Nano-Glo® Luciferase Assay System (Promega).
2. Multichannel pipette and tray (reservoir).
3. Microplate, 384-well, PS, F-bottom, med. Binding, black, 10 pcs/bag (Greiner).
4. Plate reader.

2.4 Insulin Measurement

1. Rat/mouse insulin ELISA kit (Mercodia).
2. Multichannel pipette and tray (reservoir).
3. ThermoMixer.
4. Plate reader.

3 Methods

3.1 Plate the β-Cell Line

1. Cultivate 1.1E7 INS_{vesc}-A12-A15 cells in 12 mL of complete RPMI (supplemented with 10% FBS and 1% penicillin–streptomycin) at 37 °C in an atmosphere of 5% CO_2.

2. After the cells reach 70–80% confluence, aspirate the medium and wash the cells with PBS. Aspirate the PBS and expose the cells to 1 mL of 0.05% (v/v) trypsin–EDTA solution for 5–10 min at 37 °C in the incubator.
3. Remove the trypsinized cells from the incubator, and add 5 mL of complete RPMI to the cells. Detach the cells from the plate by gentle tapping.
4. Harvest the cells from the plate, transfer them into a 15 mL conical tube, and centrifuge for 5 min at 200× *g*.
5. Discard the supernatant, and resuspend the cells in 10 mL of complete RPMI.
6. Count the cells using the cell counting device.
7. Prepare 12 mL of seeding cell suspension in a 15 mL tube at a concentration of 6.2×10^4 cells/well in a 24-well.
8. Transfer 500 μL of the cells to each well of a 24-well plate.
9. Incubate the plate with the seeded β-cells at 37 °C under 5% CO_2 for 12–24 h prior to transfection.

3.2 Transfection of Light-Sensitive Receptor into β-Cell Line

1. Prepare a master mix for 3 wells of a 24-well plate with pHY42 and pDF145 (200 ng: 200 ng for each well) and for 3 wells of a 24-well plate with pFOX12 and pDF145 (200 ng: 200 ng for each well). Per well, add 20 μL of Opti-MEM™ and 1 μL of Lipofectamine 3000 in a tube (tube A) and 20 μL of Opti-MEM™ solution with 1 μL of P3000 in a separate tube (tube B) (*see* **Notes 2** and **3**).
2. Mix the two tubes (A and B) thoroughly and incubate them at room temperature for 5 min.
3. Add 20 μL from Tube A to Tube B, pipette up and down, and incubate the mixture for another 20 min at room temperature.
4. Transfer 40 μL of Lipofectamine 3000/DNA/Opti-MEM™ mixture to the desired wells.
5. Incubate the plate with the seeded β-cells at 37 °C under 5% CO_2 for 12–24 h prior to transfection (*See* **Note 4**).
6. Centrifuge the plate for 1 min at 200× *g*.
7. Incubate transfected cells for 12 h at 37 °C under 5% CO_2 (*see* **Note 5**).
8. Dispose of the supernatant, and repeat steps described in Subheading 3.1, **steps 1–4**.
9. Resuspend the transfected cells in 2 mL of complete RPMI.
10. Count the cells using the cell counting device and prepare 2 mL of seeding cell suspension at a concentration of 3.2×10^4 cells/96-well.

11. Transfer 120 μL of the cells to each well of a 96-well plate (two rows: one for melanopsin-expressing cells and the other one for eGFP-expressing cells).
12. Incubate the plate with the seeded β-cells at 37 °C under 5% CO_2 for 12–24 h prior to transfection.

3.3 Illumination Experiment

1. Program the LED platform using the Arduino™ microcontroller to 475 nm (10 s ON and 5 s OFF) at 300 μW cm^{-2}.
2. Take 12 mL of prewarmed colorless medium (+10% FBS), add 5 μM ATR, and transfer into a pipetting reservoir (*see* **Note 6**).
3. Gently remove the medium from the transfected cell plate by tapping it upside–down onto a stack of paper towels in the dark.
4. Gently add 120 μL of the medium supplemented with ATR to each well of the 96-well plate using the multichannel pipette in the dark.
5. Repeat **steps 3** and **4** four times more (all in the dark).
6. Add 120 μL of the fresh medium supplemented with ATR to each well of the 96-well plate in the dark. Incubate the plates at 37 °C under 5% CO_2 for 15 min.
7. Gently remove the medium, store the samples at 4 °C, and label them as "Dark."
8. Add 120 μL of the fresh medium supplemented with ATR to each well of the 96-well plate. Put the LED array on top of the 96-well plate and connect it to the power supply. Incubate the plates at 37 °C under 5% CO_2 for 15 min.
9. Gently remove the medium and store the samples at 4 °C and label them as "Light."
10. Add 120 μL of the fresh medium supplemented with 40 mM KCl to each well of the 96-well plate. Incubate the plates at 37 °C under 5% CO_2 for 15 min.
11. Gently remove the medium and store the samples at 4 °C and label them as "KCl."

3.4 nLuc Secretion Measurement

1. Transfer 7.5 μL of the nLuc-containing cell supernatant from the Dark, Light, and KCl samples to a 384-well assay plate using a multichannel pipette (*See* **Note** 7).
2. Add 7.5 μL of the substrate/buffer (1:50) mixture to each well using a multichannel pipette.
3. Centrifuge the plate for 1 min at 200× *g*.
4. Incubate the plate for 3 min at room temperature.
5. Measure the bioluminescence for 1 s with a plate reader.

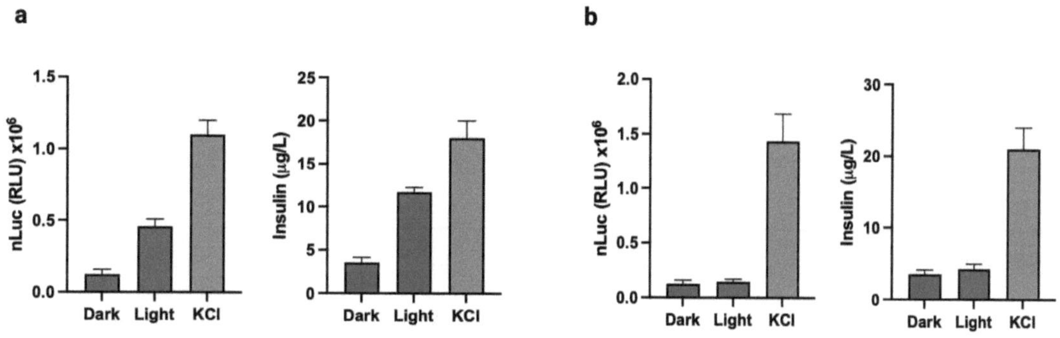

Fig. 2 Light-controlled rapid insulin–nLuc secretion. (**a**). Quantifications of nLuc (left) and insulin (right) released from melanopsin-transfected β-cells in the presence and absence of light. KCl was used as a potent membrane-depolarizing agent (positive control). (**b**) nLuc (left) and insulin (right) levels secreted from eGFP-transfected β-cells in the presence and absence of trigger illumination. eGFP was used as a control to confirm the illumination itself did not cause depolarization. KCl was used as a potent membrane-depolarizing agent

3.5 Insulin Secretion Measurement

1. Transfer 10 μL of the insulin-containing cell supernatant from the Dark, Light, and KCl samples to a 96-well assay plate provided with the ELISA kit using a multichannel pipette.
2. Follow the protocol of the commercial ELISA kit and measure the absorbance at 450 nm with a plate reader (*See* **Note 8**).

Illumination with blue light activates the melanopsin receptor on the cell surface and triggers calcium influx, leading to depolarization of the membrane and the release of the stored nLuc and insulin (Fig. 1). The levels of secreted nLuc and insulin are quantified in Fig. 2.

4 Notes

1. The gene encoding the reporter nLuc can in principle be exchanged for any other desired gene, encoding either a therapeutic protein or other reporter (e.g., a fluorescent protein).
2. The amount of insulin–nLuc release can be modulated by varying the transfection ratio of melanopsin DNA.
3. 1.1E7 cells show 10–20% transfection efficiency. Transduction with viruses (e.g., lentiviruses) or electroporation can increase the efficiency of gene delivery to these cells.
4. The generation of a monoclonal stable cell line expressing melanopsin (called iβ cell line) helps to increase fold induction and reduce leakiness (*See* also [12]).
5. After transfection, cells should be handled in the dark. Alternatively, a hood with a safe red light can be used.
6. ATR stock should be freshly prepared.

7. Increasing the number of plated cells can increase the fold induction. *See also* [12], Supplementary Fig. 6.
8. The level of nLuc–insulin release can be fine-tuned by applying various intensities and cycles (ON/OFF) of light induction. *See also* [12], Fig. 2.

Acknowledgments

Work in the laboratory of M.F. is financially supported in part through a European Research Council advanced grant (ElectroGene, no. 785800) and in part by the National Centre of Competence in Research (NCCR) for Molecular Systems Engineering as well as the EC Horizon 2020 Framework Programme ENLIGHT.

References

1. Xie M, Fussenegger M (2018) Designing cell function: assembly of synthetic gene circuits for cell biology applications. Nat Rev Mol Cell Biol 19:507–525
2. Mansouri M, Fussenegger M (2021) Therapeutic cell engineering: designing programmable synthetic genetic circuits in mammalian cells. Protein Cell 13:476. https://doi.org/10.1007/S13238-021-00876-1
3. Rippe RA, Stefanovic B (2005) Methods for assessing the molecular mechanisms controlling gene regulation. Methods Mol Med 117:141–160. https://doi.org/10.1385/1-59259-940-0:141
4. Mansouri M, Ray PG, Franko N, Xue S, Fussenegger M (2013) Design of programmable post-translational switch control platform for on-demand protein secretion in mammalian cells. Nucleic Acids Res 1:13–14. https://doi.org/10.1093/NAR/GKAC916
5. Mansouri M, Fussenegger M (2021) Remote control of mammalian therapeutic designer cells. 53–67. https://doi.org/10.1007/978-3-030-79871-0_2
6. Mansouri M, Fussenegger M (2021) Synthetic biology-based optogenetic approaches to control therapeutic designer cells. Curr Opin Syst Biol 28:100396. https://doi.org/10.1016/J.COISB.2021.100396
7. Mansouri M, Fussenegger M (2022) Electrogenetics: bridging synthetic biology and electronics to remotely control the behavior of mammalian designer cells. Curr Opin Chem Biol 68:102151. https://doi.org/10.1016/J.CBPA.2022.102151
8. Krawczyk K, Xue S, Buchmann P, Charpin-El-Hamri G, Saxena P, Hussherr MD, Shao J, Ye H, Xie M, Fussenegger M (2020) Electrogenetic cellular insulin release for real-time glycemic control in type 1 diabetic mice. Science 368:993–1001. https://doi.org/10.1126/science.aau7187
9. McCluskey JT, Hamid M, Guo-Parke H, McClenaghan NH, Gomis R, Flatt PR (2011) Development and functional characterization of insulin-releasing human pancreatic beta cell lines produced by electrofusion. J Biol Chem 286:21982–21992. https://doi.org/10.1074/jbc.M111.226795
10. Burns SM, Vetere A, Walpita D, Dančík V, Khodier C, Perez J, Clemons PA, Wagner BK, Altshuler D (2015) High-throughput luminescent reporter of insulin secretion for discovering regulators of pancreatic beta-cell function. Cell Metab 21:126–137. https://doi.org/10.1016/j.cmet.2014.12.010
11. Hou JC, Min L, Pessin JE (2009) Insulin granule biogenesis, trafficking and exocytosis. Vitam Horm 80:473. https://doi.org/10.1016/S0083-6729(08)00616-X
12. Mansouri M, Xue S, Hussherr M-D, Strittmatter T, Camenisch G, Fussenegger M (2021) Smartphone-flashlight-mediated remote control of rapid insulin secretion restores glucose homeostasis in experimental Type-1 diabetes. Small 2101939. https://doi.org/10.1002/SMLL.202101939
13. Stefanov B-A, Mansouri M, Charpin-El Hamri G, Fussenegger M, Stefanov B, Mansouri M, Fussenegger M, Charpin-El Hamri G (2022) Sunlight-controllable biopharmaceutical production for remote

emergency supply of directly injectable therapeutic proteins. Small 18:2202566. https://doi.org/10.1002/SMLL.202202566

14. Ye H, El BMD, Peng RW, Fussenegger M (2011) A synthetic optogenetic transcription device enhances blood-glucose homeostasis in mice. Science 332:1565–1568. https://doi.org/10.1126/science.1203535

15. Ausländer S, Fuchs D, Hürlemann S, Ausländer D, Fussenegger M (2016) Engineering a ribozyme cleavage-induced split fluorescent aptamer complementation assay. Nucleic Acids Res 44:94. https://doi.org/10.1093/nar/gkw117

16. Mansouri M, Lichtenstein S, Strittmatter T, Buchmann P, Fussenegger M (2020) Construction of a multiwell light-induction platform for traceless control of gene expression in mammalian cells. In: Methods in molecular biology. Humana Press Inc., pp 189–199

Chapter 16

Detection of MicroRNAs Using Synthetic Toehold Switch in Mammalian Cells

Yuwen Zhao, Pratima Poudel, and Shue Wang

Abstract

Engineering synthetic gene circuits to control cellular functions has a broad application in the field of synthetic biology. Synthetic RNA-based switches that can operate at the transcriptional and posttranscriptional level have also drawn significant interest for the application of next-generation therapeutics and diagnostics. Thus, RNA-based switchable platforms are needed to report dynamic cellular mechanisms which play an important role in cell development and diseases. Recently, several RNA-based switches have been designed and utilized for biosensing and molecular diagnostics. However, miRNA-based switches have not been well established or characterized, especially for eukaryotic translational control. Here, we designed a novel synthetic toehold switch for detection of exogenously and endogenously expressed miRNAs in CHO, HeLa, HEK 293, and MDA-MB-231 breast cancer cells. Multiplex detection of miR-155 and miR-21 was tested using two toehold switches to evaluate the orthogonality and programmability of this synthetic platform.

Key words MicroRNA, Synthetic toehold switch, miRNA-based switch, Synthetic biology

1 Introduction

MicroRNAs (miRNAs) are short, long noncoding RNA molecules that have been demonstrated to play critical roles in different biological processes including cell proliferation, cell apoptosis, and development. miRNAs regulate gene expression by mediating translational repression through targeting message RNA (mRNA) [1–6]. Increasing evidence has shown that miRNAs play an important role in the development of human diseases, including angiogenesis [7–9], cancer [10, 11], cardiovascular diseases [12], and infectious diseases [13]. Recent studies have shown that several miRNAs were directly functioning as oncogenes or tumor suppressors in human cancers, including breast, lung, brain, liver, and colon cancers and leukemia [14–16]. Aberrantly expressed miRNAs could modulate the epigenetic status of the genome, leading to the changes in cancer-associated genes that consequently affect

Francesca Ceroni and Karen Polizzi (eds.), *Mammalian Synthetic Systems*, Methods in Molecular Biology, vol. 2774, https://doi.org/10.1007/978-1-0716-3718-0_16,

initiation, progression, and metastasis of human cancers [17]. For example, miR-21 and miR-155 are two multifunctional miRNAs that play critical roles in various cancers. Thus, miRNA detection and visualization are critical for studying miRNAs' regulated pathways to identify new biomarkers for anticancer therapy. Although current techniques such as microarrays, Northern blot, and RT-qPCR are capable of measuring aggregated miRNA expression levels, the studies of miRNA expression dynamics at the single cell level are limited. Molecular probes have been utilized to detect mRNA, miRNA, and protein in mammalian cells [18–21]; however, the promise of mammalian synthetic biology is to couple an RNA input to regulate gene expression. The current challenge in the field of miRNA biology is to obtain spatiotemporal miRNA expression information under pathophysiological conditions, which are required to identify multimodal miRNAs, characterize expression variability, and enable accurate investigation of cancer.

In the past decade, engineering synthetic gene circuits has gained significant interest due to their capacity of regulating cell behaviors for basic research and biomedical applications [22]. Engineering transcriptional regulatory circuits depends on the availability of programmable, sequence-specific, and powerful synthetic transcription factors. Recently, synthetic RNA-based gene switches have gained researchers' interests due to the ability of engineering cellular functions in mammalian cells [23–25]. For example, Saito et al. constructed a set of RNA-based logic circuits using miRNA- and protein-responsive mRNAs as decision-making controllers to control cellular response [24]. They demonstrated that an apoptosis–AND gate could selectively eliminate target cells by sensing two different miRNAs. Li et al. developed a toehold-initiated rolling circle amplification for visualizing individual miRNAs in single cells [26]. Hu et al. designed an RNA-based synthetic switch to minimize CRISPR off-target effects in mammalian cells [27]. Thus, RNA-based switches have been constructed in diverse contexts to regulate or control gene expressions. Recently, several groups have developed synthetic gene circuits for miRNA sensing in mammalian cells. Kei et al. developed multiple microRNA-responsive synthetic mRNAs for high-resolution identification and separation of living cells [28]. The same group designed several synthetic miRNA-based switches to isolate and separate human pluripotent stem cell (hPSC)-derived cardiomyocytes [29]. Saito et al. developed a miRNA-responsive CRISPR-Cas9 switch using a miRNA-complementary sequence in the 5'-UTR of mRNA encoding *Streptococcus pyogenes* Cas9 [30]. They reported miR-21-Cap9 or miR-302-Cas9 switches respond to miRNAs and posttranscriptionally attenuated the Cas9 activity.

Recently, Green et al. created de novo-designed riboregulators called toehold switches to activate protein translation in response to trigger RNAs [31]. Pardee et al. have developed different types of toehold switches for various applications including detection of Zika virus and coronavirus and disease diagnosis [32–34]. Recently, we reported a novel toehold switch for miRNA detection in mammalian cells [35]. Here, we discussed the basic principles and design strategies of toehold switches, including working mechanisms and optimization of the stem-loop structure. We next introduced the characterization of the toehold switch by comparing it with a non-toehold sensor. Taking miR-155 and miR-21 as examples, multiplex detection of miRNAs in several mammalian cells lines was tested by using two different miRNA switches. Lastly, we discussed how to detect exogenously and endogenously expressed miRNAs in CHO, HeLa, HEK 293, and MDA-MB-231 breast cancer cells. Together with advancements in design strategies and performance, these miRNA-based toehold switches will provide opportunities for the application of novel therapeutic strategies.

2 Materials

2.1 Plasmids

1. pcDNA 3.0-GFP.
2. pcDNA-str-miR155-GFP.
3. pcDNA-toehold-miR155-GFP.
4. pcDNA-toehold-control-GFP.
5. pcDNA 3.0-mRFP.
6. pcDNA-str-miR21-mRFP.
7. pcDNA-toehold-miR21-mRFP.
8. pcDNA-toehold-control-mRFP.

2.2 Materials

2.2.1 Cloning Materials

1. Agarose powder.
2. 1× TAE buffer.
3. DNA gel loading dye, red (6×).
4. 1 kb DNA ladder.
5. LB Broth Miller powder.
6. Q5 site-directed mutagenesis kit.
7. NEB 5-alpha competent *E. coli* cells.
8. Wizard SV Gel and PCR Clean-Up System.
9. Wizard Plus SV Minipreps DNA purification system.
10. T4 DNA ligase.
11. pcDNA 3.0 mRFP.
12. pcDNA 3.0 RFP,
13. Q5 site-directed mutagenesis kit.

2.2.2 Cell Culture Material

1. CHO cells (ATCC).
2. HEK 293 cells (ATCC).
3. HeLa cells (ATCC).
4. MDA-MB-231 cells (ATCC).
5. DMEM, high glucose.
6. Fetal bovine serum (FBS).
7. Penicillin–streptomycin–glutamine (100X).
8. RPMI 1640 medium.
9. Gentamycin solution.
10. Trypsin–EDTA (0.25%), phenol red.
11. Bovine serum albumin (BSA).
12. Human transforming growth factor-beta (TGF-β).

2.2.3 Transfection

1. Lipofectamine™ 2000 transfection reagent (Invitrogen™, Cat no. 11668019).
2. Opti-MEM I Reduced Serum Media (Gibco™, Cat no. 31985062).
3. Lipofectamine RNAiMax transfection reagent (Invitrogen™, Cat no. 13778030).
4. miRNA-155 mimics 5′-UUAAU GCUAA UUGUG AUAG GGGU-3′ (Assay ID. MC28440; Cat no. 4464066, Thermo Fisher Scientific).
5. miRNA-21 mimics: 5′-UAGCU UAUC AGACU GAUG UUGA-3′ (Assay ID. MC10206; Cat no. 4464066, Thermo Fisher Scientific).
6. Negative control mimics (mirVana™ miRNA Mimic, Negative Control #1, Cat no. 4464058, Thermo Fisher Scientific).
7. Negative control inhibitor (mirVana™ miRNA Inhibitor, Negative Control #1, Cat no. 4464079, Thermo Fisher Scientific).
8. miRNA-155 inhibitor, mirVana® miRNA inhibitor (Assay ID. MC28440; Cat no. 4464084, Thermo Fisher Scientific).
9. miR-21 inhibitor, mirVana® miRNA inhibitor (Assay ID. MC10206; Cat no. 4464084, Thermo Fisher Scientific).

2.2.4 Reverse Transcription and RT-PCR

1. TaqMan MicroRNA Cells-to-CT Kit (Applied Biosystems™).
2. TaqMan MicroRNA Reverse Transcription Kit (Applied Biosystems™).
3. TaqMan MiRNA Assays (Applied Biosystems™).

2.3 Equipment

1. Nanodrop (ND-2000).
2. Guava EasyCyte Flow Cytometer (Merck Millipore).
3. Real-time PCR instruments (Bio-Rad real-time PCR system).
4. T100 thermal cycler (Bio-Rad).
5. Electrophoresis tank and power (Bio-Rad).

3 Methods

3.1 Construction of Toehold Structures

1. Identify the target miRNAs of interest and find the relative sequences from miRBase (https://www.mirbase.org/). For miR-155 and miR-21 detection in human mammalian cells, we first find the miRNA sequences for *Homo sapiens*, hsa-miR-155, and hsa-miR-21. Next, choose 5p-arm mature miRNA strands, including has-miR-155-5p and has-miR-21-5p.
   - hsa-miR-155-5p (5′-3′): UUAAUGCUAAUCGU GAUAGGGGUU,
   - has-miR-21-5p (5′-3′): UAGCUUAUCAGACUGAU GUUGA,
2. Write down the relative DNA sequences for miR-155 and miR-21
   - hsa-miR-155-5p (5′-3′): TTAAT GCTAA TCGTG ATAGG GGTT,
   - has-miR-21-5p (5′-3′): TAGCTTATCAGACT GATGTTGA.
3. Design sensing region of toehold switch based on the miRNA sequences. The detection regions are DNA sequences that are complementary to miRNA sequences.
   - Detection region of miR-155 (5′-3′): AACCC CTATC ACGAT TAGCA TTAA.
   - Detection region of miR-21 (5′-3′): TCAAC ATCAG TCTGA TAAGC TA.
4. Construct hairpin toehold structure

The hairpin toehold structure includes a sensing region, the Kozak sequence (GCCACC), start codon (ATG), and a repressed reporter gene. The target sensing region is employed at the hairpin stem region (labeled in red color), and Kozak sequence and start codon are located in the loop of the hairpin structure (Fig. 1). The other parts of the toehold switch sequences are designed based on the sensing region, Kozak sequence, and start codon. The hair stem and loop length can be adjusted and optimized using RNAfold WebServer. The designed toehold switch sequences should only have one secondary structure. Otherwise, this toehold switch

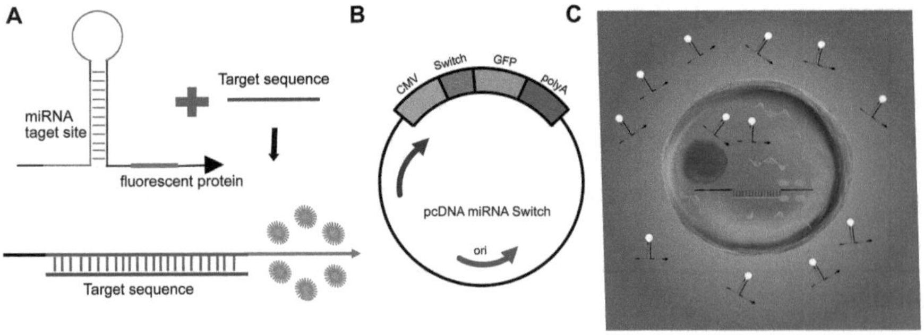

Fig. 1 Illustration of synthetic miRNA toehold switch

sequence will form a different secondary structure and will not be able to detect miRNAs. For the unwound region, add several randomly generated sequences in front of the sensing region. For miRNA detection, this length could be 5–15 nts. A fluorescent reporter gene is added at the end of the toehold structure. There are 10–30 nts between the stem region and the repressed reporter gene (*see* **Note 1**).

For the control switch, a toehold structure was constructed with a random non-sensing region. The random sequences have the same length as the miRNA target sequence. The sequence could be generated using IDT or other web servers.

For miR-155, the toehold structure sequence is:

5′- AAAAC CCCTA TCACG ATTAG CATTA ATTTT CCATC AAGAA CAGGC CACCA TGGAA AATTA ATGCT AAAAA CCTG GCGGC AGCGC AAAAG –3′.

For miR-21, the toehold structure sequence is:

5′- AAATT TTCAA CATCA GTCTG ATAAG CTATT TTCCA TCAAG AACAG GCCAC CATGG AAAAT AGCTT ATCAG AAATA TATAA AAAA – 3′.

For control switch, the toehold structure sequence is:

5′- GGGGT ATGTA ATTGA TTTGG CTTCT GTTAG TTTCA TACAA GAACA AGCCA CCTAT GAAAT GAACA GAAGC AACCT GGCGG CAGCG CAAAA G – 3’.

3.2 Plasmid Construction (See Note 2)

The pcDNA-toehold-miR-155-switch plasmid was generated by cloning a DNA fragment encoding miR-155 target sequence and reporter gene green fluorescence protein (GFP) between XhoI and BamHI restriction sites in pcDNA 3.0-GFP.

3.2.1 Construct Toehold Switch for miR-155 Detection

1. Amplify the miR-155-switch coding region by PCR, using XhoI forward primer and BamHI reverse primer (Table 1). The template DNA fragment can be ordered as gene synthesis, e.g., from a company such as IDT.

Table 1
Forward and reverse primers for cloning

DNA plasmid name	Forward primer (5′-3′)	Reverse primer (5′-3′)
pcDNA-toehold-miR-155-switch	TAAGC TCTCG AGAAA ACCCC TATCA CGATT	CGTCA GGATCC TTAT TAAAC TGATG CAGCG
pcDNA-str-miR-155-GFP	TCCAT CAAGA ACAGG CCACC TCGCC ACCAT GTTAT TAATA GTAAT CAATT ACGGG G	AAATT AATGC TAATC GTGAT AGGGG TTTTG GTCTA ATGCA TGGCG GTAAT AC
pcDNA-toehold-miR-21-mRFP	AGGCC ACCAT GGAAA ATAGC TTATC AGAAA TATAT AAAAA AACAC TGGCG GCCGC TCGA	TTCTT GATGG AAAAT AGCTT ATCAG ACTGA TGTTG AAAAT TTGAT GGATA TCTGC AGAAT TCCAG CACAC TGG
pcDNA-str-miR-21-mRFP	TATTT TCCAT CAAGA ACAGG CCACC ACGGA TCGGG AGATC TCC	GCTTA TCAGA CTGAT GTTGA AAATT TCGAC GTCAG GTGGC ACTT
pcDNA-control	TAAGCT CTCGAG GGGTA TGTAA TTGAT TTGGC TTCTGT	CGTCA GGATC CTTAT TAAAC TGATG CAGCG

Table 2
Assembly of digestion solution

	Insert	Backbone
DNA	3 μg	1 μg
CutSmart buffer	2 μL	2 μL
BamHI	2 μL	2 μL
Xbal	2 μL	2 μL
ddH20	Variable	Variable
Total	20 μL	20 μL

2. Digest PCR product with XhoI and BamHI. Assemble the solution based on Table 2.

First, add all components to small centrifuge tubes on ice and mix carefully. Next, incubate at 37 °C in an incubator for 90 min. After incubation, prepare 1% agarose gel in a flask by adding 0.3 g agarose powder to 30 mL 1xTAE buffer. Microwave for 30 s, and stop. Repeat until agarose is fully dissolved. Allow gel to solidify during restriction digestion. Add 3 μL DNA gel loading dye to each sample and load 5 μL 1 kb DNA ladder to the first well. Next, load complete sample to

Table 3
Assembly of ligation solution

	Concentration 1	Concentration 2
Insert	5 μL	5 μL
Backbone	5 μL	3 μL
T4 ligase buffer (5x)	3 μL	3 μL
ddH_2O	1 μL	3 μL
Mix and wait 10 min. Keep sample on ice		
T4 DNA ligase	1 μL	1 μL
Total	15 μL	15 μL

two different wells, respectively. After loading, run the gel at 100 V for 30 min. After 30 min, turn off the power, disconnect the electrodes from the power source, and then carefully remove the gel from the gel box.

3. Gel Extraction.

Cut out bands using a razor blaze under UV light and weigh the bands on a sensitive scale and record *mg* mass. After weighing, melt the bands by heating the bands at 65 °C for 10 min. Next, follow the Gel Extraction Kit protocol to isolate DNA from each gel slice. For the final step of the protocol, elute the final product in 30 μL double-distilled water (ddH_2O).

4. Ligation.

First, assemble the ligation solution based on Table 3. Prepare two different concentrations for insert and backbone, 1:1 and 2:1 volume ratio, respectively. Next, mix all components on ice and incubate the reaction at room temperature for 1.5–2 h. Once finished, stop the reaction by incubating samples at 65 °C for 5 min.

5. Transformation.

Thaw DH5*α*-competent cell tubes on ice. At the same time, set heat block at 42 °C. Once the stopping reaction has cooled, add each sample to their respective DH5*α* cells and mix gently. After mixing, incubate on ice for 10 min and afterward, heat shock the cells for 90 s at 42 °C. After heat shocking, put back the tube on ice for 4 min and add 150 μL non-contaminated LB medium without antibiotics to each tube of cells. Incubate and shake at 37 °C for 1 h. Next, plate each reaction (200 μL) on AMP plate, upside down, and incubate at 37 °C overnight.

6. Sequence the constructed plasmids using appropriate primers by Sanger sequencing for verification.

The pcDNA-str-miR-155-GFP plasmid can be constructed using Q5 site-directed mutagenesis kit.

3.2.2 Construct Toehold Switch for miR-21 Detection

The pcDNA-toehold-miR-21-mRFP switch could be constructed using site-directed mutagenesis method.

1. Design miR-21 toehold switch sequence which includes stem region and loop region.
2. Encode the miR-21 detection sequence in the stem region. Insert the custom-designed miR-21 toehold switch sequence into pcDNA 3.0-mRFP using Q5 site-directed mutagenesis kit with a pair of back-to-back orientated primers (*see* **Note 3**).
3. Amplify backbone plasmid (pcDNA 3.0-mRFP) using a pair of standard primers and a master mix formulation of A5 Hot Start High-Fidelity DNA Polymerase.
4. Analyze the PCR product using agarose gel electrophoresis before enzyme reaction.
5. Incubate the PCR product with a unique enzyme mix (KLD enzyme mix) which allows for rapid circulation of the PCR product and removes the template DNA. The enzyme mix contains a kinase, a ligase, and Dpnl.
6. Transform into *E. coli* as described in Subheading 3.2.1.
7. Construct pcDNA-str-miR-21-mRFP plasmid using Q5 site-directed mutagenesis kit following the same procedures.

3.2.3 Construct Control Toehold Switch

1. Construct control switch plasmid by cloning the DNA fragment encoding control switch sequence and GFP protein between XhoI and BamHI restriction sites in pcDNA 3.0.
2. Amplify the coding region of control switch through PCR using XhoI forward primer and BamHI reverse primer (Table 1).
3. Digest, amplify, and clone the DNA fragment into the vector downstream of the CMV promoter between the XhoI and BamHI cloning sites to result in pcDNA-control plasmid.

3.3 Cell Culture and Transfection

1. Culture the cells in tissue culture dish at 37 °C in a humidified incubator with 5% CO_2. Culture Chinese hamster ovary (CHO), human embryonic kidney (HEK) 293, and HeLa cells in DMEM, high glucose, supplemented with 10% FBS and 1% penicillin/streptomycin solution. Culture MDA-MB-231 breast cancer cells in RPMI-1640 media with 25 mM HEPES and L-glutamine, supplemented with 10% FBS, 1% penicillin/streptomycin solution, and 5 μg/mL Gibco

gentamycin. Change media every 2–3 days and trypsinize cells using 0.25% trypsin–EDTA when needed.

2. For DNA plasmids, transfect the cells using Lipofectamine™ 2000 transfection reagent in 6-well plates. Seed cells at the concentration of 3 x 10^5 cells per well with 1.5 mL culture medium. Transfect once the cells reach 70–80% confluency.
3. For each well, follow the procedures below. For multiple wells of cells, multiply the amount of solution by the number of wells.
4. For miRNA mimics or inhibitors, transfect using Lipofectamine™ RNAiMax transfection reagent according to manufacturer's protocol. For the control group, transfect cells with equal amount of negative control mimics or inhibitors for comparison. The concentrations of miRNA mimics, inhibitors, and negative control are set to 25 nM.
5. For co-transfection of DNA plasmids and miRNA mimics/inhibitors, use Lipofectamine™ 2000 transfection reagent. Briefly, dilute DNA (500 ng/mL) and miRNA mimics/inhibitors (25 nM) in Opti-MEM and mix with Lipofectamine 2000 to form DNA–miRNA mimics/inhibitors–Lipofectamine™ 2000 complex. The complexes can be added to the cells after 20 min of incubation at room temperature.

3.4 Reverse Transcription and RT-PCR

For miRNA quantification, use TaqMan™ MicroRNA RT Kit and TaqMan™ MiRNA Assays to generate cDNA and to quantitatively detect mature miRNAs, respectively. In addition, to quantify the levels of mature miRNAs in cells, TaqMan™ MicroRNA Cells-to-CT Kit can be used.

3.4.1 Cell Lysis

1. Rinse the cells three times with cold 1x phosphate-buffered saline (PBS) and aspirate.
2. Add 50 μL of lysis solution to each sample and mix the lysis reaction by pipetting up and down five times.
3. Incubate for 8 min at room temperature (19–25 °C). During this incubation, cells were lysed, and RNA was released into a lysis solution containing reagents to inactivate endogenous RNases.
4. Add 5 μL stop solution and mix five times. In this step, a stop solution is used to inactivate the lysis reagents to avoid inhibition of reverse transcription.
5. Incubate for 2 min at room temperature.

3.4.2 Reverse Transcription

1. Program thermal cycle for reverse transcription.
2. Assemble RT master mix and distribute to reaction (Table 4).
3. Mix master mix solution gently or centrifuge briefly.

Table 4
Assembly of RT master mix solution

10X RT buffer	1.5 μL
dNTP mix	0.15 μL
RNase inhibitor	0.19 μL
MultiScribe TM RT	1 μL
Water	4.16 μL
Total (master mix solution)	7 μL

4. Distribute RT master mix to PCR tubes.
5. Add 3 μL 5x RT primers.
6. Add 5 μL cell lysate solution.
7. Run the RT thermal cycle program.

3.4.3 Real-Time PCR

1. Program the real-time PCR.
2. Assemble PCR cocktail and aliquot into reaction. PCR cocktail setup: 10 μL of TaqMan Master Mix (2x), 1 μL of TaqMan MicroRNA Assay (20x), and 7.67 μL of water. Total 18.67 μL. Distribute PCR cocktail into PCR tubes.
3. Add RT product. The reverse transcription product (1.33 μL) was added to the PCR cocktail mix.
4. Run the PCRs in real-time PCR instruments. In this step, the RT product was amplified by real-time PCR using TaqMan™ Universal PCR Master Mix and TaqMan™ MicroRNA Assay. The data was collected and analyzed after the quantitative PCR was done.

3.4.4 Data Analysis

1. Obtain C_T values for all samples in order to compare relative concentration.
2. ΔC_T is a relative gene expression level which was calculated by subtracting the C_T value of miR-155 from the C_T value of the same sample.
3. For comparing fold change, convert this number using $\log(2^{-(\Delta CT)})$.

3.5 Flow Cytometry and Toehold Switch Characterization

1. For each experiment, add a control group by using non-treated cells to adjust the setting to minimize autofluorescence signal.
2. Culture cells at a density of 3×10^5 cells per well in 6-well plates. Once cells reach 70–80% confluency, transfect cells following the procedures in Subheading 3.4.
3. After 48 h of transfection, harvest cells using 0.25% EDTA–trypsin and resuspend in 500 μL of cold 1% BSA/PBS solution.

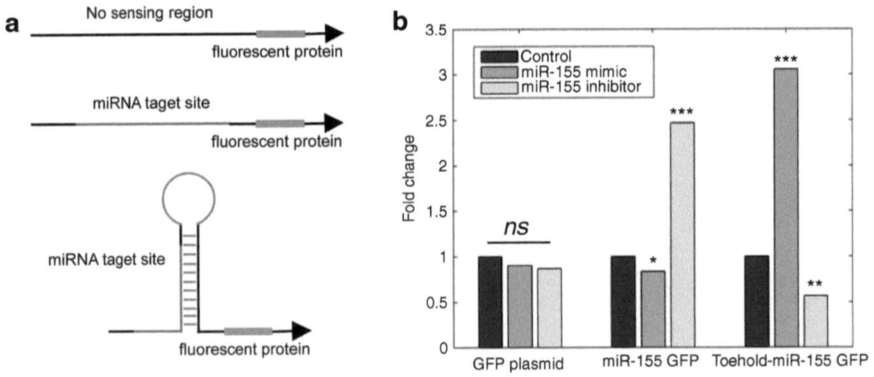

Fig. 2 Representative results of three different plasmids for comparison

4. Transfer the cells into a centrifuge tube and keep on ice.
5. Analyze the cells using a Guava EasyCyte Flow Cytometer. A total number of 10,000 cells is recommended for each sample. A 488 nm excitation laser and a Green-B 525/30 nm filter to detect GFP. A 587 nm excitation laser and a Red-B 695/50 nm filter to detect mRFP. Adjust the threshold gate to avoid the background signal.
6. Collect the data and analyze the data using the Incyte software. Calculate the mean fluorescent intensities for comparison of samples.
7. Characterize the performance of toehold switch comparing it with a non-toehold miRNA sensor and a positive control plasmid (Fig. 2; *see* **Note 4**).

3.6 Detection of Exogenously Expressed miRNAs

The miRNA toehold switch is capable of detecting exogenously expressed miRNA in mammalian cells (Fig. 3).

1. Use RT-PCR in the presence of miRNA mimics or inhibitors to confirm that the mimics or inhibitors can change the expression of miRNAs.
2. Once confirmed, co-transect the cell line of interest with toehold switch plasmid and miRNA mimics or inhibitors. Use a negative control and a toehold control plasmid for comparison (*see* **Note 5**).
3. After transfection, perform flow cytometry to acquire the fluorescent intensity of GFP (Fig. 3a). Acquire the mean intensity of each sample for comparison (Fig. 3b).

3.7 Detection of Simulated Endogenous miRNAs

The miRNA toehold switch can be utilized to detect simulated endogenously expressed miRNA in mammalian cells.

1. First, stimulate the endogenous miR-155 expression in HeLa and MDA-MB-231 cells using transforming growth factor-

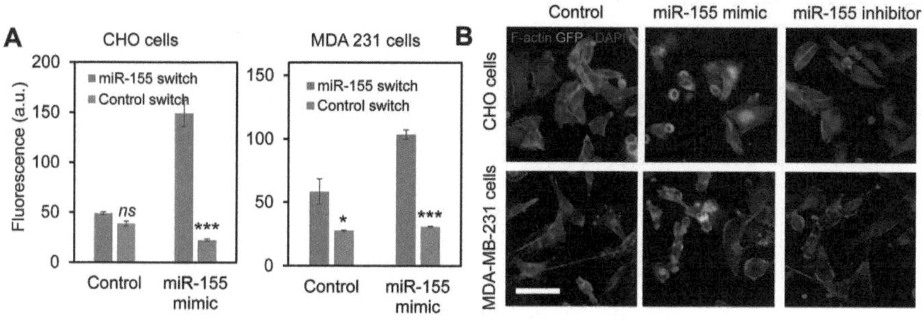

Fig. 3 Representative results of detecting miRNAs in CHO and MDA-MB-231 cells

beta (TGF-β) (*see* **Note 6**) [36–39]. First, treat the cells with different concentrations of TGF-β (0–2 ng/mL) to simulate miR-155 expression.

2. After 48 h of incubation, quantify the miR-155 expression levels using RT-PCR (*see* **Note 7**).
3. Transfect the cells with pcDNA-toehold-miR-155 switch and treat with TGF-β at three different concentrations, 0.5, 1, and 2 ng/mL. Measure the GFP expression by flow cytometry.

3.8 Multiplex Detection of miRNAs

The toehold switch can be designed for multiplex detection of miRNAs in mammalian cells (*see* **Note 8**).

1. Design a miR-21 toehold switch according to the same principle as miR-155 toehold switch. Optimize the toehold structure to make sure the Kozak sequence and start codon are located in the loop and stem region, respectively. Use an mRFP as a repressed protein (*see* **Note 9**).
2. Test and characterize the miR-21 toehold switch by measuring exogenously expressed miR-21 in HEK 293 cells. This step is similar to procedure 3.5. Briefly, miR-21 expression levels can be modulated using miR-21 mimics or inhibitors. Use a control construct without a sensing region. Measure and analyze the mRFP expression using flow cytometry. For the miR-21 toehold switch, the mRFP expression in HEK 293 cells can be increased in the presence of miR-21 mimics, while it can be decreased in the presence of miR-21 inhibitors. For the control switch, the mRFP expression levels should not have a significant difference in the presence of miR-21 mimics or inhibitors.
3. After optimization, both miR-155 toehold switch and miR-21 toehold switch can be utilized for multiplex detection of miR-155 and miR-21 in mammalian cells. The GFP expression in mammalian cells indicates miR-155 level, while the mRFP expression indicates miR-21 level.

4 Notes

1. The design of miRNA toehold switch has two important parameters, the toehold length and stem unwound length. Since miRNAs are shorter at 22 nt, the ratio of toehold length and stem unwound length should be in the range of 0.8–1.2. This ratio could also be adjusted based on the switch ON/OFF ratio. This step is important since it will affect the performance of the toehold switch, including sensitivity and specificity.
2. All the toehold switch and control plasmids were constructed using standard cloning techniques unless specified. Competent *E. coli* strain DH5α was grown in Luria broth (LB) Miller medium with appropriate antibiotics (ampicillin, 100 μg/mL; kanamycin, 50 μg/mL) and used for cloning purposes. All the DNA fragments and primer sequences were synthesized by IDT. Digestion products and PCR products were purified using Wizard SV Gel and PCR Clean-Up System. All ligations were performed using T4 DNA ligase with adjusted incubation time and temperature. The ligation products were then transformed into chemically competent *E. coli* DH5α and plated on LB agar plates with appropriate antibiotics. All the plasmid DNA was prepared from *E. coli* using Wizard Plus SV Minipreps DNA purification system. After plasmid purification, the DNA amounts were quantified using Nanodrop.
3. Site-directed mutagenesis can only be utilized if the inserted DNA sequences are less than 100 bp. Larger DNA fragment inserts should be done using traditional cloning methods or Golden Gate DNA assembly approach.
4. Figure 2 is an example of the characterization process for miR-155 toehold switch. Three plasmids were used for toehold switch, non-toehold sensor, and positive control, respectively. These are pcDNA-Toehold-miR-155-GFP, pcDNA-miR-155-GFP, and pcDNA-GFP. For each plasmid, cells were co-transfected with plasmid and miRNA mimics (or inhibitors, negative control). After transfection, flow cytometry was performed to acquire the fluorescence intensity of GFP. The mean intensity of each sample was acquired and compared. In this step, a cell line with high transfection efficiency should be chosen, i.e., HEK 293 or HeLa. For positive control plasmid (pcDNA-GFP), the GFP intensity should not have a significant difference in the presence of miRNA mimics or inhibitors. For non-toehold miRNA sensors, GFP intensity should decrease in the presence of miRNA mimics and increase in the presence of miRNA inhibitors. For the toehold switch sensor, GFP intensity should increase in the presence of

miRNA mimics and decrease in the presence of miRNA inhibitors.

5. For the toehold control plasmid, there is a toehold structure without a sensing region. Thus, the GFP intensity should be minimum for the groups of control plasmid.
6. TGF-β is a cytokine that is known to have important effects on processes such as angiogenesis, fibrosis, and cancer metastasis [37, 40]. Previous studies have shown that TGF-β could induce miR-155 expression in breast cancer cells and promote collective cell migration and invasion [36–39]. It is noted that different types of cells line may have different results due to their transfection efficiency and ability to respond to TGF-β.
7. It is noted that TGF-β can only induce the expression of miR-155 in specific cell lines, including MDA-MB-231. For HeLa cells, the miR-155 expression level has no difference with or without TGF-β treatment.
8. It is noted that the performance of toehold switches for multiplex detection of miRNA is dependent on the transfection efficiency of mammalian cells, types of cells, and the amount of endogenously expressed miRNA levels.
9. For multiplex detection, different reporter genes (i.e., RFP, YFP, mCherry) should be chosen.

References

1. Lin S, Gregory RI (2015) MicroRNA biogenesis pathways in cancer. Nat Rev Cancer 15: 321–333
2. Hayes J, Peruzzi PP, Lawler S (2014) MicroRNAs in cancer: biomarkers, functions and therapy. Trends Mol Med 20:460–469
3. Jansson MD, Lund AH (2012) MicroRNA and cancer. Mol Oncol 6:590–610
4. Suzuki HI, Katsura A et al (2015) MicroRNA regulons in tumor microenvironment. Oncogene 34:3085–3094
5. Forterre A, Komuro H et al (2020) A comprehensive review of cancer microRNA therapeutic delivery strategies. Cancers 12:1852
6. Wilczynska A, Bushell M (2015) The complexity of miRNA-mediated repression. Cell Death Differ 22:22–33
7. Suarez Y, Sessa WC (2009) MicroRNAs as novel regulators of angiogenesis. Circ Res 104:442–454
8. Urbich C, Kuehbacher A, Dimmeler S (2008) Role of microRNAs in vascular diseases, inflammation and angiogenesis. Cardiovasc Res 79: 581
9. Khurana R, Simons M et al (2005) Role of angiogenesis in cardiovascular disease a critical appraisal. Circulation 112:1813–1824
10. Anastasiadou E, Jacob LS, Slack FJ (2018) Non-coding RNA networks in cancer. Nat Rev Cancer 18:5
11. Rupaimoole R, Slack FJ (2017) MicroRNA therapeutics: towards a new era for the management of cancer and other diseases. Nat Rev Drug Discov 16:203–222
12. Wojciechowska A, Braniewska A, Kozar-Kamińska K (2017) MicroRNA in cardiovascular biology and disease. Adv Clin Exp Med 26: 865–874
13. Tribolet L, Kerr E et al (2020) MicroRNA biomarkers for infectious diseases: from basic research to biosensing. Front Microbiol 11: 1197
14. Hu S, Zhu W et al (2014) MicroRNA-155 broadly orchestrates inflammation-induced changes of microRNA expression in breast cancer. Cell Res 24:254–257
15. Eis PS, Tam W et al (2005) Accumulation of MiR-155 and BIC RNA in human b cell

lymphomas. Proc Natl Acad Sci U S A 102: 3627–3632

16. Takahashi RU, Miyazaki H, Ochiya T (2015) The roles of microRNAs in breast cancer. Cancers (Basel) 7:598–616
17. Nana-Sinkam SP, Croce CM (2011) MicroRNAs as therapeutic targets in cancer. Transl Res 157:216–225
18. Wang S, Xiao Y et al (2018) A gapmer aptamer nanobiosensor for real-time monitoring of transcription and translation in single cells. Biomaterials 156:56–64
19. Wang S, Majumder S et al (2018) Simultaneous monitoring of transcription and translation in mammalian cell-free expression in bulk and in cell-sized droplets. Synth Biol 3:ysy005
20. Wang S, Sun J et al (2016) A nanobiosensor for dynamic single cell analysis during microvascular self-organization. Nanoscale 8:16894–16901
21. Wang S, Riahi R et al (2015) Single cell nanobiosensors for dynamic gene expression profiling in native tissue microenvironments. Adv Mater 27:6034–6038
22. Jusiak B, Cleto S et al (2016) Engineering synthetic gene circuits in living cells with CRISPR technology. Trends Biotechnol 34: 535–547
23. Ausländer S, Fussenegger M (2017) Synthetic RNA-based switches for mammalian gene expression control. Curr Opin Biotechnol 48: 54–60
24. Matsuura S, Ono H et al (2018) Synthetic RNA-based logic computation in mammalian cells. Nat Commun 9:1–8
25. Rossetti M, Del Grosso E et al (2019) Programmable RNA-based systems for sensing and diagnostic applications. Anal Bioanal Chem 411:4293–4302
26. Deng R, Tang L et al (2014) Toehold-initiated rolling circle amplification for visualizing individual microRNAs in situ in single cells. Angew Chem 126:2421–2425
27. Shen C-C, Hsu M-N et al (2019) Synthetic switch to minimize CRISPR off-target effects by self-restricting cas9 transcription and translation. Nucleic Acids Res 47:e13–e13
28. Endo K, Hayashi K, Saito H (2016) High-resolution identification and separation of living cell types by multiple microRNA-responsive synthetic mRNAs. Sci Rep 6:1–8
29. Miki K, Endo K et al (2015) Efficient detection and purification of cell populations using synthetic microRNA switches. Cell Stem Cell 16: 699–711
30. Hirosawa M, Fujita Y et al (2017) Cell-type-specific genome editing with a microRNA-responsive CRISPR–Cas9 switch. Nucleic Acids Res 45:e118–e118
31. Green AA, Silver PA et al (2014) Toehold switches: De-novo-designed regulators of gene expression. Cell 159:925–939
32. Mousavi PS, Smith SJ et al (2020) A multiplexed, electrochemical interface for gene-circuit-based sensors. Nat Chem 12:48–55
33. Pardee K, Green AA et al (2016) Rapid, low-cost detection of zika virus using programmable biomolecular components. Cell 165: 1255–1266
34. Da Silva SJR, Silva CT, a. D. et al (2020) Clinical and laboratory diagnosis of sars-cov-2, the virus causing covid-19. ACS Infect Dis 6: 2319–2336
35. Wang S, Emery NJ, Liu AP (2019) A novel synthetic toehold switch for microRNA detection in mammalian cells. ACS Synth Biol 8: 1079–1088
36. Cui W, Meng W et al (2019) TGF-β-induced long non-coding RNA MiR155hg promotes the progression and EMT of laryngeal squamous cell carcinoma by regulating the MiR-155-5p/sox10 axis. Int J Oncol 54: 2005–2018
37. Xie F, Ling L et al (2018) TGF-β signaling in cancer metastasis. Acta Biochim Biophys Sin 50:121–132
38. Adams JM, Cory S (2018) The BCL-2 arbiters of apoptosis and their growing role as cancer targets. Cell Death Differ 25:27
39. Li D-P, Fan J et al (2017) MiR-155 up-regulated by TGF-β promotes epithelial-mesenchymal transition, invasion and metastasis of human hepatocellular carcinoma cells in vitro. Am J Transl Res 9:2956
40. Louafi F, Martinez-Nunez RT, Sanchez-Elsner T (2010) MicroRNA-155 targets smad2 and modulates the response of macrophages to transforming growth factor-β. J Biol Chem 285:41328–41336

Chapter 17

Imaging *S*-Adenosyl Methionine Dynamics in Living Cells Using an RNA-Based Fluorescent Sensor

Jared D. Moon, Kevin Yusko, Lindsey Nassimos, and Jiahui Wu

Abstract

S-Adenosyl methionine (SAM) is a critical metabolite involved in numerous cellular processes, including DNA methylation and gene expression regulation. Understanding the spatiotemporal dynamics of SAM within living cells is essential for deciphering its roles in maintaining cell homeostasis and in disease development. Here, we describe a protocol based on a recently reported SAM sensor exploiting a fluorogenic RNA and an RNA three-way junction for visualizing SAM dynamics in cultured mammalian cells.

Key words RNA-based fluorescent sensor, Fluorescence imaging, Fluorogenic aptamer, Live cell imaging, RNA aptamer

1 Introduction

Metabolites are small molecules that play important roles in cellular processes such as biosynthesis, signaling, and gene expression. Proper regulation of metabolites is crucial for maintaining cell homeostasis. One important metabolite is *S*-adenosyl methionine (SAM), which is involved in DNA methylation and gene expression regulation [1]. During DNA methylation, SAM serves as a methyl donor to a cytosine base in DNA, resulting in 5-methylcytosine, which is associated with silencing gene expression by altering the accessibility of DNA to the transcription machinery [2]. Importantly, dysregulation of SAM levels is highly associated with cancer development [3–5]. Therefore, visualizing the spatiotemporal dynamics of SAM in living cells could provide valuable insights into how cells regulate SAM levels to maintain homeostasis and how dysregulation of SAM levels contributes to cancer progression.

Traditional methods to image metabolites in living cells rely on protein-based fluorescent sensors [6, 7]. Fluorescent sensors are a class of fluorescent molecules whose fluorescence intensity is proportional to the levels of a ligand of interest. A fluorescent sensor

Francesca Ceroni and Karen Polizzi (eds.), *Mammalian Synthetic Systems*, Methods in Molecular Biology, vol. 2774, https://doi.org/10.1007/978-1-0716-3718-0_17,

comprises two domains: (1) a fluorescence output domain and (2) a ligand-binding domain. These two domains are connected so that ligand binding at the ligand-binding domain will proportionally change the fluorescence intensity from the fluorescence output domain [8]. This change of fluorescence intensity reflecting the change of ligand concentration can be captured using fluorescence microscopy.

Despite tremendous development, using protein-based fluorescent sensors to image SAM is challenging. This is because there is a lack of SAM-binding protein domains that undergo conformational changes upon binding SAM for constructing protein-based fluorescent sensors. To address this, researchers have developed a series of SAM sensors based on RNA aptamers [9–11]. RNA aptamers are single-stranded RNA sequences that can fold into structures to specifically bind to target molecules. RNA-based sensors utilize nongenetically encoded fluorophores for their fluorescence output. These fluorophores, such as DFHBI (3,5-difluoro-4-hydroxybenzylidene-imidazolinone), have minimal fluorescence by themselves. However, when bound to the RNA-based fluorescence output domain, such as Broccoli [12] and Squash [10], these fluorophores become highly fluorescent.

Here, we demonstrate the application of a recently reported RNA-based fluorescent sensor for SAM in living cells. This SAM sensor is based on an RNA three-way junction, F30 [13]. In this SAM sensor, a SAM-binding RNA aptamer is connected to a critical stem of Broccoli through an RNA three-way junction [9]. In the absence of SAM, the SAM-binding aptamer is unfolded, which disrupts the folding of Broccoli, resulting in minimal fluorescence output. However, binding to SAM induces the folding of Broccoli, resulting in an increase of Broccoli fluorescence (Fig. 1). This SAM sensor can reliably detect SAM level changes in living cells.

Fig. 1 Design strategy of SAM biosensor using the F30 RNA three-way junction. In this SAM sensor, a SAM-binding RNA aptamer is connected to a critical stem of Broccoli through an RNA three-way junction, F30. In the absence of SAM, the SAM-binding aptamer is unfolded, which disrupts the folding of Broccoli, resulting in minimal fluorescence output. However, binding to SAM induces the folding of Broccoli, resulting in an increase of Broccoli fluorescence

2 Materials

2.1 Cell Culture Reagents

1. HEK293T cells.
2. Cell culture medium: Dulbecco's Modified Eagle Medium (DMEM) supplemented with 10% fetal bovine serum (FBS), 100 U mL^{-1} penicillin, and 100 μg mL^{-1} of streptomycin. Store at 4 °C.
3. T75 cell culture flask.
4. 1× phosphate-buffered saline (PBS), premade by manufacturer. Store at room temperature.
5. 1× TrypLE Express (no phenol red), premade by manufacturer. Store at room temperature.
6. 15 mL tubes.
7. 50 mL tubes.
8. Serological pipettes and pipette controller.
9. Water bath set to 37 °C.
10. Centrifuge with rotors for 15 mL and 50 mL tubes.
11. Trypan blue solution, 0.4%, premade by manufacturer.
12. Hemocytometer.
13. 24-well plates with glass bottom (*see* **Note 1**).
14. Poly-D-lysine solution. Store at 4 °C.
15. Mouse laminin I. Store at −20 °C.

2.2 Transfection Reagents

1. Opti-MEM, premade by manufacturer (*see* **Note 2**). Store at 4 °C.
2. DNA plasmid: pAV-U6 + 27-Tornado-F30-SAM sensor (*see* **Notes 3** and **4**).
3. FuGENE® HD transfection reagent or other equivalent transfection reagents, premade by manufacturer. Store at 4 °C.
4. 1.7 mL microcentrifuge tubes.

2.3 Imaging Reagents

1. 100× GlutaMAX™, premade by manufacturer.
2. 100 mM sodium pyruvate, premade by manufacturer.
3. Imaging medium: FluoroBrite DMEM supplemented with 10% FBS, 1× GlutaMAX-I, and 1 mM sodium pyruvate.
4. Hoechst 33342 dye solution, premade by manufacturer.
5. BI ((Z)-3-((1H-benzo[d]imadazol-4-yl)methyl)-5-(3,5-difluoro-4-hydroxybenzylidene)-2-methyl-3,5-dihydro-4H-imidazol-4-one), prepared by dissolving BI powder in DMSO to a final concentration of 10 mM.

6. Cycloleucine 500 mM solution, prepared by dissolving cycloleucine powder in imaging medium before use.
7. An inverted fluorescence microscope equipped with a white light source, a DAPI filter cube, a FITC filter cube, a 40× air objective, a camera, and an imaging stage with an adaptor for 24-well plate.

3 Methods

3.1 Culturing and Seeding Cells

1. Maintain HEK293T cells with cell culture medium in a T75 cell culture flask at 37 °C with 5% CO_2 until cell confluency reaches ~80%.
2. Prewarm cell culture medium, 1× PBS, 1× TrypLE Express in a 37 °C water bath.
3. Carefully aspirate cell culture medium from the T75 cell culture flask without detaching HEK293T cells in the flask, then rinse the HEK293T cells with ~10 mL prewarmed 1× PBS once, and aspirate the 1× PBS.
4. After aspirating the 1× PBS, add 2 mL 1× TrypLE Express to the same T75 cell culture flask with HEK293T cells, and then place the flask back to the cell incubator at 37 °C with 5% CO_2 for up to 5 min to let cells detach from the flask.
5. After incubation, take the T75 cell culture flask out of the incubator, and add ~8 mL of prewarmed cell culture medium to the flask. Gently pipette the 1× TrypLE Express and cell culture medium mixture using a serological pipette to detach cells from the flask.
6. Then transfer the detached cells into a 15 mL tube and centrifuge this 15 mL tube at 300*g* for 3 min.
7. After centrifugation, cells should be pelleted at the bottom of the 15 mL tube. Aspirate supernatant carefully without disrupting the cell pellet.
8. Resuspend cells using ~10 mL prewarmed cell culture medium. Pipette gently to ensure cells are fully resuspended.
9. Pipette 10 μL of the resuspended cells and mix them with 10 μL of trypan blue solution. Then pipette 10 μL of this mixture into a hemocytometer and count the number of cells.
10. Based on the cell density calculated from **step 9**, dilute HEK293T cells to a density of 2×10^5 per 5 mL using cell culture medium. Then, seed 0.4×10^5 HEK293T cells in 1 mL of cell culture medium to each well of a 24-well plate, shake gently to ensure cells are evenly distributed in the well, and put the 24-well plate back to an incubator at 37 °C with 5% CO_2 and culture overnight (*see* **Note 5**).

3.2 Transfection

1. Transfect HEK293T cells for 18–24 h after seeding them on a 24-well plate. To transfect HEK293T cells in one well of a 24-well plate, mix the following reagents in a 1.7 mL tube: 25 μL of Opti-MEM, 0.5 μg of pAV-U6+27-Tornado-F30-SAM Sensor DNA plasmid, and 2 μL of FuGENE® HD reagent.
2. Thoroughly mix the above mixture by gently pipetting, and then incubate at room temperature for 15 min.
3. Pipette dropwise the above transfection mixture to HEK293T cells in a well of a 24-well plate. Gently shake the 24-well plate to ensure transfection mixture is well-mixed with the cell culture medium. Then move the 24-well plate back to the incubator at 37 °C with 5% CO_2 to culture for 36–48 h.

3.3 Imaging

1. 6 h prior to imaging, prewarm 1× PBS and imaging medium in water bath to 37 °C.
2. Aspirate cell culture medium from a 24-well plate. Gently rinse the transfected HEK293T cells with 0.5 mL prewarmed 1× PBS. Aspirate 1× PBS, and then add 0.5 mL fresh prewarmed imaging medium supplemented with 10 μM BI to each well of HEK293T cells. Incubate cells at 37 °C with 5% CO_2.
3. 30–60 min prior to imaging, add Hoechst 33342 to the cells to a final concentration of 2 μg/mL. Incubate cells at 37 °C with 5% CO_2.
4. Prior to imaging, transfer the 24-well plate with transfected HEK293T cells to prewarmed microscope imaging stage at 37 °C with 5% CO_2. Adjust the X,Y position of the microscope imaging stage such that the objective is under the well with transfected cells. Then, turn on bright-field illumination and adjust the height of the objective (the Z position) to focus on adhered HEK293T cells (*see* **Note 6**).
5. Turn off bright-field illumination. Switch to the DAPI channel (equipped with a filter cube containing an excitation filter of 350 ± 25 nm, a dichroic mirror of 400 nm with long pass, and an emission filter of 460 ± 25 nm) and turn on fluorescence acquisition. In the DAPI channel, adjust the height of the objective to fine-tune the focus of HEK293T cells.
6. After adjusting the focus, switch to the FITC channel (equipped with a filter cube containing an excitation filter 470 ± 20 nm, a dichroic mirror of 495 nm with long pass, and an emission filter of 525 ± 25 nm) to image SAM sensor. Adjust exposure time in the microscope software to acquire sufficient fluorescence signals from HEK293T cells expressing SAM sensor (*see* **Note 7**).

7. Set time-lapse imaging protocol in the microscope software for image acquisition every 5 min for a total imaging time of 6 h (*see* **Note 8**).
8. Start time-lapse imaging protocol and acquire images for 5 min. Pipette 55.55 μL of cycloleucine solution (500 mM) dropwise onto each well of the 24-well plate, taking care to not disrupt the cells and the position of the plate on the stage. Continue imaging for 2 h.
9. Pause image acquisition on the microscope software, and then carefully aspirate imaging medium containing cycloleucine without disrupting the cells or moving the 24-well plate.
10. Add 0.5 mL imaging medium supplemented with 10 μM BI to HEK293T cells and resume image acquisition for another 3 to 4 h (Fig. 2).

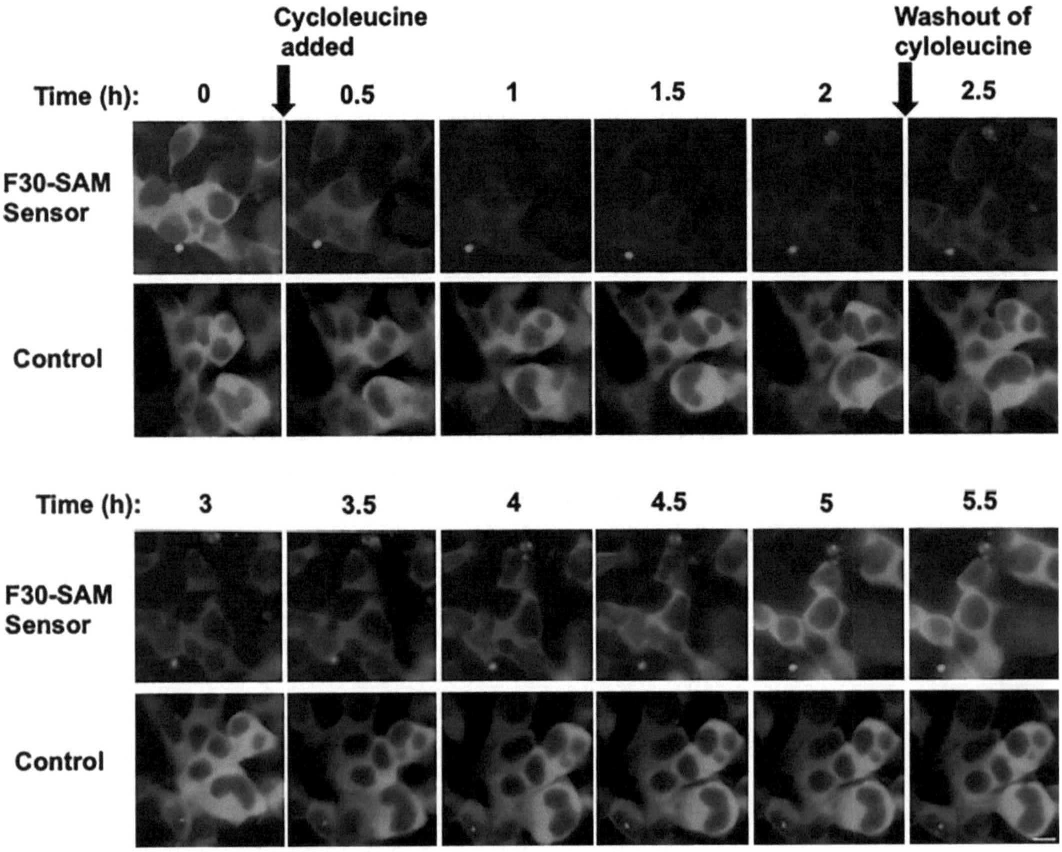

Fig. 2 Representative images of HEK293T cells expressing circular SAM sensor and circular Broccoli (control), respectively. Upon inhibition of SAM synthesis by cycloleucine, the fluorescence of the SAM sensor decreased, reflecting the depletion of SAM. When cycloleucine is removed, SAM sensor fluorescence was restored. In cells expressing Broccoli as a control, Broccoli fluorescence remains at approximately the same level

11. Once imaging is finished, discard 24-well plate containing HEK293T cells, and turn off the microscope software and then all the components of the microscope.

3.4 Data Analysis

1. After data acquisition, export time-lapse images to a series of files with TIF format. Open this image series by an image processing software, such as ImageJ or Fiji, via "File → Import → Image Sequence."
2. In ImageJ, open "ROI Manager" (Analyze → Tools → ROI Manager), and then use the "Freehand selections" tool to define and add ROIs (regions of interest) in different cells for measuring fluorescence intensity changes over time. Be sure to use the "Freehand selections" tool to draw a few ROIs reflecting background fluorescence intensity for background subtraction.
3. After finish drawing ROIs, check the box in "Analyze → Set measurements → Mean gray value." Then, measure the fluorescence intensity of the ROIs in the time-lapse image series (In the ROI Manager tab, More → Multi Measure). Export data.
4. Open data with a data visualization and analysis software, such as Microsoft Excel. Subtract each ROI to the averaged of ROIs reflecting background fluorescence. After this, plot the fluorescence intensity of each ROI against time.

4 Notes

1. For cell types that are prone to detach from glass surface, such as HEK293T cells used in this protocol, we recommend coating the 24-well imaging plate with poly-D-lysine and mouse laminin I. For cell types that are easy to adhere to glass surface, such as HeLa cells and U2OS cells, no extra coating is needed.
2. For transfection using FuGENE® HD, Opti-MEM can be replaced by double-distilled water.
3. pAV-U6 + 27-Tornado-F30-SAM Sensor can be obtained by digesting pAV-U6 + 27-Tornado-F30-Pepper(TAR Variant-2) [14] (Plasmid #129405) by SalI and SacII and then ligating to the sensor insert sequence (listed in Table 1) digested by the same restriction enzymes.
4. DNA plasmids can be purified using silica membrane-based miniprep columns. However, if high cytotoxicity is observed after transfection, consider using anion exchange-based maxiprep columns to minimize the levels of endotoxin from bacteria used for propagating the DNA plasmids.

Table 1
Insert sequence for constructing a DNA plasmid expressing a circular SAM sensor based on the F30 RNA three-way junction

Name	Sequence
Sensor insert sequence	ATACTAGTCGACGGGCCGCACTCGCCGGTCCCAAGCCCGGATAAAA TGGGAGGGGGCGGGAAACCGCCTAACCATGCCGAG TGCGGCCGCAGGTTGCCATGTGTTCTGTCGAGTAGAGTGTGGGCTC TTCGGAGACGGTCGGGTCCAGAACTCTGATCCGAAAGGA TGGCGGAAACGCCAGATGCCTTGTAACCGAAAGGGGATTCA TGGCAACCGCGGTCGGCG

5. We typically seed 0.4×10^5 HEK293T cells in each well of a 24-well plate. However, the number of cells seeded in each well should be adjusted according to cell type and optimized accordingly to each experimental need.
6. The Broccoli-based F30 SAM sensor is prone to rapid, reversible photobleaching. We recommend using transmitted light microscopy to search for areas of interest for imaging. Using the FITC channel to search for cells may cause photobleaching of the SAM sensor. Should this happen, close the excitation shutter, and leave the cells in the dark for a few minutes to let BI bind to the SAM sensor again.
7. To acquire sufficient fluorescence signals, decreasing excitation light intensity and increasing exposure time are typically recommended.
8. It is important to set the microscope to close the shutter while not imaging to minimize photobleaching.

Acknowledgments

This work was supported by the National Institutes of Health—National Institute of Mental Health (R25MH08646) and the Leon Levy Foundation (J.D.M.) and Binghamton University startup funds (K.Y., L.N., and J.W.).

References

1. Chiang PK, Gordon RK, Tal J et al (1996) S-Adenosylmethionine and methylation. FASEB J 10:471–480
2. Moore LD, Le T, Fan G (2013) DNA methylation and its basic function. Neuropsychopharmacology 38:23–38
3. Kottakis F, Nicolay BN, Roumane A et al (2016) LKB1 loss links serine metabolism to DNA methylation and tumorigenesis. Nature 539:390–395
4. Mody HR, Hung SW, AlSaggar M et al (2016) Inhibition of S-adenosylmethionine-dependent methyltransferase attenuates TGFβ1-induced EMT and metastasis in pancreatic cancer: putative roles of miR-663a and miR-4787-5p. Mol Cancer Res 14:1124–1135

5. Murakami S, Nemazanyy I, White SM et al (2019) A Yap-Myc-Sox2-p53 regulatory network dictates metabolic homeostasis and differentiation in Kras-driven pancreatic ductal adenocarcinomas. Dev Cell 51:113–128.e9
6. Koveal D, Díaz-García CM, Yellen G (2020) Fluorescent sensors for neuronal metabolism and the challenges of quantitation. Curr Opin Neurobiol 63:111–121
7. Choe M, Titov DV (2022) Genetically encoded tools for measuring and manipulating metabolism. Nat Chem Biol 18:451–460
8. Nasu Y, Shen Y, Kramer L, Campbell RE (2021) Structure- and mechanism-guided design of single fluorescent protein-based sensors. Nat Chem Biol 17:509–518
9. Moon JD, Wu J, Dey SK et al (2021) Naturally occurring three-way junctions can be repurposed as genetically encoded RNA-based sensors. Cell Chem Biol 28:1569–1580.e4
10. Dey SK, Filonov GS, Olarerin-George AO et al (2021) Repurposing an adenine riboswitch into a fluorogenic imaging and sensing tag. Nat Chem Biol. https://doi.org/10.1038/s41589-021-00925-0
11. Li X, Mo L, Litke JL et al (2020) Imaging intracellular S-adenosyl methionine dynamics in live mammalian cells with a genetically encoded red fluorescent RNA-based sensor. J Am Chem Soc 142:14117–14124
12. Filonov GS, Moon JD, Svensen N, Jaffrey SR (2014) Broccoli: rapid selection of an RNA mimic of green fluorescent protein by fluorescence-based selection and directed evolution. J Am Chem Soc 136:16299–16308
13. Filonov GS, Kam CW, Song W, Jaffrey SR (2015) In-gel imaging of RNA processing using broccoli reveals optimal aptamer expression strategies. Chem Biol 22:649–660
14. Wu J, Zaccara S, Khuperkar D et al (2019) Live imaging of mRNA using RNA-stabilized fluorogenic proteins. Nat Methods 16:862–865

Chapter 18

High-Throughput Spectroscopic Analysis of mRNA Capping Level

Chileab Redwood-Sawyerr, Rochelle Aw, Roberto Di Blasi, Ignacio Moya-Ramírez, Cleo Kontoravdi, Francesca Ceroni, and Karen Polizzi

Abstract

Eukaryotic mRNAs are characterized by terminal 5′ cap structures and 3′ polyadenylation sites, which are essential for posttranscriptional processing, translation initiation, and stability. Here, we describe a novel biosensor method designed to detect the presence of both cap structures and polyadenylation sites on mRNA molecules. This novel biosensor is sensitive to mRNA degradation and can quantitatively determine capping levels of mRNA molecules within a mixture of capped and uncapped mRNA molecules. The biosensor displays a constant dynamic range between 254 nt and 6507 nt with reproducible sensitivity to increases in capping level of at least 20% and a limit of detection of 2.4 pmol of mRNA. Overall, the biosensor can provide key information about mRNA quality before mammalian cell transfection.

Key words Biosensing techniques, RNA cap, RNA stability, RNA isolation and purification, Transcription

1 Introduction

The choice of quality control method for RNA analysis is dependent on the sample type and the characteristics of interest. RNA studies typically require sample and/or signal amplification to overcome low mRNA abundance, RNA instability at room temperature, and RNase activity [1]. Sample amplification using polymerase-based methods improves the sensitivity of RNA detection methods; however, sample amplification introduces sequence biases potentially increasing the occurrence of false-positive results [2].

Alternatives to sample amplification strategies include sample enrichment processes employing complementary base pairing to capture poly(A) motifs on mRNA molecules [3]. Additionally, mRNA molecules can be detected and captured using labels such

Francesca Ceroni and Karen Polizzi (eds.), *Mammalian Synthetic Systems*, Methods in Molecular Biology, vol. 2774, https://doi.org/10.1007/978-1-0716-3718-0_18,

as 3DNA dendrimers or labeled via splinted ligation and quantified by qPCR methods probing the presence and abundance of specific regions of RNA [4, 5]. Unfortunately, these techniques are subject to enzymatic stalling and incomplete detection of mRNA degradation products.

High-resolution LC–MS methods to estimate mRNA integrity independent from rRNA abundance, such as the Agilent 6545XT AdvanceBio LC/Q-TOF and Acquity Ultra-Performance Liquid Chromatography instruments, offer methods that are sequence independent and have high precision [6–8]. These capillary electrophoresis instruments rapidly analyze mRNA up to 3800 nt long for 5′ capping and polyadenylation using ARCA cap 0 and ARCA cap 1 analogues and poly(A) standards [9, 10]. However, these LC–MS methods are optimized for shorter sequences; hence, e.g., capping level analysis of long mRNAs typically requires pre-treatment with site-directed cleavage by RNAse H.

The protocol described here offers a method of analyzing mRNA capping level and integrity in a single step decoupled from rRNA abundance. It includes the in vitro transcription of mRNA, analysis of the integrity using a high-throughput biosensor assay, and transient transfection into mammalian cells (Fig. 1). The biosensor provides a one-step measurement of the amount of functional (translatable) mRNA in a sample. The unique advantage of the biosensor is the ability to analyze RNA directly from in vitro transcription (IVT) and RNA extraction processes and with low cost. Additionally, this biosensor has been validated for mRNA products of lengths up to 6507 nt as well as for heterogenous total cell mRNA [11].

Finally, while state-of-the-art LC–MS systems are well-suited for detailing the characteristics of mRNA capping structures and poly(A) tail motifs, as well as for identifying degraded product, this biosensor is adept for rapid, high-throughput determination of capping and integrity of full-length mRNA transcripts and can be operated without specialist training.

2 Materials and Reagents

Prepare all solutions using UltraPure DNase/RNase-Free Distilled Water and analytical-grade reagents. Prepare all reagents at room temperature and store prepared reagents at 4 °C. Store RNA samples at −80 °C, keeping them on ice when in use and ensure that pipettes used for RNA are regularly cleaned with 70% ethanol and specialist chemicals to prevent RNAse contamination such as RNase Away ™ (Thermo). Take care to achieve accurate pH readings for prepared reagents to ensure proper functioning of all proteins. Adhere to waste disposal regulations when disposing of waste materials, particularly where contaminated waste is generated.

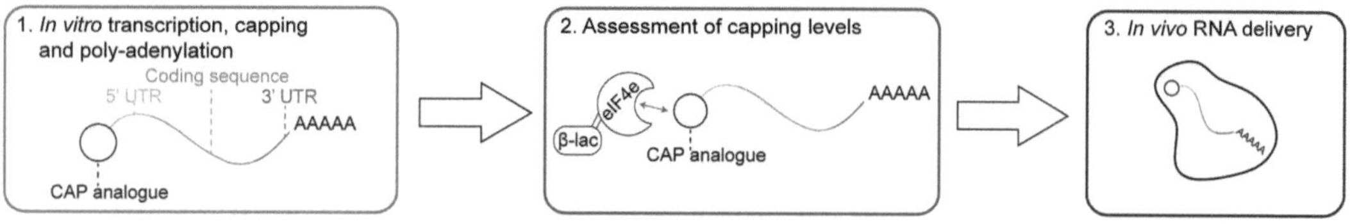

Fig. 1 Diagrammatic workflow of the protocol. mRNAs of interest are first transcribed, capped, and polyadenylated in vitro. mRNA quality and integrity are assessed using the biosensor assay. Once capping levels are verified, the mRNAs are purified and transfected in mammalian cells to quantify expression

2.1 In Vitro mRNA Synthesis, Capping, and Polyadenylation

1. HiScribe® T7 ARCA mRNA Kit (NEB) or equivalent (*see* **Note 1**).
2. RNA Clean & Concentrator-25 (Zymo Research) or equivalent.

2.2 mRNA Substrates and Reporter Enzymes

For the quantification of the percentage of capped, full-length mRNA molecules in a sample, a standard curve is used. This requires mRNA molecules of a similar length to the target in capped and uncapped (or truncated) forms. Accordingly, the synthesis of mRNA molecules using an appropriately linearized template DNA by in vitro transcription (IVT) is required as described in Subheading 3.1.

Standard mRNA solutions with varied mRNA capping levels should then be prepared using mixtures of capped and uncapped mRNA at the required percentages maintaining the required total mRNA concentration.

A reporter β-lactamase eIF4E (B4E) protein catalyzes the chemical reaction in the final step of the assay described here. The pET28a-B4E plasmid encoding the B4E protein can be obtained as an *E. coli* DH5α cell bacterial stab from Addgene (plasmid #162067). Once obtained, single colonies can be prepared by streaking LB agar plates from these bacterial stabs using 50 μg/mL kanamycin as a working concentration. Typical DNA extraction protocols can be performed using overnight growths from these single colonies with 50 μg/mL kanamycin. Extracted plasmid DNA can be used to transform bacterial cells expressing T7 polymerase (e.g., Invitrogen One Shot BL21 DE3™) for ITPG-inducible expression of B4E and His-tag purification.

2.2.1 Materials

1. Streptavidin-conjugated magnetic beads (e.g., Dynabeads T1™, Thermo).
2. Magnetic separation rack for 200 μL tubes.
3. Thin-wall 0.2 mL PCR tubes.
4. Microplate reader.
5. Thermocycler.
6. Flow cytometer.

2.2.2 Reagents

1. 50 mM $MgCl_2$.
2. 2× Buffer A: 100 mM HEPES, 100 mM KCl, pH 9.
3. Buffer B: 33 mM HEPES, 50 mM KCl, 1% (w/v) BSA, 0.1% (v/v) Tween 20, 10 mM DTT, pH 7.56.
4. Oxoid™ nitrocefin (Thermo Fisher Scientific) rehydrated provided with phosphate buffer to 0.5 mg/mL (1 mM) (*see* **Note 2**).
5. Biotinylated poly-deoxythymidine oligonucleotide (pdT_{25}).

2.3 Cell Culture

2.3.1 Materials

1. Mammalian cell culture facility.
2. Stripettes.
3. Sterile T75 flasks.
4. 15 mL and 50 mL Falcon tubes.
5. 1.5 mL microcentrifuge tubes.
6. Sterile tips.
7. Automated cell counter.

2.3.2 Reagents

1. HEK293T cell line (ATCC; *see* **Note 3**).
2. Culture media (DMEM high glucose, GlutaMAX™, pyruvate supplement) (Thermo Fisher).
3. FBS.
4. Trypsin.

2.4 Transfection of mRNAs into HEK293T Mammalian Cells

Once the quality of the mRNA is established, the mRNA can be transfected to assess functionality and protein expression in mammalian cells.

2.4.1 Reagents

1. TransIT®-mRNA Transfection Kit (Mirus).
2. Opti-MEM™ serum-free medium (Thermo Fisher).

2.5 Assessment of Expression by Flow Cytometry

2.5.1 Materials

1. Flow cytometer.
2. 5 mL tubes with cell strainer cap.

2.5.2 Reagents

1. DPBS.

3 Methods

3.1 In Vitro mRNA Transcription

1. To transcribe your template DNA into mRNA and to perform capping and poly-adenylation, use the HiScribe® T7 ARCA mRNA Kit as per manufacturer's instructions.

3.2 mRNA Purification

1. To isolate your mRNA of interest, use a spin column purification kit such as the RNA Clean & Concentrator-25 as per manufacturer's instructions. Store samples at −80 °C until use.

3.3 Rehydrating Biotinylated Poly-deoxythymidine (pdT_{25})

1. Centrifuge the tube containing lyophilized biotinylated deoxythymidine oligonucleotide pdT_{25} and make up to 100 μM using water.
2. Dilute 5 μL of 100 μM biotinylated poly(pdT_{25}) stock to a final concentration of 1 μM in an Eppendorf tube to a final volume of 500 μL using 495 μL water.

3.4 Equilibration of Streptavidin-Coated Magnetic Beads and Immobilization with pdT_{25}

1. Pipette 10 μL of resuspended streptavidin-coated magnetic beads into a thin-walled 200 μL PCR tube. Use one tube per reaction condition.
2. Pellet the magnetic beads by transferring the PCR tubes to a magnetic separation rack for 200 μL tubes and allow the beads to collect the base of the tube for 1 min.
3. Mix the magnetic beads with 10 μL 2× Buffer A and 10 μL water before collecting beads using the magnetic rack and discarding the supernatant (*see* **Note 4**). Repeat this wash twice more, discarding the supernatant each time.
4. Resuspend the beads with 20 μL of 2× Buffer A.
5. Mix the beads with 20 μL of 1 μM biotinylated pdT_{25} and incubate at room temperature for 10 min on a rotatory platform at 60 rpm.
6. Collect beads using the magnetic separator rack and discard the supernatant. Repeat the 20 μL wash from **step 3**, discarding the supernatant.

3.5 mRNA Capture by pdT_{25}

1. Refold mRNA samples by first defrosting the mRNA on ice. Prepare mRNA solutions by appropriately diluting mRNA as follows. Calculate the necessary volume of mRNA stock solutions required to prepare 0.6 μM mRNA solutions in 19.6 μL. Additionally, determine the volume of water to prepare these solutions.
2. Use a thermocycler for the following incubation steps. Incubate the mixture at 80 °C for 2 min, followed by 60 °C for 2 min.
3. Pipette 0.4 μL of concentrated 50 mM stock $MgCl_2$ solution to the mixture to a final volume of 20 μL and final concentration of 1 mM. Finally, incubate the mixture at 37 °C for 30 min.
4. Mix 20 μL of the diluted, refolded mRNA with the collected beads and incubate for 20 min.

5. Mix 100 µL of Buffer B with the beads and incubate for 5 min at room temperature. Collect the beads using a magnetic separation rack and dispose of the supernatant.

3.6 B4E Binding to Capped mRNA

1. Dilute His-tag purified B4E to 0.2 µM with Buffer B.
2. Mix the collected beads with 50 µL of the 0.2 µM B4E solution and incubate for 1 h at room temperature on a rotatory platform at 60 rpm.
3. Post incubation, collect the beads using a magnetic separation rack and discard the supernatant.
4. Wash the beads by mixing them with 200 µL Buffer B and collect using the magnetic rack. Discard the supernatant to remove unbound B4E. Repeat this wash step twice more, discarding the supernatant each time.
5. Resuspend collected beads by mixing with 100 µL of 2× Buffer A and 100 µL water.

3.7 Nitrocefin Colorimetric Assay

1. Set up a plate reader to record absorbance at 492 nm (A_{492}) of the biosensor over a 90-min reaction period, measuring absorbance every minute.
2. Pipette 172 µL phosphate buffer and 20 µL 1 mM nitrocefin solution into the reaction wells of a 96-well plate (*see* **Note 5**).
3. Pipette 8 µL of the resuspended beads (approximately 4 µg beads) into reaction wells and mix well.
4. Begin A_{492} measurements (*see* **Note 6**).
5. For each point on the standard curve, determine the linear region of the A_{492} measurement. Extract the value at the end of the linear region and average the measurements for each technical replicate of the standard curve to generate the transfer function graph. The A_{492} measurements of the biosensor of test samples can be interpolated using the linear transfer function for mRNA standards of the same length (as shown in Fig. 2) to determine the capping and integrity of the test sample.

3.8 Cell Culture

1. Maintain HEK293T cells in culture (37 °C, 5% CO_2) by seeding them in T75 flasks at the cell data bank recommended concentration (ATCC).
2. When cells reach confluence, remove the spent culture medium, wash in DPBS, and detach them by adding 3 mL of trypsin. Incubate at 37 °C for 5–10 min.
3. Add 5 mL of fresh culture medium to inactivate trypsin and collect the cells in a 15 mL Falcon tube.

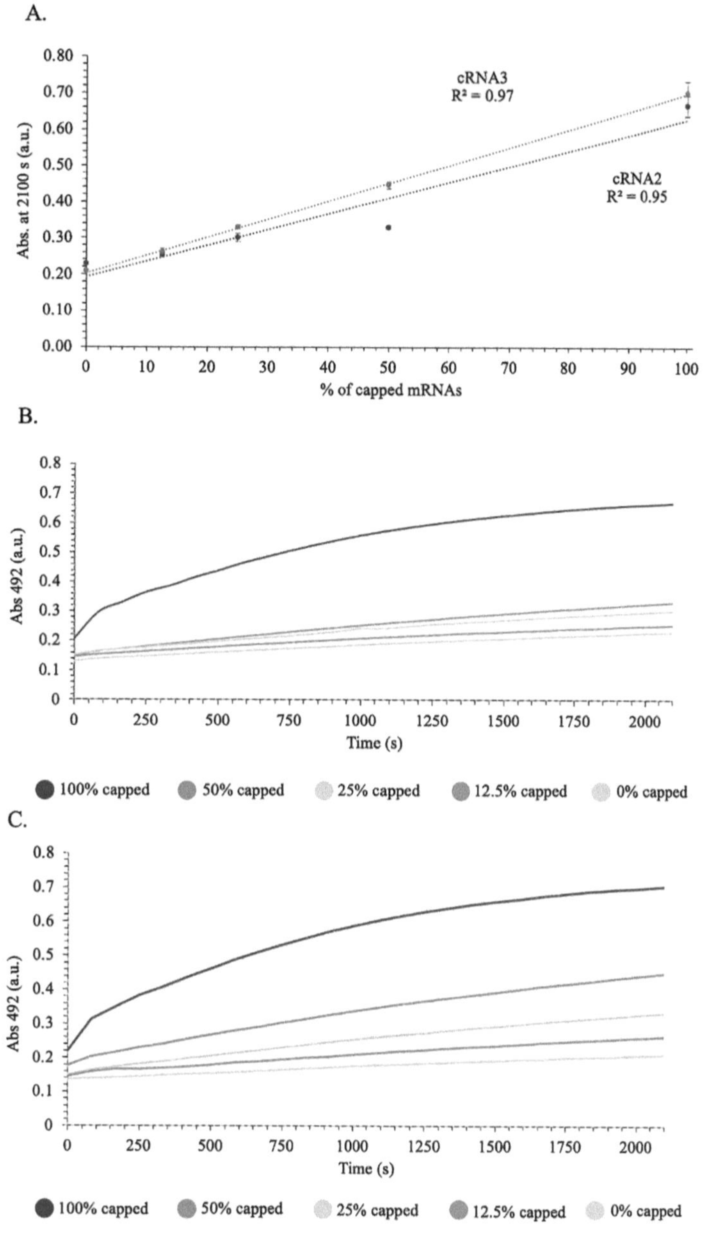

Fig. 2 Representative results from the biosensor assay. (**a**) Transfer function of the response of the sensor against different percentages of full-length, capped cRNA2 (1051 nt) and cRNA3 (2699 nt) at 2100 s using response curves (**b**, **c**), where the absorbance value for nitrocefin hydrolysis was taken at the end of the linear first-order (linear) reaction phase. Biosensor responses at different percentages of full-length, capped mRNAs for cRNA2 (**b**) and cRNA3 (**c**) used in [11]. The R^2 values for transfer functions should be above 0.9 in all the cases ensuring high reproducibility of the assay and the linearity of the biosensor response during the chosen time range. Error bars represent the standard deviation of three technical replicates

4. Count the cells using an automated cell counter as per manufacturer instructions.
5. Transfer the required number of cells into a fresh 15 mL Falcon tube.
6. Centrifuge at 1000 rpm for 5 min.
7. Remove the supernatant, resuspend the cells in fresh medium, and add them to a new T75 flask.
8. Repeat every 2–3 days.

3.9 mRNA Transfection

1. Seed 10^5 HEK293T per well in a 24-well plate.
2. One day after seeding, prepare the transfection mix using 0.25 μg of in vitro transcribed mRNA, 1 μL of TransIT®--mRNA transfection reagent, and 1 μL of TransIT®-mRNA transfection boost in a final volume of 50 μL in Opti-MEM™ serum-free medium (volumes are per well, scale up accordingly based on your needs) (*see* **Notes** 7 and **8**).
3. Incubate at room temperature for 5 min.
4. Add the transfection mix dropwise to the cells, gently rock the plate back and forth, and incubate at 37 °C, 5% CO_2.

3.10 Assessment of Protein Expression by Flow Cytometry

1. Two days after transfection, detach the cells, wash in DPBS, and resuspend in 500 μL DPBS.
2. Filter the cell suspension using 5 mL tubes with cell strainer caps to disrupt any cell clumps.
3. Analyze the cell samples using a flow cytometer. Figure 3 shows representative results from the transfection of a mRNA encoding the green fluorescent protein.

4 Notes

1. Production of capped mRNA can be achieved either co-transcriptionally using a reaction mixture containing a cap analogue or post-transcriptionally using enzymes. Either work with the biosensor described here. We have obtained high yields of mRNA with the HiScribe ® kit from NEB and versions of this containing enzymes for simultaneous polyadenylation are available. To use the kit for in vitro transcription, the template DNA must be previously prepared to include the T7 promoter upstream of the coding region. Primer design is advised in the HiScribe® T7 ARCA mRNA Kit as per manufacturer instructions.
2. Aliquot prepared nitrocefin substrates into 1 mL aliquots, and protect each aliquot from light degradation and oxygen exposure by wrapping aliquots in aluminum foil at −20 °C. Avoid

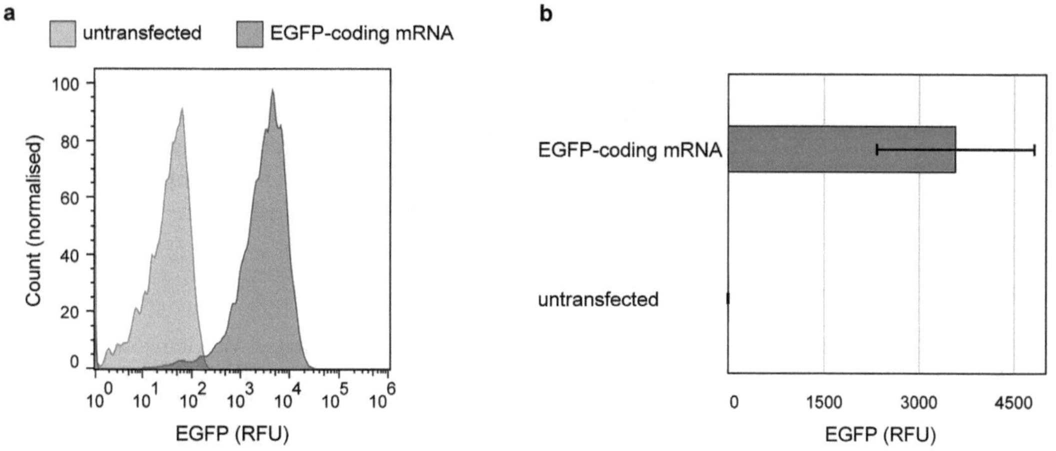

Fig. 3 Representative transfection results. (**a**) Flow cytometry histograms for HEK293T cells transfected with an EGFP-encoding mRNA compared to the untransfected control. (**b**) Bars show the EGFP quantification from flow cytometry in (**a**). EGFP mean and standard deviation of two independent biological replicates are shown

multiple freeze–thaw cycles of prepared nitrocefin substrate solutions for best assay performance.

3. Any mammalian cell line of choice can be used to test the mRNA functionality. Follow the transfection protocol optimized for your specific cell type.
4. Leaving approximately 20 μL of liquid in the PCR tube when washing the magnetic beads helps to maximize retention. On the final wash, the supernatant can be completely removed by ensuring that the PCR tube is tilted, and the supernatant is aspirated away from the magnetic beads.
5. The number of wells needed will depend on the number of samples being tested and the number of points on the standard curve. Subheadings 3.3–3.6 provide sufficient material for multiple technical replicates of the assay from the same reaction tube.
6. If the β-lactamase reporter does not produce a signal within the first 5 min of analysis, possible explanations include:
   - The mRNA under investigation has insufficient capping or polyadenylation.
   - The yellow substrate is already hydrolyzed to a red–brown color.
   - The buffer is not sufficiently reducing, potentially lacking DTT.
   - Insufficient β-lactamase reporter may have been used.
7. Briefly vortex the TransIT®-mRNA transfection and TransIT®-mRNA transfection boost reagents before use, to ensure homogeneous mixing.

8. The transfection protocol has been optimized for use in HEK293T. When using other cell lines, additional experiments might be needed to identify the conditions needed to obtain high transfection efficiencies. Variables include cell seeding density, the ratio of mRNA to transfection reagents (including the boost reagent), incubation time for complex formation, etc., similar to any standard DNA transfection.

References

1. Carrascosa LG, Huertas CS, Lechuga LM (2016) Prospects of optical biosensors for emerging label-free RNA analysis. TrAC – Trends Anal, Chem, p 80
2. Degrelle SA, Hennequet-Antier C, Chiapello H, Piot-Kaminski K, Piumi F, Robin S, Renard JP, Hue I (2008) Amplification biases: possible differences among deviating gene expressions. BMC Genomics 9. https://doi.org/10.1186/1471-2164-9-46
3. Anderson AJ, Culver HR, Prieto TR, Martinez PJ, Sinha J, Bryant SJ, Bowman CN (2020) Messenger RNA enrichment using synthetic oligo(T) click nucleic acids. Chem Commun 56. https://doi.org/10.1039/d0cc05815g
4. Mora JR, Zielinski TL, Nelson BP, Getts RC (2008) Protein detection enhanced by 3DNA dendrimer signal amplification. BioTechniques 44. https://doi.org/10.2144/000112733
5. Blewett N, Coller J, Goldstrohm A (2011) A quantitative assay for measuring mRNA decapping by splinted ligation reverse transcription polymerase chain reaction: qSL-RT-PCR. RNA 17. https://doi.org/10.1261/rna.2436411
6. Schroeder A, Mueller O, Stocker S, Salowsky R, Leiber M, Gassmann M, Lightfoot S, Menzel W, Granzow M, Ragg T (2006) The RIN: an RNA integrity number for assigning integrity values to RNA measurements. BMC Mol Biol 7. https://doi.org/10.1186/1471-2199-7-3
7. Palmer M, E P (2014) Assessing RNA quality. Tech. Ref. Libr. – RNA Tech. Resour. from Ambion 2014
8. Liau B (2021) Rapid analysis of mRNA 5′ capping with high resolution LC/MS. Agil Technol Inc
9. Beverly M, Dell A, Parmar P, Houghton L (2016) Label-free analysis of mRNA capping efficiency using RNase H probes and LC-MS. Anal Bioanal Chem 408. https://doi.org/10.1007/s00216-016-9605-x
10. Beverly M, Hagen C, Slack O (2018) Poly A tail length analysis of in vitro transcribed mRNA by LC-MS. Anal Bioanal Chem 410. https://doi.org/10.1007/s00216-017-0840-6
11. Moya-Ramírez I, Bouton C, Kontoravdi C, Polizzi K (2020) High resolution biosensor to test the capping level and integrity of mRNAs. Nucleic Acids Res 48:e129. https://doi.org/10.1093/NAR/GKAA955

Chapter 19

In Vitro Generation of Megakaryocytes from Engineered Mouse Embryonic Stem Cells

Mitchell R. Lewis and Tara L. Deans

Abstract

The in vitro differentiation of pluripotent stem cells into desired lineages enables mechanistic studies of cell transitions into more mature states that can provide insights into the design principles governing cell fate control. We are interested in reprogramming pluripotent stem cells with synthetic gene circuits to drive mouse embryonic stem cells (mESCs) down the hematopoietic lineage for the production of megakaryocytes, the progenitor cells for platelets. Here, we describe the methodology for growing and differentiating mESCs, in addition to inserting a transgene to observe its expression throughout differentiation. This entails four key methods: (1) growing and preparing mouse embryonic fibroblasts for supporting mESC growth and expansion, (2) growing and preparing OP9 feeder cells to support the differentiation of mESCs, (3) the differentiation of mESCs into megakaryocytes, and (4) utilizing an integrase-mediated docking site to insert transgenes for their stable integration and expression throughout differentiation. Altogether, this approach demonstrates a streamline differentiation protocol that emphasizes the reprogramming potential of mESCs that can be used for future mechanistic and therapeutic studies of controlling cell fate outcomes.

Key words Megakaryocytes, Differentiation, Pluripotent stem cells, Synthetic gene circuits, Synthetic biology

1 Introduction

Pluripotent stem cells have the potential to produce the three primary germ layers that make up the mammalian body. During development, embryonic stem cells (ESCs) undergo specialized decision-making to yield tissue-specific characteristics that allow them to perform particular functions. Understanding how these cells tightly control their spatial and temporal gene expression of lineage-specific transcription factors will provide new insights into the design principles governing how cells transition from one cell state to another. Megakaryocytes (MKs) are a rare population of cells that develop from hematopoietic stem cells (HSCs) in the bone marrow and function to produce platelets that circulate

Francesca Ceroni and Karen Polizzi (eds.), *Mammalian Synthetic Systems*, Methods in Molecular Biology, vol. 2774, https://doi.org/10.1007/978-1-0716-3718-0_19,

throughout the body [1]. The misguided differentiation cues and incomplete maturation of MKs have been shown to cause a number of blood disorders [2]. In a healthy individual, roughly 1 in 10,000 bone marrow cells are MKs, making their development and maturation difficult to study. Another major challenge in studying MK development has been the identification, classification, and enrichment of MK progenitor cells that are produced during hematopoiesis, the process of making all cells of the blood system. We and others have recently identified, isolated, and expanded an important MK progenitor cell population that marks a critical transition state for stem cells in their journey to become mature MKs for the production of platelets [3–5]. In order to study the cell fate transitions and establish the molecular rules governing hematopoietic commitment, cell fate transitions, and the dynamic processes guiding stem cell differentiation during the development and maturation of MKs, it is critical to have a consistent and well-established protocol for the differentiation of pluripotent stem cells into MKs. Mouse embryonic stem cells (mESCs) are particularly interesting because of the large number of mouse disease models that exist for future studies of abnormal MK development and maturation, in addition to testing the function of platelets made in vitro [6] for mitigating diseases and better understanding their role during infection in vivo.

Here, we describe a new approach for differentiating mESCs in vitro for the production of MKs. Indeed, protocols exist in the literature for differentiating mESCs into MKs in vitro; however, many protocols contradict each other and give inconsistent results, and their MK production is minimal. Taking what we learned from attempting many protocols, we have established a new protocol for the consistent and robust differentiation of mESCs into MKs in vitro. Next, we show that mESCs can be engineered to express exogenous transgenes throughout differentiation, demonstrating the potential for reprogramming pluripotent stem cells with exogenous transgenes and synthetic gene circuits [7–13] to be used for driving their cell fate. In this protocol, we describe differentiating mESCs on a support layer of OP9 cells, a bone marrow-derived stromal cell line, in the presence of thrombopoietin (TPO) that leads to the in vitro differentiation of mESCs into MKs. We also describe how to use phage integrases to enable site-specific genome editing to insert desired DNA into the genome. Specifically, we use ϕC31 integrase, which catalyzes the irreversible recombination between appropriate *attB* and *attP* sites to insert desired DNA sequences [14, 15]. To accomplish this, we use mESCs from the TARGATT mouse line that have an *attP* site integrated in the Hipp11 chromosome [16]. We transfected mESCs with plasmids containing the ϕC31 integrase and GFP with *attB* sites flanking a promoter and GFP to enable its integration into the Hipp11 chromosome. After confirming integration, these mESCs were differentiated into MKs. Using the protocols outlined here will enable the growth and expansion of mESCs, their differentiation into MKs in vitro, and studies for inserting transgenes and synthetic gene circuits into the Hipp11 chromosome of mESCs. These methods will facilitate studying the molecular rules and dynamic processes governing cell fate transitions into MKs, in addition to the mechanisms of transgene silencing, the challenge of losing expression of the inserted transgene(s) over time [17, 18].

2 Materials

2.1 Expanding and Mitomycin C Treating Mouse Embryonic Fibroblasts (MEFs)

1. Primary mouse embryonic fibroblasts, neomycin resistant, not treated, P3 (MilliporeSigma, PMEF-NL).
2. MEF growth medium: 500 mL high glucose Dulbecco's Modified Eagle's Medium (DMEM, Thermo Fisher Scientific, 11965-092), 50 mL fetal bovine serum (FBS; 10% final concentration, Thermo Fisher Scientific, 10437028), 5 mL penicillin/streptomycin (1% final concentration, Thermo Fisher Scientific, 15140-122), 5 mL L-glutamine (2 mM final concentration, Thermo Fisher Scientific, 25030-081), 5 mL nonessential amino acids (NEAA; 1× final concentration, Thermo Fisher Scientific, 11140050).
3. Gelatin (Sigma, G1890).
4. Mitomycin C (MMC) (Fisher Scientific, BP2531-2).
5. DMSO (Fisher Scientific, MT-25950CQC).
6. 0.25% trypsin–EDTA, phenol red (Thermo Fisher Scientific, 25200056).
7. PBS (Thermo Fisher Scientific, 10010-023).
8. Milli-Q water.
9. T75 flasks, vented cap.
10. T175 flasks, vented cap.
11. 15 mL conical tubes.
12. 50 mL conical tubes.
13. Sterile disposable filter unit, 250 mL (Fisher Scientific, FB12566502).
14. 70% ethanol in a spray bottle.
15. Centrifuge.
16. Hemocytometer.
17. Cryovials.
18. Magnetic stir bar.
19. Magnetic stirrer hot plate.

20. Cell culture incubator capable of regulating temperature (37 °C), humidity, and carbon dioxide (5%).
21. Biosafety hood suitable for growing cells aseptically.

2.2 Expanding Mouse Embryonic Stem Cells (mESCs)

1. Mouse embryonic stem cells (harvested from a TARGATT mouse, Applied StemCell) (*see* **Note 1**).
2. Confluent T25 flask of mitomycin C treated MEFs.
3. mESC growth medium: 500 mL high glucose knockout Dulbecco's Modified Eagle's Medium (DMEM, Thermo Fisher Scientific 10829-018), 75 mL fetal bovine serum (FBS), embryonic stem cell certified (FBS, ES certified; 15% final concentration, Thermo Fisher Scientific, 10439024), 5 mL penicillin/streptomycin (1% final concentration, Thermo Fisher Scientific, 15140-122), 5 mL L-glutamine (2 mM final concentration, Thermo Fisher Scientific, 25030-081), 5 mL nonessential amino acids (NEAA; 1× final concentration, Thermo Fisher Scientific, 11140050), 500 μL β-mercaptoethanol (21985-023), leukemia inhibitory factor (LIF, Thermo Fisher Scientific, PMC9484; *see* **Note 2**).
4. 0.25% trypsin–EDTA, phenol red (Thermo Fisher Scientific, 25200056).
5. T25 flasks, vented cap.
6. 15 mL conical tubes.
7. Centrifuge.
8. Hemocytometer.
9. PBS.
10. Cell culture incubator capable of regulating temperature (37 °C), humidity, and carbon dioxide (5%).
11. Water bath at 37 °C.
12. Biosafety hood suitable for growing cells aseptically.

2.3 Expanding OP9 Cells

1. OP9 cells (ATCC, CRL-2749).
2. OP9 growth medium: 500 mL minimum essential medium α, no nucleosides (Thermo Fisher Scientific, 12561-056), 100 mL fetal bovine serum (FBS; 20% final concentration, Thermo Fisher Scientific, 10437028), 5 mL penicillin/streptomycin (1% final concentration, Thermo Fisher Scientific, 15140-122).
3. 0.25% trypsin–EDTA, phenol red (Thermo Fisher Scientific, 25200056).
4. T25 flasks, vented cap.
5. 12-well plates.

6. Cell culture incubator capable of regulating temperature (37 °C), humidity, and carbon dioxide (5%).
7. Biosafety hood suitable for growing cells aseptically.

2.4 Differentiating mESCs

1. 12-well plate with OP9 cells ~80% confluent in each well.
2. 5000–7500 mESCs.
3. 0.25% trypsin–EDTA, phenol red (Thermo Fisher Scientific, 25200056).
4. Recombinant murine thrombopoietin (TPO) (VWR, 10770-952).
5. PBS.
6. Bovine serum albumin (Fisher Scientific, BP1600-100).
7. Cell culture incubator capable of regulating temperature (37 °C), humidity, and carbon dioxide (5%).
8. Biosafety hood suitable for growing cells aseptically.

2.5 Docking Transgenes into the Genome of mESCs Using ϕC31 Integrase

1. Eight to twelve wells in a 12-well plate with confluent mESCs growing on mitomycin C-treated MEFs (this can be scaled up or down, depending on the experimental design).
2. Lipofectamine™ 2000 (Thermo Fisher Scientific, 11668-019).
3. Opti-MEM™ (Thermo Fisher Scientific, 31985070).
4. Plasmid for inserting transgene into mouse genome: pBT378_pattB-pCA-GFP-pA-attB plasmid (Addgene, 52554).
5. Plasmid with ϕC31 integrase to enable recognition and integration at the *attP* location in the Hipp11 chromosome (Addgene, 13795).
6. Microcentrifuge tubes.
7. Biosafety hood suitable for growing cells aseptically.

2.6 Flow Cytometry Analysis of Differentiation and Loaded Transgene Markers

1. PBS.
2. Bovine serum albumin (BSA) (Fisher Scientific, BP1600-100).
3. Hoechst 33342 solution for identifying live cells (Fisher Scientific, BDB561908).
4. CD117 (c-Kit) clone 2B8 for identifying hematopoietic stem cell progenitors, PerCP conjugated (VWR, 105821).
5. CD45 for identifying nucleated hematopoietic cells, APC conjugated (30-F11) (VWR, 103111).
6. CD41 for identifying MKs and platelets, PE conjugated (Fisher Scientific, BDB561850).
7. Flow cytometry: Beckman Coulter Life Sciences CytoFLEX S cytometer.

2.7 Imaging for Analysis of Differentiation and Loaded Transgene Markers

1. 12-well plates.
2. PBS.
3. 4% formaldehyde (Thermo Fisher Scientific, FB002).
4. Permeabilizing blocking solution: 5% normal donkey serum (Jackson ImmunoResearch Laboratories, 017-000-121), 1% bovine serum albumin, IgG-free, protease-free (Jackson ImmunoResearch Laboratories, 001-000-162), 0.25% triton (Fisher Scientific, BP151-500).
5. Blocking solution: 5% normal donkey serum (Jackson ImmunoResearch Laboratories, 017-000-121), 1% bovine serum albumin, IgG-free, protease-free (Jackson ImmunoResearch Laboratories, 001-000-162).
6. Glycerol (Fisher Scientific, G31-1).
7. DAPI to stain nuclei (Fisher Scientific, PI62247).
8. Phalloidin to stain actin filaments (Thermo Fisher Scientific, A30106).
9. Rat anti-CD41 primary antibody, unconjugated to stain MKs and platelets (Thermo Fisher Scientific, MA516875).
10. Donkey anti-rat secondary antibody, Alexa Fluor® 647 AffiniPure Donkey Anti-Rat IgG (H + L) (Jackson Labs, 712-605-150).
11. Mouse anti-GFP primary antibody, unconjugated to stain for docked GFP (Thermo Fisher Scientific, MA5-15256).
12. Donkey anti-mouse secondary antibody, Alexa Fluor® 647 AffiniPure Donkey Anti-Mouse IgG (H + L) (Jackson Labs, 715-605-150).
13. Zeiss Axio Observer 7 live cell imaging inverted fluorescence microscope.
14. Nikon TS100 inverted microscope.

3 Methods

An overview of the entire differentiation process described in this protocol is shown in Fig. 1.

3.1 Expanding and Mitomycin C Treating MEFs

1. Make a 0.1% gelatin solution (0.1 g/100 mL Milli-Q water) by weighing out 0.1 g of gelatin and adding it to a beaker of 100 mL Milli-Q water with a magnetic stirrer and placing it on a magnetic stirrer hot plate (*see* **Note 3**). Once the gelatin is completely dissolved, let it cool to room temperature, and then filter-sterilize by running through a sterile disposable filter unit. Coat a T75 flask with 5 mL of 0.1% gelatin. Rotate flask to

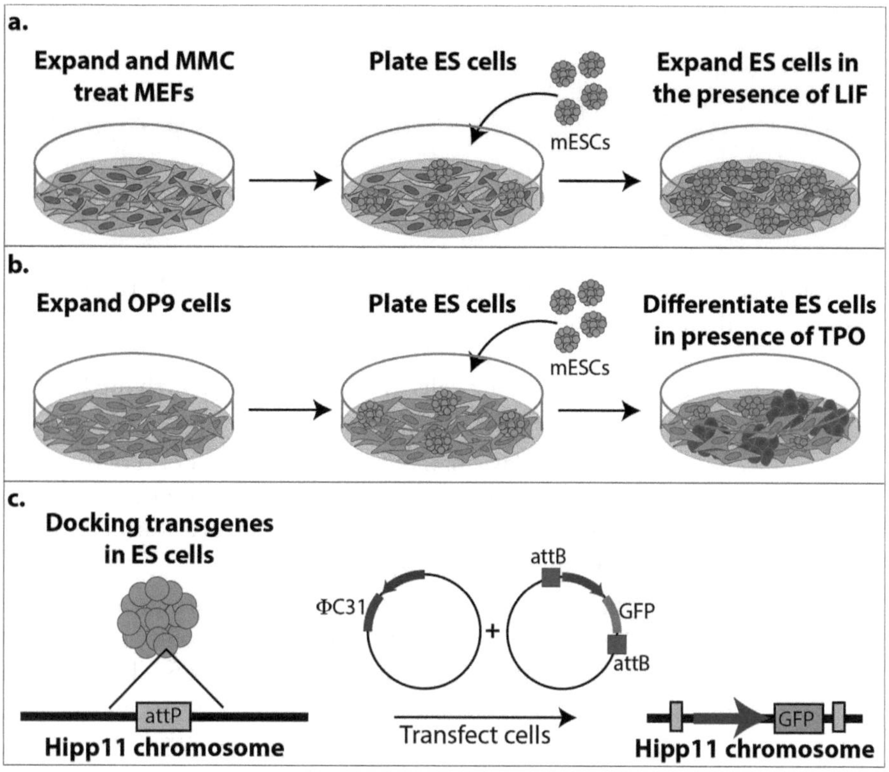

Fig. 1 Schematic of the differentiation process. (**a**) Expansion and mitomycin C (MMC) treating mouse embryonic fibroblasts (MEFs, gray cells) for supporting the growth and expansion of mouse embryonic stem cells (mESCs, turquoise cells). (**b**) Expansion of OP9 cells (orange cells) for differentiating mESCs into megakaryocytes (MKs, purple cells) in the presence of thrombopoietin (TPO). (**c**) Docking transgenes in the Hipp11 chromosome of mESCs using ϕC31 integrase. The mESCs were harvested from a TARGATT mouse line that has an *attP* site located in the Hipp11 chromosome for ϕC31 integrase to catalyze the recombination and insertion of between *attB* and *attP* sites to insert a promoter and GFP into the Hipp11 chromosome

ensure the entire bottom of the flask is covered. Let it sit at room temperature in the hood for 20–30 min.

2. Thaw a vial of the neomycin-resistant MEFs by moving a tube around in a 37 °C water bath (*see* **Note 4**).
3. Once thawed, wipe water from the tube and spray with 70% ethanol before placing in hood. Transfer cells to 7 mL of pre-warmed MEF growth media in a 15 mL conical tube (*see* **Note 5**).
4. Centrifuge cells at 300 × *g* for 5 min.
5. Aspirate medium from the cell pellet, being careful not to disturb the pellet.
6. Aspirate the gelatin from the T75 flask (no need to wash the flask).

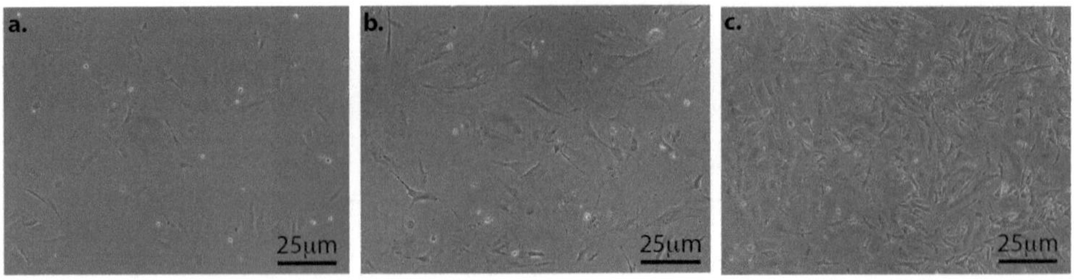

Fig. 2 Bright-field images of MEFs grown on gelatin-coated plates. (**a**) and (**b**) are MEFs growing and are not ready to be passed. (**c**) The cells have reached ~90% confluency and are ready to be passed

7. Resuspend the cells in 10 mL MEF growth medium and transfer to a T75 flask.
8. The next day, change the medium to get rid of the cells that did not survive the freeze–thaw by aspirating the medium from the cells and adding 10 mL of fresh pre-warmed MEF growth medium (*see* **Note 5**).
9. Check on cells daily to ensure they are growing (Fig. 2a, b). About 1–2 days after thawing, when cells are about 90% confluent (Fig. 2c), pass into 2x-T175 (or 5x-T75) gelatin-coated flasks (*see* **Note 6**).
10. Once the cells are ready for passage (Fig. 2c), pass cells into 8x-T175 gelatin-coated flasks (*see* **Note 6**).
11. When cells are about 90% confluent (4–5 days), mitomycin C (MMC) treat the cells. MMC comes in 2 mg of powder. Resuspend MMC in 2 mL sterile PBS to make it 1 mg/mL (*see* **Note 7**). Add 100 μL of 1 mg/mL to 10 mL MEF growth medium, so the final concentration is 10 μg/mL of MMC on the cells (*see* **Note 8**). Incubate the cells in the incubator for 2–4 h (*see* **Note 9**).
12. Wash the cells with PBS by aspirating the medium with MMC off of the cells and adding 15 mL of PBS. Rotate the flask to ensure PBS is washed over all of the cells.
13. Aspirate the PBS and add 7 mL trypsin to each flask. Return to the incubator for about 5 min.
14. Check that the cells have detached in all flasks and once they have, add 14 mL MEF growth medium and transfer all cells to 50 mL conical tubes.
15. Count the number of cells using a hemocytometer.
16. Centrifuge cells at 300 × *g* for 5 min.
17. Aspirate the medium from the cell pellet, being careful not to disturb the pellet.

18. Resuspend cells in freeze medium (90% FBS, 10% DMSO) to freeze cells at 750,000–900,000 cells per cryovial (*see* **Note 10**).
19. Store at −80 °C overnight and transfer to liquid nitrogen the next day.

3.2 Expanding Mouse Embryonic Stem Cells (mESCs)

1. The day before thawing mESCs, coat the bottom of a T25 flask with 2 mL of 0.1% gelatin. Rotate the flask to ensure the entire bottom of the flask is covered. Let it sit at room temperature in the hood for 20–30 min.
2. Thaw one vial of MMC-treated MEFs by moving a tube around in a 37 °C water bath.
3. Once thawed, wipe water from the tube and spray with 70% ethanol before placing in hood. Transfer cells to 9 mL of pre-warmed MEF growth media in a 15 mL conical tube (*see* **Note 5**).
4. Centrifuge cells at 300 × *g* for 5 min.
5. Aspirate the medium from the cell pellet, being careful not to disturb the pellet.
6. Aspirate the gelatin from the T25 flask (no need to wash the flask).
7. Resuspend the MEFs in 5 mL MEF growth medium and transfer to the T25 flask.
8. Place the flask in an incubator overnight.
9. The next day, thaw a vial of the TARGATT mESC line by moving the tube around in a 37 °C water bath.
10. Resuspend the mESCs in 5 mL mESC growth medium with LIF and transfer to the T25 flask with the MMC-treated MEFs that were thawed the previous day (*see* **Note 11**).
11. mESCs need fresh medium every day (*see* **Note 12**). To change the medium, aspirate the medium out of the T25 flask and replace with 5 mL pre-warmed ES medium containing LIF.
12. Carefully watch the growth of the mESCs daily (Fig. 3). The mESC colonies will grow over days, and the day before they become very confluent (Fig. 3b), prepare a T25 flask of MMC-treated MEFs (**step 1**) (*see* **Note 13**). The next day, the mESCs will be confluent (Fig. 3c) and need to be passed. Pass the cells 1:3 if you are actively running experiments or up to 1:25 if you need more time between passages (*see* **Note 6**).

3.3 Expanding OP9 Cells and Plating for mESC Differentiation Studies

1. Thaw a vial of OP9 cells by moving a tube around in a 37 °C water bath.
2. Once thawed, wipe water from the tube and spray with 70% ethanol before placing in hood. Transfer cells to 7 mL of pre-warmed MEF growth medium in a 15 mL conical tube (*see* **Note 5**).

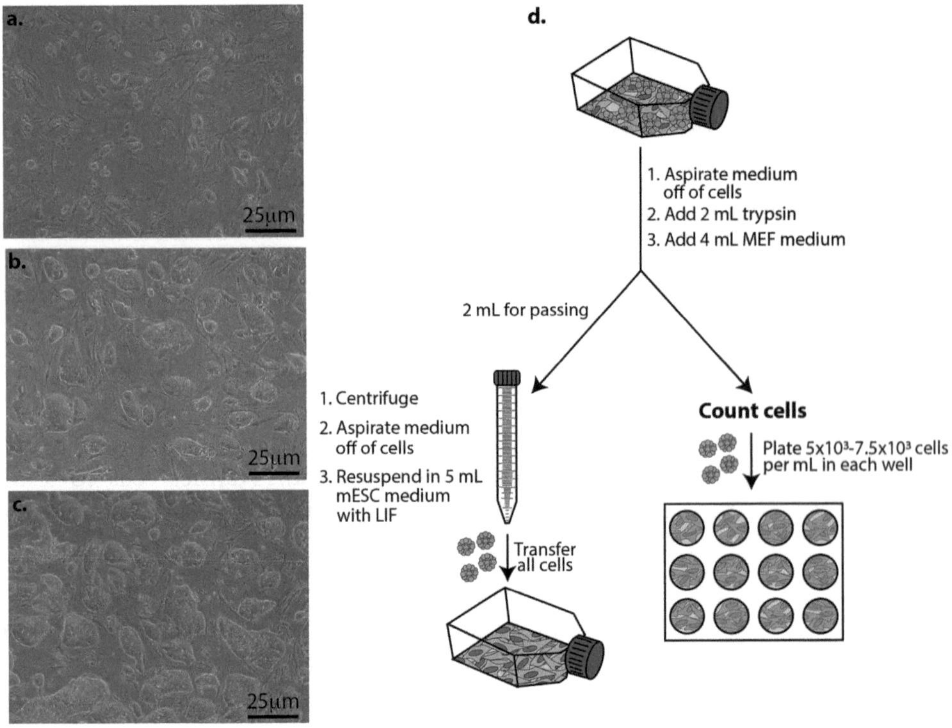

Fig. 3 Growing mESCs on a layer of mitomycin C-treated MEFs and plating mESCs for differentiation studies. (**a**) Representative bright-field image 2 days after plating mESCs on MEFs. (**b**) Representative bright-field image 4 days after plating mESCs on MEFs. (**c**) Representative bright-field image 5 days after plating mESCs on MEFs. (**d**) mESCs (blue cells) are grown in a T25 flask until they are ready to pass (Fig. 3c). Pass cells into a T25 flask with MMC-treated MEFs (gray cells) for expansion (left) and plate mESCs into a 12-well plate containing OP9 cells (orange cells) for differentiation studies (right)

3. Centrifuge cells at 300 × *g* for 5 min.
4. Aspirate the medium from the cell pellet, being careful not to disturb the pellet.
5. Resuspend the OP9 cell in 5 mL OP9 growth medium and transfer to the T25 flask.
6. Place the flask in an incubator overnight.
7. Pass the OP9 cells when they reach about 80% confluency (Fig. 4b), at a ratio of 1:4 or 1:5. It is critical to watch the growth of these cells because the cell density is important. If the cells are too sparse (e.g., below 4×10^3 cells/cm^2), they will senesce and never reach confluency. It is also critical to not let the OP9 cells become over 80% confluent because the cells will start to differentiate into adipocytes and deposit lipid droplets in their cytoplasm (Fig. 4c). When OP9 cells start to deposit lipid droplets, they are no longer able to support the maintenance or differentiation of hematopoietic cells. Do not use them for differentiation studies (*see* **Note 14**).

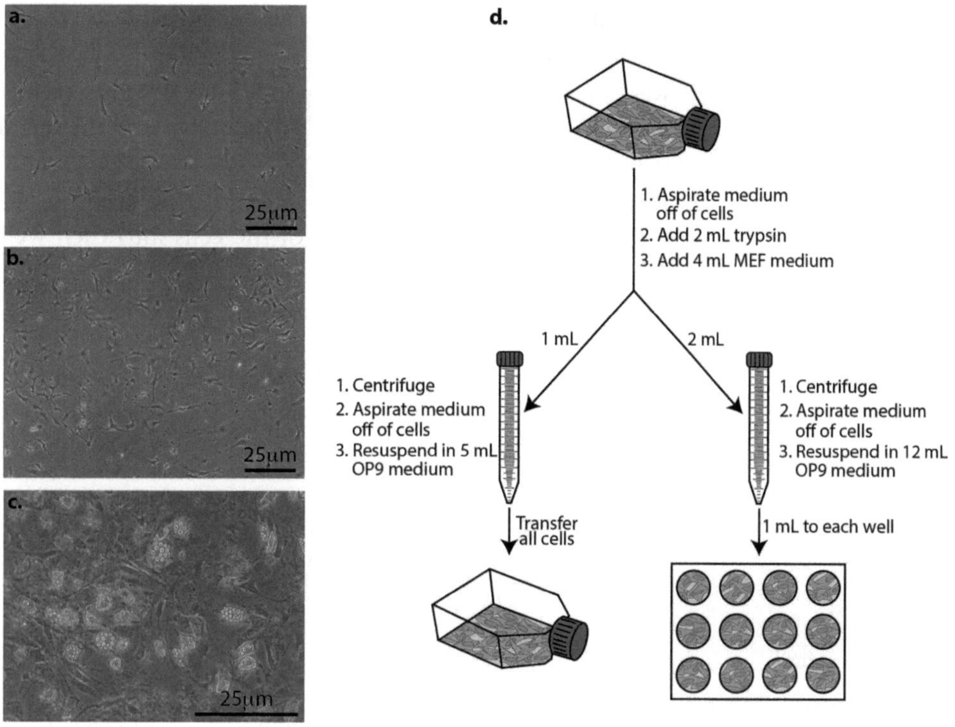

Fig. 4 Expanding OP9 cells and preparing them for differentiation studies. (**a**) The day after passing OP9 cells. (**b**) Cells at ~80% confluency and are either ready for mESCs to be seeded on them for differentiation studies, or they need to be passed. (**c**) When cells become too confluent, they start to differentiate into adipocytes and deposit lipid droplets in their cytoplasm. At this stage, they will no longer support the maintenance and differentiation of hematopoietic cells and they should no longer be used for differentiation studies. (**d**) OP9 cells (orange cells) are grown in a T25 flask until about 80% confluent (Fig. 4b). At this stage, the OP9 cells need to be passed into a T25 flask for expansion (left) and/or into a 12-well plate for differentiation studies (right)

3.4 Differentiating mESCs

3.4.1 Preparing OP9 Cells (Fig. 4)

1. You will need to prepare OP9 cells at least 1 day before the differentiation study can begin. When OP9 cells are ready to pass in a T25 flask (Fig. 4b), aspirate the medium out of the flask and add 2 mL of trypsin. Incubate at 37 °C until the cells have detached from the flask, about 5 min.
2. Once the cells are detached, add 4 mL of complete MEF medium to deactivate the trypsin (*see* **Note 15**).
3. Add 1 mL of the resuspended cells to a 15 mL conical tube for passing (Fig. 4d, left) and 2 mL to a 15 mL conical tube for plating in a 12-well plate for differentiation studies (Fig. 4d, right). This can be scaled up or down depending on the scale of your differentiation studies. Discard the remaining cells.
4. Centrifuge cells at 300 × *g* for 5 min.

5. Aspirate the medium from the cell pellet, being careful not to disturb the pellet.
6. Resuspend the pellet for passing in 5 mL OP9 growth medium and transfer to a T25 flask (Fig. 4d, left). Discard the remaining resuspended cells.
7. Resuspend the pellet for the differentiation study in 12 mL OP9 medium and plate 1 mL of the cells into each well of a 12-well plate (Fig. 4d, right).
8. Place the flask and plate in incubator.

3.4.2 Differentiating mESCs (Fig. 5)

Day 0: Plate ES cells on prepared OP9 cells (Fig. 3d, right).

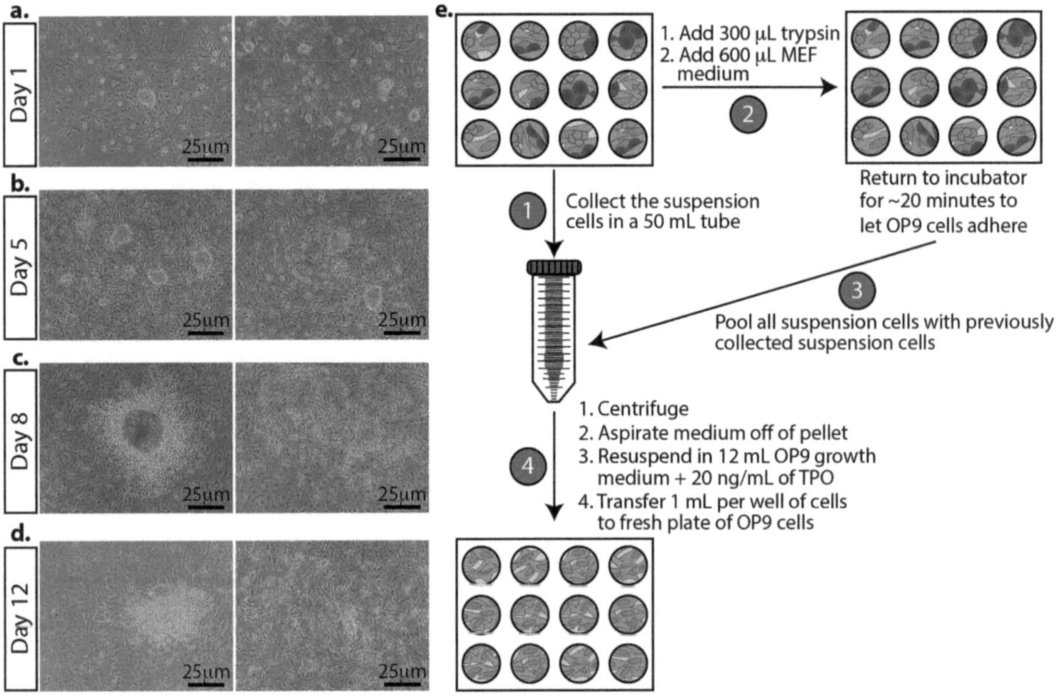

Fig. 5 Morphological changes in mESCs during differentiation and passing differentiating mESCs. Representative bright-field images. (**a**) The day after plating mESCs on OP9 cells. (**b**) mESCs grown on OP9 cells for 5 days. TPO is added at this stage. (**c**) Three days after adding TPO to the cultures. (**d**) The last day of differentiation. The cells are ready to be analyzed for differentiation markers. (**e**) Schematic for differentiating mESCs. First, collect and pool the suspension cells from each well (left, labeled **step 1**). Second, add trypsin to the adherent cells. Once the cells are detached and the trypsin is neutralized with the medium, return the plate to the incubator to remove the majority of the OP9 cells (they will attach first) for about 20 min (top, labeled **step 2**). After the majority of the OP9 cells have attached, take the cells in suspension and pool them with the previously collected suspension cells (right, labeled **step 3**). Plate onto a fresh plate of OP9 cells growing in a 12-well plate (bottom, labeled **step 4**)

1. Aspirate the medium from a flask of confluent mESCs. Add 2 mL of trypsin and incubate at 37 °C until the cells have detached from the flask, about 5 min.
2. Once the cells are detached, add 4 mL of complete MEF medium to deactivate the trypsin and transfer 1 mL of cells to a 15 mL conical tube (Fig. 3, left) (*see* **Note 16**).
3. With the remaining mESCs, count the number of cells using a hemocytometer.
4. Aspirate the OP9 medium from the cells in the 12-well plate and add 1 mL of fresh OP9 medium to each well.
5. Plate between 5×10^3 and 7.5×10^3 mESCs per well in a 12-well plate. Because the volume of mESCs is so low (usually under 25 μL), you can add this small volume to each well without centrifuging. Indeed, some MEFs will be a part of this count; however, the number of mESCs is significantly higher than the number of MEFs during this passage so the MEFs are negligible. Going above the given range of mESCs will likely result in a failed differentiation study (*see* **Note 17**). Plating all 12 wells usually gives enough samples for assessing differentiation at different stages during the process, in addition to samples for imaging. You can scale up or down depending on the needs of your experiment.
6. Gently mix the mESCs in the fresh OP9 medium by rotating the plate front and back, and then side to side, avoiding a circular pattern to prevent cells from pooling in the center of the wells.
7. Return the plate and T25 flask to the incubator until the next step.

Day 3: Half medium swap on differentiating cells.

1. Remove half of the medium from each well (500 μL) of the 12-well plate and add fresh OP9 growth medium (500 μL).
2. Return plates to the incubator until the next step.

Day 4: Pass OP9 cells (Fig. 4d).

1. Pass the OP9 cells according to your experimental setup and how many plates you will need for your differentiation study (see Subheading 3.4.1).

Day 5: Pass differentiating mESCs (Fig. 5).

1. Transfer the medium growing on the cells to a 50 mL conical tube. This contains many semi- and non-adherent hematopoietic stem cells that you want to continue differentiating. It's okay to pool all the medium and cells from each well. Set aside (Fig. 5e, **step 1**).
2. To the remaining adherent cells in the 12-well plate, add 300 μL of trypsin to each well and return it to the incubator until the cells have detached (~5 min).
3. Once the cells have detached, add 600 μL MEF medium to each well to deactivate the trypsin. Gently pipette up and down each sample to separate any cell clumps. Return the plate to the incubator for ~20 min to allow the majority of the OP9 cells to adhere to the plate (Fig. 5, **step 2**).
4. Collect the remaining suspension cells and pool them with the 50 mL tube of the semi- and non-adherent hematopoietic stem cells that you already collected in **step 1** (Fig. 5, **step 3**).
5. Centrifuge cells at 300 × *g* for 5 min.
6. Aspirate the medium from the cell pellet, being careful not to disturb the pellet.
7. Resuspend the pellet with 12 mL OP9 growth medium and add 20 ng/mL of TPO (*see* **Note 18**).
8. Take the plate of OP9 cells that was prepared the previous day and aspirate the medium from the OP9 cells. Transfer 1 mL of differentiating mESCs in OP9 growth medium containing TPO onto the fresh plate of OP9s (Fig. 5, **step 4**).
9. Return the plate to the incubator until the next step.

Day 7: Pass OP9 cells.

1. Pass your OP9 cells according to your experimental setup and how many plates you will need for your differentiation study (see Subheading 3.4.1) (Fig. 4d).

Day 8: Pass differentiating mESCs (Fig. 3d).

1. Repeat steps in day 5 for passing differentiating mESCs (*see* **Note 19**).

Day 12: Assess suspension and adherent cells for differentiation (Fig. 6).

1. Assess differentiation using flow cytometry (Subheading 3.6).
2. Assess differentiation using fluorescent microscopy (Subheading 3.7).

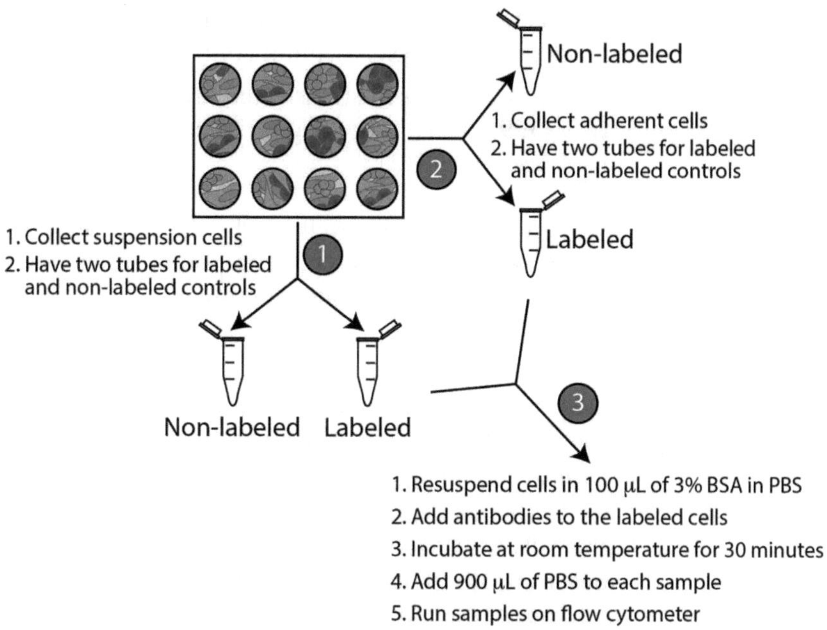

Fig. 6 Schematic of collecting cells for flow cytometry. Collect the suspension cells and split them into two tubes (labeled as **step 1**). Next, collect the adherent cells (labeled as **step 2**). Lastly, resuspend cells in 3% BSA and label the appropriate tubes with antibodies (labeled as **step 3**)

3.5 Docking Transgenes into the Genome of mESCs Using ϕC31 Integrase

1. Two days before starting the differentiation of transfected mESCs, prepare a 12-well plate of MMC-treated MEFs (see Subheading 3.2) (*see* **Note 20**).
2. The day before transfecting mESCs, when passing mESCs (Fig. 3d), count the number of mESCs and transfer 1×10^6 to 3×10^6 mESCs to a 15 mL conical tube.
3. Centrifuge cells at 300 × *g* for 5 min.
4. Aspirate the medium from the cell pellet, being careful not to disturb the pellet.
5. Resuspend the pellet with 12 mL mESC growth medium with LIF.
6. Aspirate the MEF medium from the 12-well plate.
7. Distribute 1 mL of mESCs to each well of the 12-well plate with MMC-treated MEFs.
8. Return the plate to the incubator for a few hours to allow the mESCs to settle on the MEFs (*see* **Note 21**).
9. Later in the afternoon, after the mESCs have settled onto the MMC-treated MEFs, gather all the reagents for a Lipofectamine™ 2000 transfection. We are doing a co-transfection of ϕC31 (the integrase) and attB-GFP (the transgene to be inserted). Follow the manufacturer's instructions for the

Lipofectamine 2000™ transfection. After transfecting the cells, grow them in mESC medium containing LIF overnight.

10. The next day, pass the transfected mESCs onto a prepared plate of OP9 cells to begin the differentiation process (Subheading 3.4.2).

3.6 Flow Cytometry

1. We usually take one well of the 12-well plate for assessing differentiation at different time points using flow cytometry. Adjust as needed for your experiments.
2. Collect the suspension cells by pipetting the medium from one well of the 12-well plate into two microcentrifuge tubes (Fig. 6). One tube will be used as a non-labeled control and the other tube will be labeled with antibodies.
3. Collect the adherent cells by adding 300 μL of trypsin to the well that you already collected the suspension cells from. Once the cells are detached, add 600 μL of MEF growth medium to deactivate the trypsin. Add 450 μL of adherent cells to two microcentrifuge tubes. One tube will be used as a non-labeled control and the other tube will be labeled with antibodies.
4. Add 5 μL of Hoechst per mL of growth medium to the microcentrifuge tubes that will be labeled with antibodies. Place the tubes in the incubator for 30–60 min, protected from light.
5. Centrifuge the labeled and non-labeled cells at 300 × *g* for 5 min.
6. Aspirate the medium from the cell pellet, being careful not to disturb the pellet.
7. Resuspend cells in 100 μL of 3% bovine serum albumin (BSA) in PBS (*see* **Note 22**).
8. Add 2 ng/μL of CD41 antibody, 2 ng/μL CD45 antibody, and 50 ng/μL of c-Kit antibody to the tube with the cells already labeled with Hoechst.
9. Incubate at room temperature for 30 min, protected from light.
10. Dilute samples with 900 μL of PBS.
11. Run samples on flow cytometer (Fig. 7).

3.7 Imaging

1. Aspirate the medium from the cells you will be imaging and wash with cells with 1 mL PBS and aspirate it from the cells (*see* **Note 23**).
2. Add 500 μL (enough to cover cells) of 4% formaldehyde. Incubate at room temperature for 10 min. Aspirate formaldehyde from the cells.
3. Wash cells with 1 mL PBS.

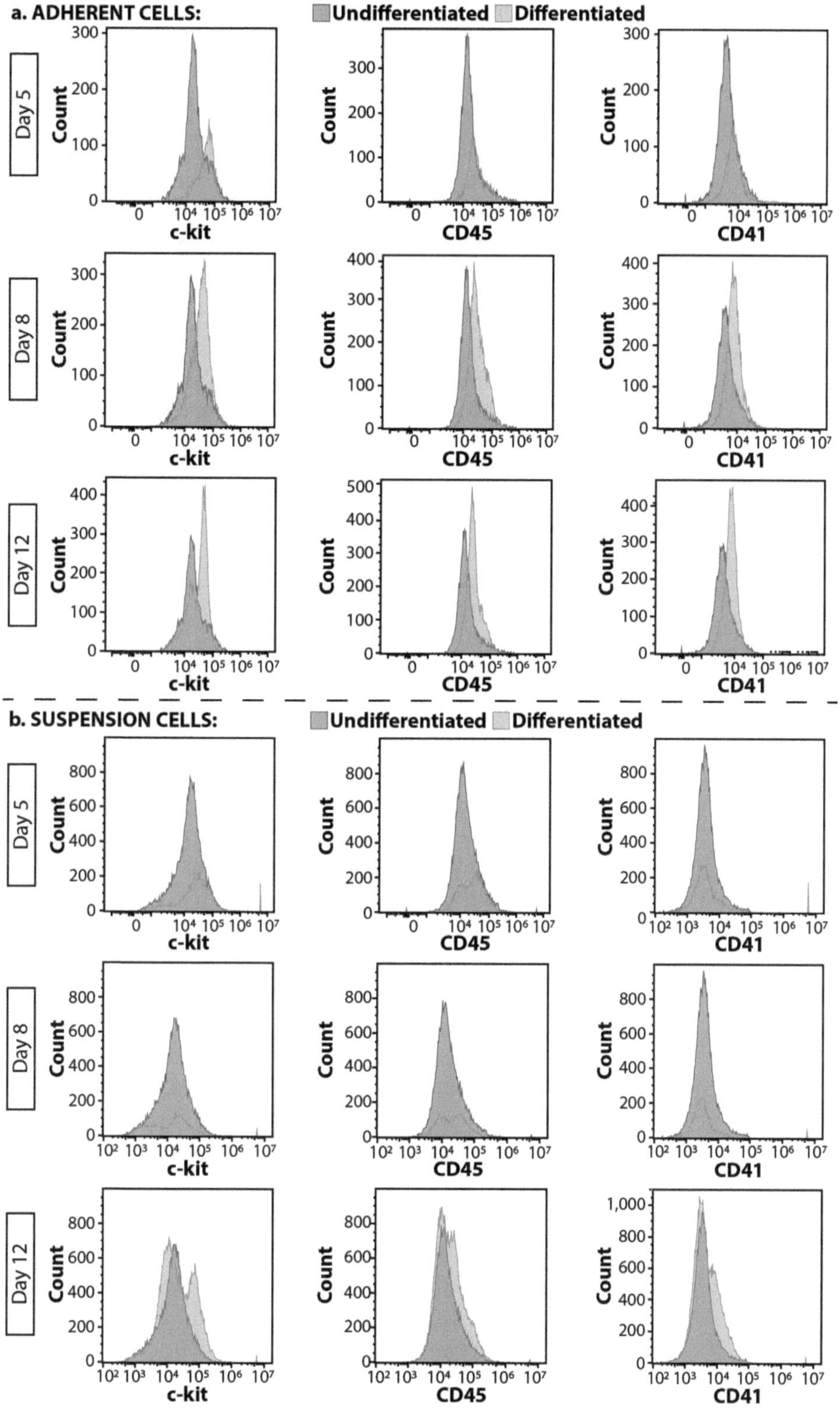

Fig. 7 Flow cytometry to assess differentiation. Outcomes of (**a**) adherent cells and (**b**) suspension cells over a 12-day differentiation period. Pink peaks are undifferentiated TARGATT cells and blue peaks are differentiated cells

4. After aspirating the PBS from the cells, add 500 μL of permeabilizing blocking solution. Incubate at room temperature for 10 min.
5. Wash cells with 1 mL PBS.
6. After aspirating the PBS from the cells, add 500 μL of blocking solution. Add GFP primary antibody between at 1:500 and 1: 2000 dilution (we typically use 1:1000) and the CD41 primary antibody at 1:50 and 1:100 dilution (we typically use 1:100). Incubate at room temperature for 30 min.
7. Gently wash cells with 1 mL PBS two times.
8. After aspirating the PBS from the cells, add 750 μL of blocking solution and add 0.5 μL of secondary antibodies and 0.75 μL of DAPI (*see* **Note 24**). Incubate at room temperature for 30 min.
9. Wash cells with 1 mL PBS and after aspirating the PBS, add another 1 mL of PBS to keep the cells from drying out.
10. Image cells (Fig. 8).

4 Notes

1. We have used this protocol with multiple mESC lines with similar results.
2. LIF comes lyophilized at 10 μg. Resuspend in 100 μL sterile PBS. After making the complete ES medium without LIF, take a 50 mL aliquot and add 5 μL of the resuspended LIF. This can be stored at 4 °C for up to a week. The LIF will degrade so only add it to smaller aliquots of mESC medium that will be used within a week.
3. Gelatin goes into solution faster with constant stirring and warming up the solution by using a magnetic stir bar with a magnetic stirrer hot plate. Larger batches of gelatin can be made and stored at 4 °C for months. Always check that the gelatin is clear before using. If the solution looks cloudy, it is likely contaminated. Throw away immediately and make fresh gelatin.
4. Neomycin-resistant MEFs can be used when selecting for stable mESC lines containing a neomycin-resistant cassette. We use these MEFs for the growth and expansion of all mESC lines that we grow, whether selecting for stable clones or not.
5. Pre-warm media in a 37 °C water bath for about 10 min before using on cells. Once the media is warm, wipe off the water and spray the entire bottle with 70% ethanol before placing it in the hood.

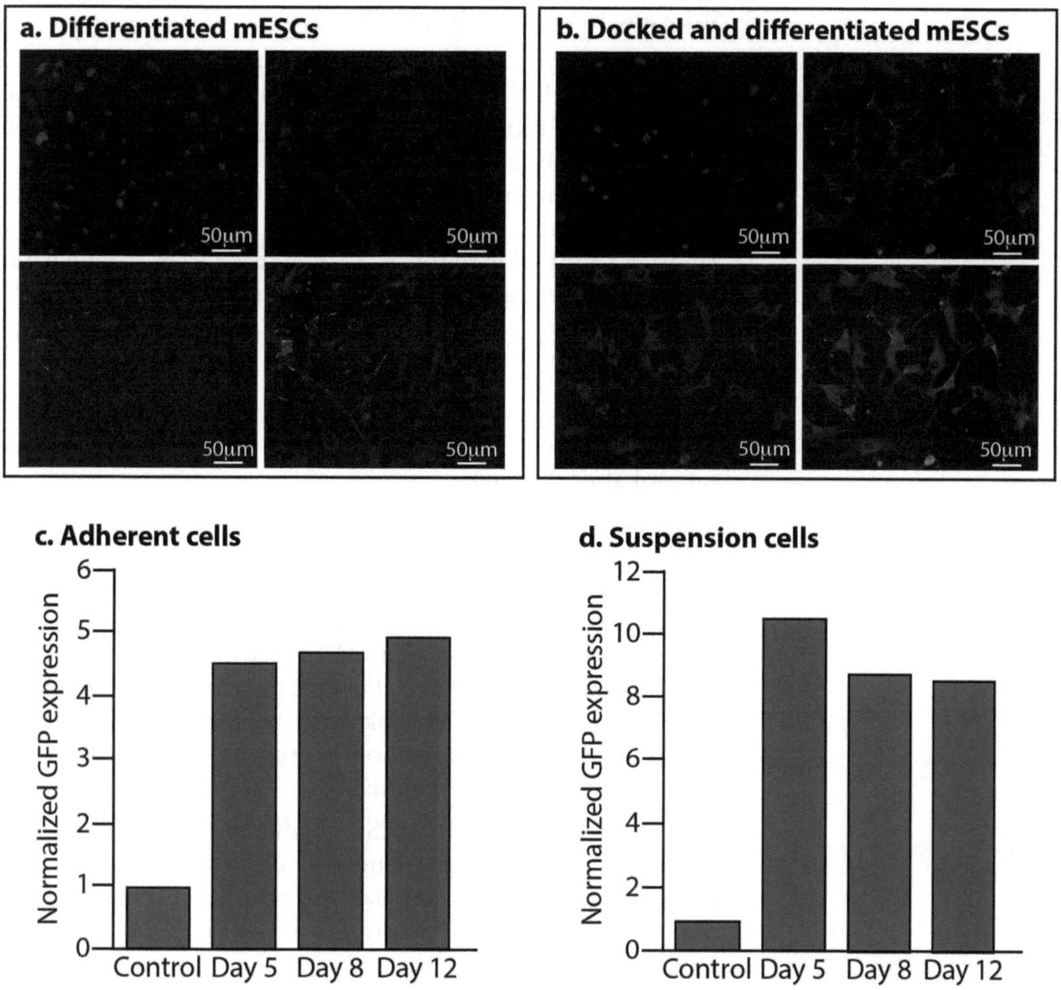

Fig. 8 Assessing docked mESCs after differentiation. (**a**) Non-docked mESCs differentiated into MKs. Blue is DAPI, red is phalloidin, and green is CD41. Bottom right is the merged image. (**b**) Docked mESCs with GFP and differentiated into MKs. Blue is DAPI, red is phalloidin, green is GFP, and the bottom right image is the merged image. (**c**) Docked GFP expression of adherent cells during the differentiation process. (**d**) Docked GFP expression of suspension cells during the differentiation process. In both (**c**) and (**d**), the control is non-docked TARGATT cells

6. Passing cells: Add pre-warmed trypsin to cells (0.3 mL to each well of a 12-well dish, 2 mL to a T25 flask, 5 mL to a T75 flask, and 7 mL to a T175 flask). Always add double the amount of medium containing FBS to deactivate the trypsin once the cells have detached. Transfer cells to a 15 mL conical tube and centrifuge at 300 × *g* for 5 min. Aspirate the medium, taking care not to disturb the cell pellet. Resuspend the cells in fresh pre-warmed medium and transfer the cells to the appropriate growth vessel (5 mL for a T25 flask, 10 mL for a T75 mL flask, and 18 mL for a T175 flask).

7. Store any leftover MMC at −20 °C.
8. To conserve the medium and FBS, transfer the medium from the growing cells (you will use about 13 mL per T175 flask) to 50 mL conical tubes (if you are growing 10x T175 flasks, you will transfer 130 mL of medium to 3× 50 mL conical tubes). Discard any remaining medium. Add 100 μL of 1 mg/mL of MMC to 10 mL media (e.g., 500 μL to 50 mL medium). Transfer 13 mL of medium + MMC to each flask. Rotate flasks to ensure all cells get covered with medium.
9. To fully MMC treat cells, incubate for no less than 2 h and no more than 6 h.
10. You can also freeze with 50% FBS, 40% MEF growth medium, and 10% DMSO; however, the higher FBS yields, the better survival after thawing.
11. ESCs grow better if they are plated on a layer of MMC-treated MEFs that have settled in the flask. These cells are capable of supporting mESC growth for up to 10 days. In unusual circumstances when you need to pass your mESCs and you do not have a plate of MMC-treated MEFs ready, you can thaw both the MEFs and the mESCs at the same time and plate them both in a gelatin-coated T25 flask with mESC growth medium containing LIF. If this method is used, it should be avoided for the next passage of the mESCs.
12. It is important that the mESC medium is changed daily to prevent mESCs from spontaneously differentiating. The only exception to this is that the day you pass the mESCs, you can double the medium (10 mL in a T25 flask) and not change the medium the next day.
13. Usually when you check on the mESCs to change the medium, the medium will start to turn orange in color when the cells are close to needing to be passed. Often the medium will turn yellow (and not cloudy) the next day when they need to be passed. It is important to take care of the cells (i.e., pass them) at this stage because the medium is more acidic and not ideal for the mESCs. If the cells are in this acidic medium too long, they will start to spontaneously differentiate.
14. As the OP9 cells undergo many passages, you will start to notice one or two cells with lipid droplets. A few of these cells present in your culture are usually okay to use in your studies, but you should consider thawing a fresh vial soon. Once you start to see the number of these cells with lipid droplets increases, you should no longer use them for your differentiation studies.
15. Trypsin can be deactivated with medium containing double the volume of serum (e.g., the OP9 medium). However, for us, the

base medium for making MEF growth medium is a lot less expensive compared to the OP9 medium. Therefore, our preference is to use MEF growth medium to deactivate the trypsin by adding medium double the volume of trypsin added (this doubles the serum).

16. We use MEF growth medium to deactivate the trypsin with the mESCs because the FBS used in MEF medium is a lot less expensive than the FBS used to grow the mESCs.
17. The range of mESCs used in a differentiation study is critical for the differentiation study to progress to completion. We have found that adding more than 7.5×10^3 of mESCs per mL of culture results in the OP9 cells becoming overwhelmed, and they start peeling from the plate, resulting in a failed differentiation attempt.
18. TPO comes in 10 μg lyophilized. Resuspend in 1 mL PBS to give 10 μg/mL stock solution. For a concentration of 10 ng/mL in the differentiation medium, add 1 μL/mL of TPO to OP9 growth medium. For a concentration of 20 ng/mL in the differentiation medium, add 2 μL/mL of TPO to OP9 growth medium.
19. We have decreased the TPO concentration from 20 ng/mL to 10 ng/mL on day 8 with the TARGATT and other mESC lines, and our differentiation outcomes are similar to those reported here.
20. The surface area of a 12-well plate is about double to that of a T25 flask. If you plan on transfecting all 12 wells, you will need two vials of the MMC-treated MEFs since they are frozen at a density for a T25 flask. If you are only using six wells, you will only need one vial. MMC-treated MEFs do not freeze–thaw well, so once you thaw a vial of MMC-treated MEFs, you should not freeze them again.
21. We usually pass the mESCs in the morning onto a plate of MMC-treated MEFs, and in the afternoon, do the transfection. You can let the mESCs grow overnight and do the transfection the next day. In both cases, the mESCs have a high rate of transfection efficiency.
22. To make 3% BSA in PBS, weigh out 3 g of BSA and dissolve in 100 mL of PBS. Store at 4 °C for short-term storage and at −20 °C for long-term storage.
23. We use a separate plate for imaging because the incubation steps are outside of the incubator and imaging the cells can take some time.

24. DAPI comes in 10 mg; add 10 mL of Milli-Q water to make 1 mg/mL working stock. This is 1000×, so add 1 μL to 1 mL of blocking solution. The secondary antibodies come lyophilized. Resuspend these in a glycerol solution where you place a 50 mL conical tube on the scale and add 10 mL of Milli-Q water. Tare the scale and drip 10 g of glycerol into the tube of water. Use this to resuspend the lyophilized secondary antibodies by adding 500 μL to each secondary antibody. Use this at 1:1500 to label cells.

Acknowledgments

We would like to thank the generous support from the National Institutes of Health Director's New Innovator Award 1DP2CA250006-01. Research reported in this publication was also supported by the Flow Cytometry Core grants from the Office of the Director of the National Institutes of Health under Award Number S10OD026959 and NCI Award Number 5P30CA042014-24.

References

1. Machlus KR, Italiano JE Jr (2013) The incredible journey: from megakaryocyte development to platelet formation. J Cell Biol 201:785–796
2. Briere J, Kiladjian JJ, Peynaud-Debayle E (1997) Megakaryocytes and platelets in myeloproliferative disorders. Baillieres Clin Haematol 10:65–88
3. Bush LM, Healy CP, Marvin JE, Deans TL (2021) High-throughput enrichment and isolation of megakaryocyte progenitor cells from the mouse bone marrow. Sci Rep 11:8268
4. Grover A, Sanjuan-Pla A, Thongjuea S, Carrelha J, Giustacchini A, Gambardella A et al (2016) Single-cell RNA sequencing reveals molecular and functional platelet bias of aged haematopoietic stem cells. Nat Commun 7:11075
5. Laurenti E, Gottgens B (2018) From haematopoietic stem cells to complex differentiation landscapes. Nature 553:418–426
6. Persson KM, Kneller PV, Livingston MW, Bush LM, Deans TL (2021) High-throughput production of platelet-like particles. Methods Mol Biol 2258:273–283
7. Deans TL, Cantor CR, Collins JJ (2007) A tunable genetic switch based on RNAi and repressor proteins for regulating gene expression in mammalian cells. Cell 130:363–372
8. Fitzgerald M, Gibbs C, Shimpi AA, Deans TL (2017) Adoption of the Q transcriptional system for regulating gene expression in stem cells. ACS Synth Biol 6:2014–2020
9. MacDonald IC, Deans TL (2016) Tools and applications in synthetic biology. Adv Drug Deliv Rev 105:20–34
10. Khalil AS, Collins JJ (2010) Synthetic biology: applications come of age. Nat Rev Genet 11: 367–379
11. Lienert F, Lohmueller JJ, Garg A, Silver PA (2014) Synthetic biology in mammalian cells: next generation research tools and therapeutics. Nat Rev Mol Cell Biol 15:95–107
12. Xie M, Haellman V, Fussenegger M (2016) Synthetic biology-application-oriented cell engineering. Curr Opin Biotechnol 40:139–148
13. Ye H, Aubel D, Fussenegger M (2013) Synthetic mammalian gene circuits for biomedical applications. Curr Opin Chem Biol 17:910–917
14. Groth AC, Calos MP (2004) Phage integrases: biology and applications. J Mol Biol 335:667–678
15. Tasic B, Hippenmeyer S, Wang C, Gamboa M, Zong H, Chen-Tsai Y et al (2011) Site-specific integrase-mediated transgenesis in mice via

pronuclear injection. Proc Natl Acad Sci U S A 108:7902–7907

16. Devine WP, Wythe JD, George M, Koshiba-Takeuchi K, Bruneau BG (2014) Early patterning and specification of cardiac progenitors in gastrulating mesoderm. eLife 3:e03848

17. Cabrera A, Edelstein HI, Glykofrydis F, Love KS, Palacios S, Tycko J et al (2022) The sound of silence: transgene silencing in mammalian cell engineering. Cell Syst 13:950–973

18. Fitzgerald M, Livingston M, Gibbs C, Deans TL (2020) Rosa26 docking sites for investigating genetic circuit silencing in stem cells. Synth Biol (Oxf) 5:ysaa014

Chapter 20

Preparation of Chromobodies for the Detection of Cell Surface Epitopes

Ugne Baronaite and Elise Cachat

Abstract

Chromobodies are nanobodies genetically fused to fluorescent proteins, which were developed to visualize endogenous intracellular antigens. These versatile bioimaging nanotools can also be used to detect cell surface epitopes, and we describe here how we use them as an alternative to conjugated antibodies. This way, we routinely test the binding efficiency of nanobodies for their cognate cell surface antigens, before integrating them as sensing domains into complex synthetic receptor architectures.

Key words Chromobody, Nanobody, Surface antigen, Synthetic receptor, Flow cytometry, Microscopy

1 Introduction

Protein binders are small protein-based affinity reagents that recognize and selectively bind a target molecule [1]. In recent decades antibodies and antibody-derived small binders, such as single-chain variable fragments (scFvs) or nanobodies (Nbs), have become indispensable research and therapeutic tools [2–9]. In research, their uses include, but are not limited to, microscopic imaging, protein–protein interaction analysis, protein function analysis, and regulation of target protein in living cells or organisms [1, 10].

Methods to generate stable, high-affinity binders have evolved over time—from animal immunization with target antigens [11] to fully *de novo* synthesized DNA libraries that are expressed and interrogated in vitro to select target-specific binders [12]. *In silico* modeling tools [13–16] can also be used to predict the sequence and structure of possible binders, which can later be generated for in vitro or in vivo screening against the selected target.

In a mammalian synthetic biology context, small binders are frequently used as detection moieties in the extracellular domains of synthetic receptors like synNotch, MESA, or GEMS receptors [17–

Francesca Ceroni and Karen Polizzi (eds.), *Mammalian Synthetic Systems*, Methods in Molecular Biology, vol. 2774,
https://doi.org/10.1007/978-1-0716-3718-0_20,

21]: nanobody or scFv sequences are incorporated into receptor sequences for the detection of customized extracellular ligands. When constructing novel synthetic receptors for the detection of specific cell surface antigens, we routinely assess the binding efficiency of these binders before embarking on the lengthy receptor assembly and testing phases. To do so, nanobodies or scFvs are genetically fused to a fluorescent protein (FP), producing "chromobodies" or scFv-FPs, respectively, used to monitor binder/antigen interactions.

In this protocol, we will explain how to produce chromobodies (or scFv-FPs) in mammalian cell medium and use them (i) to test the binding efficiency of potential synthetic receptor sensing domains or (ii) for immunofluorescence detection of cell surface epitopes as an alternative to conjugated antibodies.

2 Materials

2.1 Mammalian Cell Lines and Plasmids

1. Mammalian cells used for chromobody extracellular production were HEK293FT (ThermoFisher, #R70007), a derivative of HEK293. This cell line stably expresses SV40 large T antigen, allowing high protein expression levels from vectors containing the SV40 origin. Regular HEK293 or other protein production cell lines can also be used.
2. pUBa5001 (Fig. 1) is a chromobody extracellular expression plasmid. It contains a mammalian expression cassette with (i) a CMV promoter (strong), (ii) Igκ exportation signal derived from mouse Igκ, (iii) bacterial mRFP expression cassette as a placeholder for the GoldenGate reaction, (iv) flexible linker—

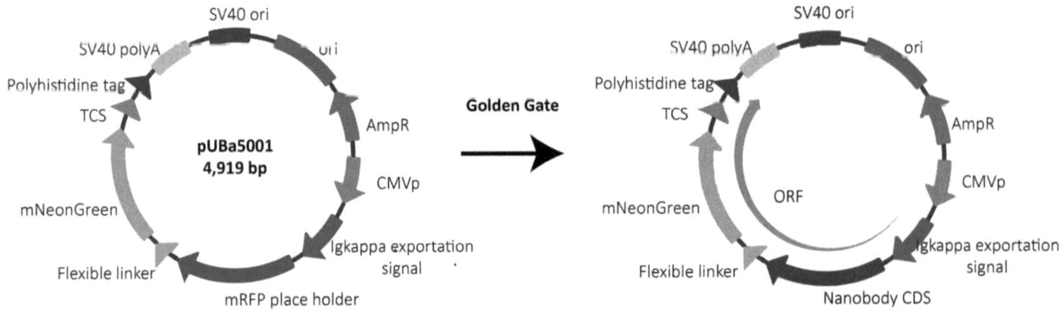

Fig. 1 Map of the plasmid used to express chromobodies extracellularly. Plasmid map pUBa5001 for extracellular nanobody-fluorescent protein fusion expression. mRFP placeholder is inserted for easier cloning and screening. The mRFP cassette contains internal BsaI restriction enzyme recognition sites and is designed to be replaced with the binder coding sequence (CDS) via the GoldenGate reaction. N-terminal Igκ signal sequence drives fusion protein exportation outside the cellular membrane. C-terminal TCS and 6xHis tags are inserted to facilitate purification if needed. SV40 origin of replication (ori) allows maintenance of the plasmid in mammalian cell lines that express SV40 large T antigen. pUBa5001 plasmid map

derivative of $(GGGS)_5$, (v) fluorescent protein—mNeonGreen in this case, (vi) TEV protease cut site (TCS), (vii) polyhistidine tag, and (viii) SV40 polyA. The plasmid also contains SV40 origin of replication (SV40 ori) for plasmid propagation in mammalian cell strains containing T antigen in their genome. TCS and polyhistidine tag are added for easier purification if deemed required. By removing the Igκ exportation signal to produce intracellular chromobodies, the plasmid can also be used for monitoring binder-intracellular antigen interactions.

2.2 Mammalian Cell Media

Dulbecco's modified Eagle's medium (DMEM) (Thermo Fisher, #41966) supplemented with 10% Fetal Bovine Serum (FBS).

1. DMEM supplemented with 10% FBS
2. DMEM supplemented with 10% FBS and GlutaMAX™ (Thermo Fisher, #35050061) (3× standard concentration) (*see* **Notes 1** and **2**)

2.3 Transfection Reagents

1. 1 mg/mL PEI in PBS (*see* **Note 3**)
2. Opti-MEM™ (Thermo Fisher, #31985062)—reduced serum medium (*see* **Note 4**)

2.4 Molecular Cloning

1. Polymerase Chain Reaction (PCR) for nanobody CDS amplification:
   (a) Q5 polymerase (New England Biolabs, #M0491S)
   (b) Q5 reaction buffer (New England Biolabs, #B9027S)
   (c) Deoxynucleoside triphosphate (dNTP) MasterMix
   (d) Primers for the nanobody CDS. For extension sequences for forward and reverse primers, *see* Table 1.
2. GoldenGate reaction for cloning nanobody CDS into expression plasmid:
   (a) BsaI-HF v2 (New England Biolabs, #R3733S)
   (b) Hi-T4™ DNA Ligase (New England Biolabs, #M2622S)
   (c) T4 DNA Ligase Reaction Buffer (New England Biolabs, #B0202S)

Table 1
Primer extension sequences for amplifying nanobody CDS (*see* Note 5)

Primer	Primer extension sequence (5′ to 3′)
Forward	nnnnGGTCTCn<u>CAGC</u>
Reverse	nnnnGGTCTCn<u>GCCT</u>**GA**

3. ReliaPrep™ DNA Clean-Up and Concentration System (promega, #A2891)
4. Thermocycler
5. Gel electrophoresis dock
6. E.Z.N.A.® Plasmid DNA Mini Kit I (Omega, #D6942-00S)

2.5 Immunofluorescence for Analysis

1. Phosphate Buffered Saline (PBS)
2. Dulbecco's Phosphate Buffered Saline (DPBS) (ThermoFisher, #14190144)
3. Paraformaldehyde (PFA) in PBS, 4%
4. StemPro™ Accutase™ Cell Dissociation Reagent (ThermoFisher, #A1110501)
5. Sorting buffer: DPBS + 1% FBS or 0.1% Bovine Serum Albumin (BSA)

3 Methods

In this section, protocols for producing small binders in mammalian cells and detecting surface epitopes using fluorescence microscopy and flow cytometry are described. The cloning of a CDS (coding sequence) for the desired binder into an expression vector, transfection of 293FT cells using PEI, and collection and clearing of the medium containing the small binder are also described.

3.1 Cloning of Nanobody CDS into Expression Plasmid

pUBa5001 plasmid is designed for an easy single-step GoldenGate assembly.

1. Design primers to amplify nanobody CDS and add extensions for GoldenGate assembly.
2. PCR amplify nanobody CDS using Q5 polymerase. Set up Q5 polymerase reaction as described in Table 2, and use the cycling condition described in Table 3.
3. Separate PCR-amplified nanobody CDS from the template by gel electrophoresis.
4. Purify the separated DNA fragment with a gel extraction kit.
5. Set up GoldenGate reaction as described in Table 4, and incubate in the following conditions: 37°C for 5 min and 60°C for 5 min.
6. Transform 2 μL of the GoldenGate reaction into desired competent *Escherichia coli* cells.
7. Pick white colonies from the plate and set up overnight cultures (*see* **Note 7**).

Table 2
Q5 PCR reaction composition as described by the manufacturer

Components	25 μL reaction	50 μL reaction	Final concentrations
5X Q5 Reaction Buffer	5 μL	10 μL	1X
10 mM dNTPs	0.5 μL	1 μL	200 μM
10 μM Forward Primer	1.25 μL	2.5 μL	0.5 μM
10 μM Reverse Primer	1.25 μL	2.5 μL	0.5 μM
Template DNA	Variable	Variable	Less than 1000 ng
Q5 High-Fidelity DNA Polymerase	0.25 μL	0.5 μL	0.02 U/μL
5X Q5 High GC Enhancer (optional)	(5 μL)	(10 μL)	(1X)
Nuclease-Free Water	to 25 μL	to 50 μL	

Table 3
Q5 PCR reaction cycling conditions as described by the manufacturer

Step	Temperature	Time
Initial denaturation	98°C	30 seconds
	98°C	5–10 seconds
25–35 cycles	*50–72°C	10–30 seconds
	72°C	20–30 seconds/kb
Final extension	72°C	2 minutes
Hold	4–10°C	

Table 4
GoldenGate reaction composition (*see* Note 6)

Components	Assembly reaction
Entry plasmid (pUBa5001 or pUBa5002)	Variable (up to 75 ng)
PCR-amplified nanobody CDS	Variable (up to 75 ng)
T4 DNA Ligase Buffer (10X)	2.5 μL
T4 DNA Ligase (2000 U/μL)	0.5 μL (1000 units)
BsaI-HFv2 (20 U/μL)	1.5 μL (30 units)
Nuclease-free H2O	to 25 μL

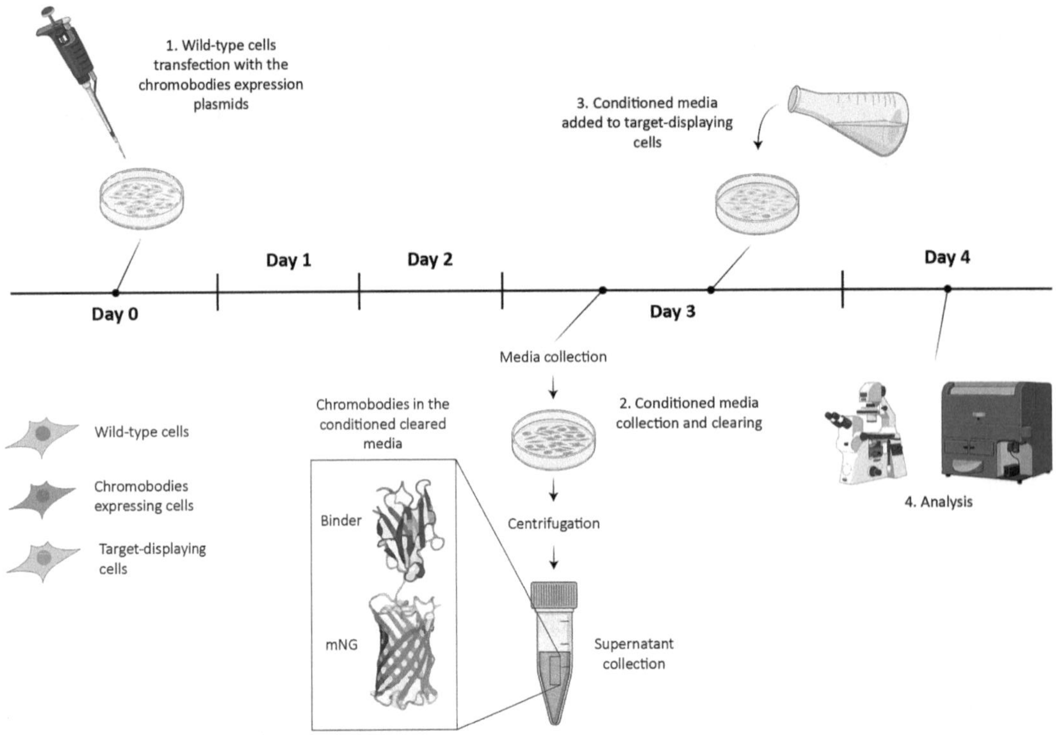

Fig. 2 Schematic of the workflow for chromobodies production in mammalian cells. Chromobodies production takes 3 days. The seeded cells are transfected on Day 0. Cells are left to grow and produce chromobodies for 3 days. The medium containing chromobodies is collected on Day 3 and cleared of cells and cell debris through centrifugation. The supernatant can be used for downstream applications. *Created with BioRender.com*

8. Extract the plasmid DNA from overnight bacterial cultures using DNA extraction kit, and confirm the sequence is correct via preferred method.

3.2 Transfection and Chromobodies Production

The general workflow of chromobodies production can be seen in Fig. 2.

1. Prepare 1 mg/mL PEI in PBS:
   (a) Dissolve 100 mg in 100 mL sterile ddH_2O
   (b) Stir while slowly adding HCl to pH 7.0
   (c) Mix for 10 minutes and then recheck pH
   (d) Filter sterilize through 0.22 μm filter
   (e) Aliquot 500 μL to 1000 μL and store in −80°C
2. Seed FT293 cells at 70% confluence a day prior transfection.
3. Assemble the transfection mixture as described in Table 5.
4. Deliver transfection mixture to mammalian cells culture vessel dropwise.

Table 5
Transfection of 293FT scaling for different culture vessels used

Culture vessel	Culture media (mL)	Opti-MEM (μL)	DNA (μg)	PEI (μg)
48-well	0.25	25	0.25	0.75
12-well	1	100	1	3
6-well	2	250	2.5	7.5
T-75	15	1500	20	60

5. The day after transfection, change the cell medium.
6. Leave cells to produce chromobodies for 3 days.
7. Collect the cell medium, spin it down, and collect the supernatant in a separate tube (*see* **Notes 8** and **9**).
8. Store collected media at 4°C in the dark.

3.3 Detecting Surface Epitopes Using Chromobodies via Microscopy

Conditioned media containing chromobodies can be used to detect cell surface epitopes via microscopy (Fig. 3).

1. Seed target-displaying adherent cells at 80% confluence into the wells containing round coverslips. Wait until cells attach, usually overnight.
2. Aspirate medium and wash cells with PBS (*see* **Note 10**).
3. Fix cells with 4% PFA for 10 min (*see* **Note 11**).
4. Aspirate 4% PFA and wash cells with PBS by incubating for 5 min. Repeat the step three times.
5. Add the conditioned medium containing anti-target chromobodies to the fixed cells.
6. Incubate at 4°C overnight in the dark.
7. Aspirate the conditioned medium and wash cells with PBS by incubating for 5 min. Repeat the step three times.
8. Carefully pick up a cover slip from its well and sandwich cells between the cover slip and a microscopy slide. Seal the coverslip to avoid evaporation.
9. Image cells immediately or store at 4°C in the dark for up to a week.

3.4 Detecting Surface Epitopes Using Chromobodies via Flow Cytometry

The conditioned medium containing chromobodies can be used to detect cell surface epitopes via flow cytometry (Fig. 4).

1. Seed target-displaying adherent cells at 80% confluence. Wait until cells attach, usually overnight.
2. Aspirate medium and wash cells with DPBS.

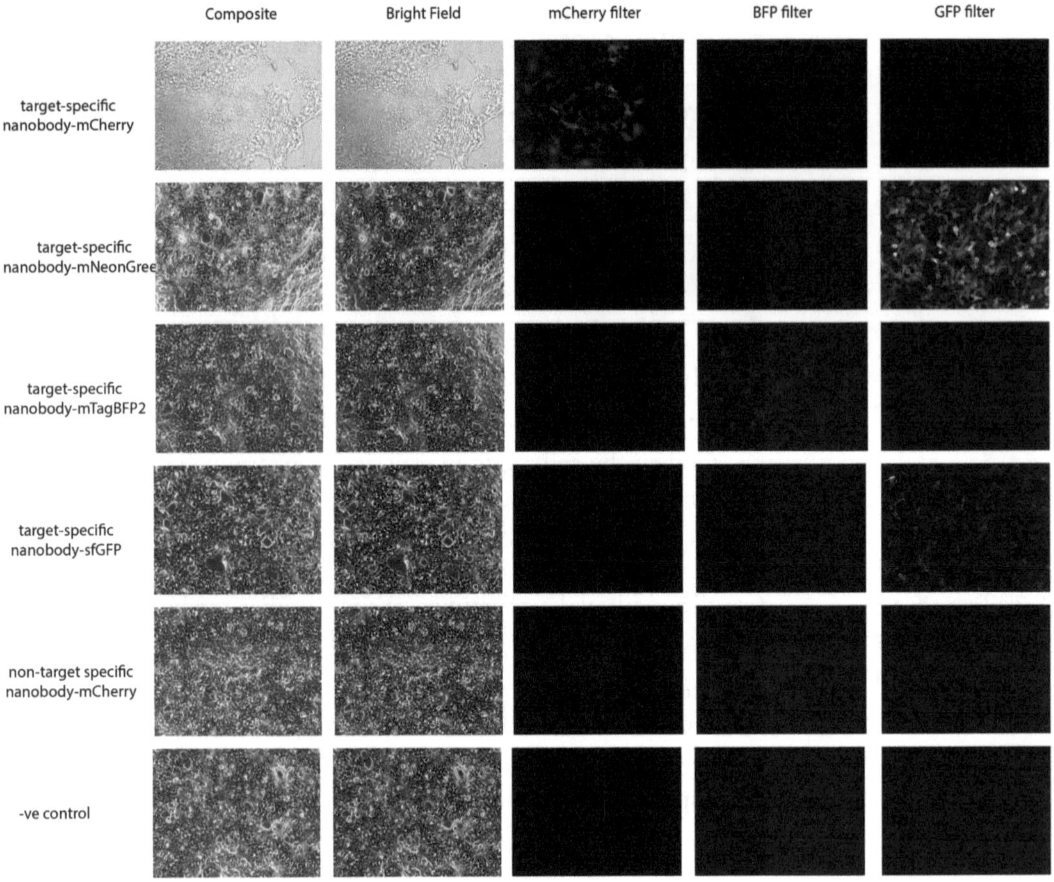

Fig. 3 Detection of a surface epitope using conditioned media containing chromobodies. The epitope-displaying cells were fixed and incubated in media, which were conditioned for 3 days by 293FT cells transfected with one of the four chromobody plasmids (*see* **Note 13**)

3. Dissociate cells using Accutase (*see* **Note 12**).
4. Collect cells and spin them down. Aspirate medium.
5. Wash cells with DPBS. Aspirate supernatant.
6. Resuspend cells in the conditioned medium containing chromobodies. Incubate at 4°C in the dark with gentle rotation for 30 min.
7. Spin the cells down. Aspirate the medium.
8. Wash the cells with DPBS by incubating for 5 min. Repeat the step three times.
9. Resuspend cells in sorting buffer.
10. Analyze using flow cytometry immediately.

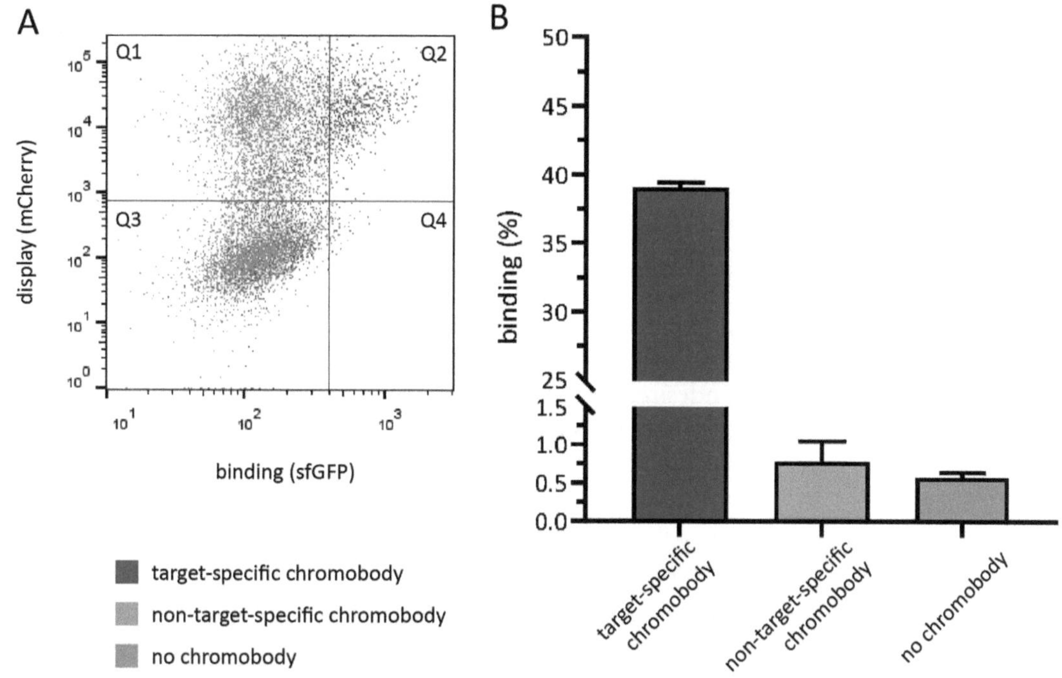

Fig. 4 Detecting surface epitopes via flow cytometry using the conditioned medium containing anti-target chromobodies. 293FT cells expressing and exporting anti-target sfGFP chromobodies (nanobody/super-folder GFP fusions) were grown for 3 days, their medium filtered and added onto cells displaying the target epitope and expressing mCherry. (**a**) Representative flow cytometry data for cell surface epitope detection, using for comparison conditioned medium containing (1) target-specific chromobodies (red population), (2) non-target-specific chromobodies (blue population), or (3) no chromobodies (orange population). Hits in the Q1 and Q2 quadrants are cells that display the target (mCherry+). Hits in the Q2 and Q4 quadrants are cells bound by chromobodies (sfGFP+). Hence, target-displaying cells bound by a detectable amount of chromobodies locate in quadrant Q2. (**b**) Percentage of target-displaying cells bound by a detectable amount of chromobodies: nQ2 x100 / (nQ1+nQ2), where nQ is the number of cells in a particular quadrant for each condition tested (*see* **Note 14**)

4 Notes

1. While HEK293 are okay with regular DMEM supplemented 10% FBS, 293FT cells require richer medium, otherwise tend to detach from the culture vessel surface. You may choose to supplement media with non-essential amino acids too.
2. It is better to use a phenol-free medium for chromobodies production as it does not change color and allows quantification of chromobodies in the conditioned medium using spectrometry (*see* **Note 8**).

3. Various transfection reagents can be used to transfect 293FT cells, such as polyethylenimine (PEI) or Lipofectamine 3000 (Thermo Fisher, #L3000001). 293FT cells are relatively easy to transfect, and hence, PEI—a cheaper and easily accessible reagent—was used.
4. DMEM can be used instead of Opti-MEM™; however, it may reduce the transfection efficiency and transfection optimization might be required.
5. The underscored nucleotides are overhangs to be produced following digestion with the BsaI restriction enzyme. *n* should be replaced with any nucleotide, taking care to avoid undesired secondary structures. The bolded nucleotides are essential for cloning the nanobody CDS in frame with the downstream fluorescent protein.
6. Although it is unlikely for the vector to religate, to minimize the risk, use 1:2 molar ratios of entry vector to nanobody CDS insert.
7. The red colonies contain the original entry plasmid. If there are many more red colonies than white ones, there is a high chance that the GoldenGate reaction was unsuccessful. Try extending the 37°C incubation step.
8. The amount of chromobodies in the medium can be quantified using a plate reader and medium from a transfection negative control used as a blank.
9. If there are a lot of debris in the cell medium, the medium containing chromobodies can be filtered through 0.45 μm or 0.22 μm filter.
10. DPBS can be used instead of PBS; however, cells tend to stay attached better if PBS is used.
11. The fixation step can be extended for up to 20 min; however, it is not recommended for fragile epitopes.
12. It is important to avoid using harsh dissociation reagents to lift off the cells, such as trypsin, as it is likely to destroy the epitope.
13. Signal strength will vary depending on the nanobody binding efficiency and/or the fluorescent protein used (mNeonGreen giving the best signal in our experience).
14. In our case, we designed the plasmid in a way that the CDS for the surface epitope is followed by an internal ribosome entry site (IRES) and the mCherry CDS. This arrangement ensures that the expression of the surface epitope is directly correlated with the expression of mCherry, which acts as a reporter. In other words, when the mCherry gene is expressed, we can also detect the expression of the surface epitope using chromobodies or scFv-FPs.

Acknowledgements

This work was supported by the School of Biological Sciences at the University of Edinburgh. Flow cytometry data were generated within the Flow Cytometry and Cell Sorting Facility in King's Buildings at the University of Edinburgh, with the help of Dr Martin Waterfall. Microscopy images were acquired on a microscope funded by the School of Biological Sciences, the Institute of Quantitative Biology, Biochemistry, and Biotechnology, and the UK Centre for Mammalian Synthetic Biology.

References

1. Harmansa S, Affolter M (2018) Protein binders and their applications in developmental biology. Development 145(2):dev148874. https://doi.org/10.1242/dev.148874. https://journals.biologists.com/dev/article/145/2/dev148874/48799/Protein-binders-and-their-applications-in
2. Sun Y, Huang T, Hammarström L, et al (2020) The immunoglobulins: new insights, implications, and applications. Annu Rev Anim Biosci 8(1):145–169. https://doi.org/10.1146/annurev-animal-021419-083720. https://www.annualreviews.org/doi/10.1146/annurev-animal-021419-083720
3. Jovčevska I, Muyldermans S (2020) The therapeutic potential of nanobodies. BioDrugs 34(1):11–26. https://doi.org/10.1007/s40259-019-00392-z. http://link.springer.com/10.1007/s40259-019-00392-z
4. Brunner JD, Schenck S (2020) Production and application of nanobodies for membrane protein structural biology. In: Methods in molecular biology, vol 2127. Humana Press Inc., Totowa, p 167–184. https://doi.org/10.1007/978-1-0716-0373-4_12. http://link.springer.com/10.1007/978-1-0716-0373-4_12
5. Tsumoto K, Isozaki Y, Yagami H, et al (2019) Future perspectives of therapeutic monoclonal antibodies. Immunotherapy 11(2):119–127. https://doi.org/10.2217/imt-2018-0130. https://www.futuremedicine.com/doi/10.2217/imt-2018-0130
6. Aoki W (2019) Engineering antibodies and alternative binders for therapeutic uses. In: Ueda M (ed) Yeast cell surface engineering. Springer, Singapore, p 123–147. https://doi.org/10.1007/978-981-13-5868-5_10. http://link.springer.com/10.1007/978-981-13-5868-5_10
7. Nordeen SA, Andersen KR, Knockenhauer KE, et al (2020) A nanobody suite for yeast scaffold nucleoporins provides details of the nuclear pore complex structure. Nat Commun 11(1):6179. https://doi.org/10.1038/s41467-020-19884-6. https://www.nature.com/articles/s41467-020-19884-6
8. Wagner TR, Rothbauer U (2020) Nanobodies right in the middle: intrabodies as toolbox to visualize and modulate antigens in the living cell. Biomolecules 10(12):1701. https://doi.org/10.3390/biom10121701. https://www.mdpi.com/2218-273X/10/12/1701
9. Kang W, Ding C, Zheng D, et al (2021) Nanobody conjugates for targeted cancer therapy and imaging. Technol Cancer Res Treat 20:153303382110,101. https://doi.org/10.1177/15330338211010117. http://journals.sagepub.com/doi/10.1177/15330338211010117
10. Beghein E, Gettemans J (2017) Nanobody technology: a versatile toolkit for microscopic imaging, protein–protein interaction analysis, and protein function exploration. Front Immunol 8(JUL):771. https://doi.org/10.3389/fimmu.2017.00771. http://journal.frontiersin.org/article/10.3389/fimmu.2017.00771/full
11. Hassanzadeh-Ghassabeh G, Saerens D, Muyldermans S (2011) Generation of Anti-infectome/Anti-proteome Nanobodies. In: Methods in Molecular Biology, vol 790. Humana Press, Totowa, p 239–259. https://doi.org/10.1007/978-1-61779-319-6_19. https://link.springer.com/10.1007/978-1-61779-319-6_19
12. Bai X, Shim H (2017) Construction of a scFv library with synthetic, non-combinatorial CDR diversity. In: Methods in Molecular Biology, vol 1575. Humana Press Inc., Totowa, p 15–29. https://doi.org/10.1007/978-1-4939-6857-2_2. http://link.springer.com/10.1007/978-1-4939-6857-2_2

13. Desautels T, Zemla A, Lau E, et al (2020) Rapid in silico design of antibodies targeting SARS-CoV-2 using machine learning and supercomputing. bioRxiv p 2020.04.03.024885. https://doi.org/10.1101/2020.04.03.024885. https://www.biorxiv.org/content/10.1101/2020.04.03.024885v1
14. Norman RA, Ambrosetti F, Bonvin AMJJ, et al (2020) Computational approaches to therapeutic antibody design: established methods and emerging trends. Brief Bioinform 21(5): 1549–1567. https://doi.org/10.1093/bib/bbz095. https://academic.oup.com/bib/article/21/5/1549/5581643
15. Sormanni P, Aprile FA, Vendruscolo M (2018) Third generation antibody discovery methods: *in silico* rational design. Chem Soc Rev 47(24): 9137–9157. https://doi.org/10.1039/C8CS00523K. http://xlink.rsc.org/?DOI=C8CS00523K
16. Zhao J, Nussinov R, Wu WJ, et al (2018) In Silico methods in antibody design. Antibodies 7(3):22. https://doi.org/10.3390/antib7030022. https://www.ncbi.nlm.nih.gov/pmc/articles/PMC6640671/
17. Scheller L (2021) Synthetic receptors for sensing soluble molecules with mammalian cells. In: Methods in molecular biology (Clifton, N.J.), vol 2312. Humana Press Inc., Totowa, p 15–33. https://doi.org/10.1007/978-1-0716-1441-9_2. https://link.springer.com/10.1007/978-1-0716-1441-9_2
18. Santorelli M, Lam C, Morsut L (2019) Synthetic development: building mammalian multicellular structures with artificial genetic programs. Curr Opin Biotechnol 59:130–140. https://doi.org/10.1016/j.copbio.2019.03.016. https://linkinghub.elsevier.com/retrieve/pii/S0958166918301617
19. Morsut L, Roybal K, Xiong X, et al (2016) Engineering customized cell sensing and response behaviors using synthetic notch receptors. Cell 164(4):780–791. https://doi.org/10.1016/j.cell.2016.01.012. https://linkinghub.elsevier.com/retrieve/pii/S0092867416000520
20. Daringer NM, Dudek RM, Schwarz KA, et al (2014) Modular extracellular sensor architecture for engineering mammalian cell-based devices. ACS Synth Biol 3(12):892–902. https://doi.org/10.1021/sb400128g. https://pubs.acs.org/doi/10.1021/sb400128g
21. Manhas J, Edelstein HI, Leonard JN, et al (2022) The evolution of synthetic receptor systems. Nat Chem Biol 18(3):244–255. https://doi.org/10.1038/s41589-021-00926-z. https://www.nature.com/articles/s41589-021-00926-z

Index

Francesca Ceroni and Karen Polizzi (eds.), *Mammalian Synthetic Systems*, Methods in Molecular Biology, vol. 2774,
https://doi.org/10.1007/978-1-0716-3718-0,